MAMMALOGY

McGraw-Hill Series in Organismic Biology

Consulting Editors

Professor Melvin S. Fuller
    Department of Botany
    University of Georgia, Athens

Professor Paul Licht
    Department of Zoology
    University of California, Berkeley

Gardiner: The Biology of Invertebrates
Gunderson: Mammalogy
Kramer: Plant and Soil Water Relationships: A Modern Synthesis
Leopold and Kriedemann: Plant Growth and Development
Patten and Carlson: Foundations of Embryology
Phillips: Introduction to the Biochemistry and Physiology of Plant Growth Hormones
Price: Molecular Approaches to Plant Physiology
Weichert and Presch: Elements of Chordate Anatomy

Harvey L. Gunderson

Professor of Zoology, School of Life Sciences
Associate Director and Curator of Zoology, University Museum
University of Nebraska

MAMMALOGY

McGraw-Hill Book Company

New York St. Louis San Francisco Auckland Düsseldorf
Johannesburg Kuala Lumpur London Mexico Montreal New Delhi Panama Paris
São Paulo Singapore Sydney Tokyo Toronto

MAMMALOGY

1234567890DODO79876

This book was set in Times Roman by Kingsport Press, Inc.
The editors were William J. Willey and Carol First;
the cover was designed by Anne Canevari Green;
the production supervisor was Charles Hess.
The drawings were done by Eric G. Hieber Associates Inc.
R. R. Donnelley & Sons Company was printer and binder.

New photographs were taken by Harvey L. Gunderson.

Library of Congress Cataloging in Publication Data

Gunderson, Harvey L
    Mammalogy.

    (McGraw-Hill series in organismic biology)
    Bibliography: p.
    1. Mammals. I. Title.
QL703.G8   599   75–26921
ISBN 0–07–025190–8

Contents

Preface

The first part of this text deals with what might be called the classical part of mammalogy—evolution, characteristics of mammals, and classification. Taxonomy is central to most biologic work. Scientific work in which the correct name of the organisms involved is unknown or incorrect is of little value. But there has been in the last quarter-century a general explosion of knowledge in all areas of mammalogy. A great variety of new and exciting information on mammals is appearing every day, gained with techniques that vary from simply making observations to using sophisticated electronic equipment.

Most of this text places its emphasis on the living mammal. How the mammals of today are related, who their ancestors were, how they outwit their enemies or survive the winter are questions researchers are trying to answer.

I have attempted to provide a thread of continuity throughout this book. That theme is how mammals have adapted to change and have survived through time. Life is not static but dynamic, continually adjusting to meet new conditions. The technical term for such adjustment is *homeostasis,* whether at the level of the individual (the species), ecosystem, or anywhere between. The concept of an internal homeostasis was developed by the French physiologist Claude Bernard in the nineteenth century. This idea led to a vastly increased understanding of the regulation of body processes through endocrine and nervous control of heart rate, respiration, temperature, growth, behavior, and reproduction. This concept has been widely applied. We now know, for instance, that at the other extreme from the individual, i.e., the ecosystem level, regulatory processes are operate for the survival of plant and animal communities.

I have included a chapter on the history of mammalogy. History is made not by robots but by human beings, whose weakness and strengths, jealousies and generosities are real.

The material in this book is intended for upper division and graduate courses. More material is included than can be covered in one semester. In writing this text two alternatives were available. One was to discuss a limited number of topics in detail. The other was to cover a wider variety of topics in less detail. I have chosen the latter and have added a rather extensive bibliography so that more information is readily available to the student or instructor who cares to pursue a subject in more detail. This should enable instructors to pick and choose the material that best fits their training, interests, the material available, and the local academic emphasis.

In preparing this manuscript I turned often to students and colleagues for suggestions and help. Former students who have read parts of the text include Clark Adams, Douglas Liesveld, Kenneth Middleton, Peter Meserve, Ray Goldstein, and Jerry Walker. Colleagues at the University of Nebraska have read chapters in their fields of specialty. Dr. Robert Hunt has read Chap. 2; Dr. Gerald D. Tharp read Chapters 8 and 9; Dr. Earl B. Barnawell read Chapter 10; Dr. Roger Sharp read Chapters 12 and 13; and Dr. Paul Johnsgard read Chapter 16. Chapter 1 was reviewed by Dr. William H. Burt, whose suggestions and encouragement were most helpful. In addition, the following professors were kind enough to provide me with useful reviews of the entire manuscript: R. E. Mumford, Paul Licht, Robert W. Seabloom, Henry McCutcheon, and Allen M. Benton. I should like to express my gratitude to all these persons for their time and help.

Few manuscripts would appear in print if it were not for secretaries and typists. I should like to express my appreciation to Mrs. Norma Wagner, Lorene Bartos, Lorraine Tuma, and Cindy Matson, who typed the greater part of this material and sometimes helped with other details.

Finally I should like to thank my family for their patience and especially my wife Erika for her constant encouragement during a long and at times tedious undertaking.

Harvey L. Gunderson

MAMMALOGY

The Evolution of Mammalogy: A History of the Science

Only in the last 100 years has mammalogy reached the status of a separate field of science. During the half century before, information on birds and mammals was usually combined into one report or publication, and most often these reports were made by persons primarily interested in ornithology. Still earlier the amount of information available on all living things was not great enough to warrant its separation into any kind of phylogenetic grouping.

For a broad understanding of mammalogy we cannot restrict ourselves to a discussion of mammals and mammalogists. The reservoir of information now available concerning mammals has been contributed not only by mammalogists or even, more broadly, by scientists, but by artists, explorers, trappers, mountain men, army surgeons, soldiers, statesmen, politicians, philosophers, and innumerable others. The history of mammalogy is not separate from the history of events and human beings.

THE BEGINNINGS OF BIOLOGY AND NATURAL HISTORY

It was imperative even for primitive people to know something of the animals around them, for they depended upon them for food and quite often for clothing. Even today there are people living under primitive conditions who must know something about bird migration, caribou *(Rangifer tarandus)* mi-

3

gration, the habits of polar bear *(Ursus maritimus)* and seals, as well as of fish, in order to survive. Others, for whom this knowledge is not so vital, may still have a great interest.

It would be difficult to say that any science, including mammalogy, began at this time or with that event. The earliest communication of information was by the spoken, not the written, word, and information was passed from generation to generation in this manner. The process of recording information was a gradual development. Much of early writing was concerned with rumor, superstition, or speculation and was not based upon observation or experiment. Furthermore, the early writing on natural history was included in works on other subjects such as theology or astronomy.

Some of the earliest preserved writings are those of the Greeks, and these in turn contain references to the "ancients." Greece, its warm waters teeming with animal life, was geographically a fine location for a marine laboratory. But the early Greeks were more philosophic than scientific. Thales, a Greek astronomer, thought life to have originated in the ocean, an idea not unfamiliar today. Anaximander believed that living things were first produced by the drying crust of the earth. Xenophanes recognized fossils as animal remains. Empedocles formulated a crude theory of evolution. These writers all lived prior to the fourth century before Christ. The early Greeks were inclined to the deductive method of thinking. They would reach a conclusion quickly, with but little evidence to support it. From this they went on to generalizations which most often led to fallacies.

One of the early Greek philosophers who stands out for his superior accomplishments is Aristotle (384–322 B.C.). He appealed directly to nature for facts and so instituted the inductive method of reasoning, although his failure to use it properly sometimes led to errors. It was he who also brought together the knowledge of his time and formulated a beginning of science. Aristotle provided the basis for a formal classification of animals.

The period of slavish acceptance of authority, known historically as the Dark Ages, lasted many centuries. Andreas Vesalius, a Belgian who lived in the sixteenth century (1514–1564), was among the first to lead the way out of the Dark Ages. Rather than rely upon the anatomic dissections, together with their errors, made by Galen in the second century, Vesalius did his own dissections. His work typifies the revival of learning which was to lead people out of the Dark Ages.

NATURAL HISTORY IN EUROPE DURING THE SEVENTEENTH AND EIGHTEENTH CENTURIES

Eighteenth-century Europe was a paradise for the amateur naturalist. It was an era of people rich enough to do whatever they liked, to do it in the grand manner, and to do it with expertise. It was also a period of great exploration and colonization. Wealthy nobles contributed data and specimens either by their own efforts or by supporting collectors and naturalists. Information increased markedly after the organization of the Royal Society of London in 1662. One of the most important of these early naturalists was Mark Catesby, who was supported by wealthy patrons and who sent back much information about the Americas to the scientists of Europe (Frick and Stearns, 1961). Catesby's birthdate is obscure, but according to Frick and Stearns (1961) he was born April 3, 1683, in Essex, England, the home also of the eminent botanist and naturalist of the time, John Ray (1627–1705). It has been speculated that Catesby's interest in natural history might have been inspired by Ray. At about age 27 Catesby left for "Virginia," where he stayed with relatives, made many friends, traveled, observed, wrote, and painted birds. This first visit lasted from 1712 to the fall of 1719. In February 1722 he began his second sojourn in the New World, this time with financial assistance and the approval of the Royal Society. In 1726, he returned to England to complete the illustrations and work for a three-volume work, *The Natural History of Carolina, Florida and the Bahama Islands,* which was finished in 1748. He had now become an elder statesman of British natural history; a year later, on December 23, 1749, he died.

Mark Catesby has usually not been given the accolades due a great naturalist; yet *The Natural History of Carolina, Florida and the Bahama Islands* was one of the finest natural history works of the eighteenth century.

Classification and natural history occupied most botanists and zoologists from the sixteenth century to the time of Carolus Linnaeus (Karl von Linné), the Swedish naturalist who lived from 1707 to 1778. Linnaeus has been called the father of taxonomy. The tenth edition of his *Systema Naturae,* published in 1758, has been arbitrarily accepted as the foundation for systematic zoology, and the synonymy of scientific names goes no further back. Here the binomial system of nomenclature was consistently used for the first time.

Linnaeus accepted species as individually created and fixed, and he characterized them by a clear-cut species description. He also gathered lower categories into higher categories, such as genera into orders and orders into classes. Although his classification was more of a filing system than a reflection of relationships, it did reflect relationships to some extent.

Specimens were sent to him from many parts of the world, and Catesby's work in North America was certainly one of the foundations upon which Linnaeus and other systematists of that day built their structures. Linnaeus is best known as a botanist, but he did describe 86 species of mammals. Today over 3600 species of mammals are described. As larger numbers of animals became known and observation of them came under the influence of Linnaean classification, which placed similar animals together, it became apparent to many that this similarity could not be mere coincidence.

The man generally credited with founding the study of comparative anatomy was Georges Cuvier, a French Huguenot who lived from 1769 to 1832. Cuvier was born poor. His first job was as a tutor to a family living in Normandy on the west coast of France. In his spare time he dissected marine animals and suggested ways of comparing their differing structures. He drew everything he studied, and some of these drawings reached Geoffrey Saint-Hilaire at the University in Paris. As a result he was appointed professor of comparative anatomy. A brilliant person, Cuvier became an indispensable authority in the fields of science and education in France, and was made a baron. He died in that first devastating epidemic of cholera that ravaged Europe in 1832. Cuvier's method was to search for bones which might fit together and to compare them with bones of what he thought were closely related living animals. Cuvier created the field of paleontology in its modern sense. He arrived at a system of classification based on a similarity of structures but did not believe in a common origin for these similarities. As a matter of fact he ridiculed the idea.

Although there was much speculation about evolution at this time, the stumbling block to the development of thought on the subject was the generally accepted belief that the earth was only 6000 years old. It was difficult for anyone to believe that changes as great as fossils seemed to indicate could take place in such a short time. There seemed to be no evidence within human memory of an animal's changing at all, much less changing so greatly that new species resulted.

Charles Lyell, the Scottish founder of modern geology, who lived from 1797 to 1875, introduced the principle of proceeding from the known to the unknown, or at least the remotely known. Before Lyell's time the doctrine of *cataclysmos,* or change brought about by catastrophes, was held; Lyell was a vigorous opponent of this view. He believed the earth's features to be the product of forces still at work and that present forces are just as effective as those in the past. From his extensive studies and travels he deduced that the world is much older than 6000 years and may even be over a million years old! Although he had at first rejected the view that animals in earlier ages were different from those living in his day, his geologic studies made the idea of slow change in plants and animals more acceptable. Lyell finally accepted the theory of his friend and pupil, Charles Darwin.

Early Ideas on Evolution   Many scientists of the eighteenth and early nineteenth centuries held various views of evolution. Buffon (1707–1788), the wealthy French nobleman who published the 15-volume *Histoire Naturelle,* held evolutionary ideas but did not champion them. The *Histoire Naturelle* was the first extensive natural history ever published.

Erasmus Darwin (1731–1802), the grandfather of Charles Darwin, published his own comprehensive ideas of evolution, but with little supporting evidence. Alexander von Humboldt (1769–1859), a brilliant personality of

German and French ancestry, spent 5 years exploring South America so thoroughly that he was called the second discoverer of America; at least tentatively he supported the idea of evolution.

Jean Baptiste Lamarck (1744–1829) was born in Picardy in Northern France, one of the youngest of a large, poor, and noble family. He started out to become a priest, but at 17 he joined the French Army and fought in the Seven Years' War. He left the army on a small pension and spent the next 15 years living in bohemian style in the Latin Quarter of Paris as a hack writer. During his garrison life on the Mediterranean he had become particularly interested in the flora of the area. Through Cuvier's influence Lamarck rose in scientific esteem. At age 50 he was appointed, during the French Revolution, to a zoology professorship in Paris. Thus he made his fame in a subject that he had never professionally followed until he was 50. In 1809 he published his theory of evolution, in which he expressed the idea that changes produced by the environment are transmitted to the offspring, a hypothesis held in low esteem by most biologists today.

NATURAL HISTORY IN THE AMERICAN COLONIES

Almost simultaneously with the appearance of vertebrate paleontology as a science in Europe, events occurred on the North American continent which triggered its emergence there. The year 1799 marked the appearance of two papers dealing with fossils in North America: one was by Dr. Caspar Wistar, the other by Thomas Jefferson.

Thomas Jefferson is best remembered as a signer of the Declaration of Independence and third President of the United States. But he was much more than this. Jefferson had a great personal love for science, and his interests included agriculture, archeology, astronomy, botany, chemistry, geology, meteorology, and paleontology.

When he arrived in Philadelphia in 1797 as Vice President of the country, his baggage included the bones of a prehistoric animal. He was also the newly elected president of the American Philosophical Institute, which met in Philadelphia and appears to have been the American counterpart of the Royal Society of England. On March 10, 1797, he read his paper on *Megalonyx* (the great claw) before the American Philosophical Society, and 2 years later it was published in their *Transactions* journal. The *Megalonyx* was later identified as a species of sloth. Jefferson arranged with the owner of Big Bone Lick, in what is now a state park in Kentucky, for Clark (of Lewis and Clark fame) to go there and dig in 1807. When Clark's cargo arrived at the presidential mansion in Washington (by then the national capital had been moved from Philadelphia), the bones were spread out in what is now the East Room. It is difficult to imagine bones of mastodons, musk oxen, bison, and deer lying around on the floor of the East Room of the White House today! Had Jefferson been artistically gifted, he might have been an American Leonardo da Vinci.

EARLY WESTERN EXPLORERS: THEIR CONTRIBUTION
TO NATURAL HISTORY

Sporadic probing of the lands beyond the Mississippi had been going on for a century and a half before the Lewis and Clark expedition, but these probings were done by adventurers, and the information gained by them was not only meager but sometimes unreliable. With the consummation of the Louisiana Purchase from France in 1803, Thomas Jefferson was concerned about the borders of British and French exploration of the West.

With the mounting of the Lewis and Clark expedition came a new era in exploration of this area, west of the Mississippi River. Future expeditions usually had the support of the federal government directly or indirectly, and they were specifically charged with bringing back an enormous quantity of information. Lewis's assignment involved the surveying and mapping of the Missouri and Columbia Rivers, studying the botany, climatology, geology, geography, mineralogy, and zoology of the area traversed, and making a detailed report on all the Indian tribes and how to deal with them peaceably! Lewis and Clark were specifically instructed to look for huge beasts, for Jefferson seemed to harbor the idea that perhaps *Megalonyx,* or similar creatures, were still living somewhere.

The actual journey started on May 14, 1804, and ended on September 23, 1806, in St. Louis, Missouri, at a total cost to the United States taxpayers of $2500! The leaders, Captain Meriwether Lewis and Captain William Clark, were men of exceptional ability and loyalty. When Lewis asked that Clark be given a captaincy, Congress saw fit to give him only a second lieutenancy. However, since none of the men in the unit, nor even President Jefferson, knew it, he was known as Captain Clark. The eminently successful expedition occupies a prominent place in American history, contributed much to our knowledge of the fauna, and probably served as a model for many succeeding expeditions.

The journals of Lewis and Clark are frequently overlooked as a source of information on early Western natural history. Anyone interested should consult Thwaites's (1904) *The Original Journals of the Lewis and Clark Expeditions.* This is the only complete and unaltered text of the diaries. It appeared exactly 100 years after the expedition started—1904. Lewis and Clark were the true discoverers, and actually the original describers, of many animals with which their names are seldom associated. They commented generously on all the vertebrates, but were generally casual in their descriptions of insects, crustaceans, and mollusks. They also had their troubles with the prairie rattlesnake.

They were of course the first to write descriptions of many of the Western fishes, reptiles, amphibians, and birds as well as mammals. Certainly Clark's nutcracker and Lewis's woodpecker come quickly to mind. Theirs were the first detailed descriptions of, among others, the white-tailed jackrabbit *(Lepus townsendi),* the eastern wood rat *(Neotoma floridana),* the bushy-tailed wood rat *(Neotoma cinerea),* the prairie dog *(Cynomys ludovicianus),* the kit fox *(Vulpes macrotis),* the grizzly bear *(Ursus horribilis),* the mule deer *(Odocoileus hemionus),* the pronghorn antelope *(Antilocapra americana),* and the mountain goat *(Oreamnos americanus).* Coues (1876; see Coues, 1875) credited Lewis and Clark with the discovery of the mountain goat. This is unlikely, for Alexander Mackenzie in 1789 referred to a "white buffalo" that was restricted to the mountains of the Northwest. Certainly George Ord based his classification of certain Western mammals and birds upon the description of Lewis and Clark.

The diaries of Lewis and Clark show that the prairie dog *(Cynomys ludovicianus),* an unusual rodent, especially piqued their curiosity. Their first encounter with this animal was about 40 km (25 mi) above the Niobrara River (in what is now Nebraska) on September 7, 1804, as reported by Clark. Lewis wrote a detailed description of this animal on the return trip, July 1, 1806. He noted that prairie dogs generally belonged to large societies and their burrows frequently occupied several hundred acres of land. He also remarked about their warning bark, which he described as "being much like that of little toy dogs." Members of the expedition must have eaten prairie dogs, for Lewis wrote that "this little animal is frequently very fat and its flesh is not unpleasant." However, their observation that it shuts up its burrow when the hard frosts commence and remains within it until spring was probably speculation or hearsay. The expedition sent four live dogs to Jefferson, and the caretaker made note of the fact that they did not seem to need drinking water.

Ord based his description of the black-tailed prairie dog *(Cynomys ludovicianus)* on the account given in the journal of Lewis and Clark and a specimen taken on the upper Missouri River, and in 1815 named it *Arcotomys ludoviciana.* This name, along with many of Ord's other scientific names, was published in Guthrie's (1815) geography.

Two skins of the prairie dog, in addition to the live specimens already mentioned, were shipped to President Jefferson from Fort Mandan. Although the explorers prepared bird and mammal skins, they lacked the facilities to protect their collections properly from insects and moisture. Those specimens that did survive the rigors of the expedition were deposited in Peale's Museum in Philadelphia. Peale's Museum was a unique institution in its day, intimately associated with the American Philosophical Society. Peale died in 1820, and Peale's Museum at Philadelphia was dissolved in 1846, when the specimens were sold at public auction. However, it had, in its short existence, served American natural history well, for many original bird and mammal descriptions by Alexander Wilson, George Ord, C. L. Bonaparte, Richard Harlan, and Thomas Say were based on its specimens.

What happened to the collections made on the Lewis and Clark expedition is a rather unfortunate bit of the history of museum collections in North

America. When Peale's collections were sold in 1846, half went to Moses Kimball, the other half to P. T. Barnum. Barnum's half was destroyed by fire in 1865. Kimball's half was deposited in the Boston Museum and after having been moved a great deal was finally transferred to the Museum of Comparative Zoology in 1914. It is probable that the mammal and other specimens from the Lewis and Clark trips may have disappeared because all of them were in the collections that P. T. Barnum bought. Though this loss was a great one, the Lewis and Clark expedition still stands as a crowning achievement in the exploration of the American West.

With the completion of the Lewis and Clark expedition, Jefferson turned his attention to discovering the source of the Mississippi. The person chosen for this task was a man whose name looms as tall as Pike's Peak in the conquest of the Far West. Lieutenant Zebulon Pike was only 26 years old when he started his northward journey in the fall of 1805. At Leech Lake, Minnesota, where Pike's Bay commemorates this adventure, Pike and the men turned back, believing thay had found the source of the Mississippi (they had not). By April 1806 they were back in St. Louis.

Pike's estimate of his accomplishments was far from modest, and evidently Jefferson and the nation agreed (although some historians think otherwise), for Congress immediately appropriated the money to send Pike to explore the country between the Arkansas River and the Red River (of the South). In July 1806 he and 23 men started up the Missouri and explored west to the Rockies. In the spring of 1807 they pressed southwest, but an army of 100 Spaniards from Santa Fe escorted them to the New Mexico capital for questioning, then to Chihuahua for further questioning. From here they were marched across northern Mexico to Natchitoches (on the Red River in present Louisiana) and onto the American side of the border. Although his maps and papers were confiscated by the Spaniards, he evidently managed to reconstruct them accurately.

It is interesting to compare Pike's description of the prairie dog *(Cynomys ludovicianus)* with that of Lewis and Clark, for neither had seen the other's description. Pike's description is longer and more detailed. Near the present Larned, Kansas, Pike and his party killed some prairie squirrels, or wishton-wishes (an Indian name), and nine "large rattlesnakes." He mentioned that their holes descended in a spiral form which made it difficult to ascertain their depth, but he did pour 140 kettles of water into one, without driving out the occupant. Evidently the men of Pike's expedition also ate them, for Pike wrote that they found them excellent but he cautioned that they should be exposed a night or two to frost to correct the rankness acquired by their subterranean dwelling! Of their numbers he wrote that there must be innumerable hosts of them, for their villages sometimes extended over "two and three miles square." Of other inhabitants of the prairie dog towns he wrote:

It is extremely dangerous to pass through their towns, as they abound with rattlesnakes, both of the yellow and black species; and strange as it may appear, I have seen the wishtonwish, the rattlesnake, the horn frog, with which the prairie abounds (termed by the Spaniards the cammellion) and a land tortoise, all take refuge in the same hole. I do not pretend to assert that it was their common place of resort; but I have witnessed the above facts more than in one instance.

The War of 1812 and the problems of reconstruction pushed interest in Western exploration into the background until 1815, when government attention was again directed to the West. But the objectives had changed. Parties were now sent out to locate military posts that would hold the Indians in check, help the fur traders, and restrain the British. Major Stephen Long in expeditions between 1817 and 1819 established Fort Smith on the Arkansas River and Fort Snelling, at the junction of the Mississippi and Minnesota Rivers in Minnesota. In 1820 Major Long undertook another expedition in search of the headwaters of the Red River of the South, and in 1823 he made his second trip up the Mississippi to ascertain the boundary between the United States and Canada, which was a "point of intersection between Red River [of the North] and the forty-nine degree of north latitude." From here they would follow the boundary east to Lake Superior and their home. He was accompanied by some highly competent men, among them Thomas Say, as zoologist, James Calhoun, as astronomer and meteorologist, and a geologist, William Keating, whose two-volume record is generally known as *Keating's Narrative* (1825). It is *Keating's Narrative* that has given us an account of the wildlife seen on this journey (Krause, 1957). From Prairie du Chien (Wisconsin) to Fort St. Anthony (Fort Snelling, Minnesota) Major Long and several of the party rode horseback while the rest went upstream by boat. In this journey of a little over 320 km (200 mi) the only game seen was a herd of elk *(Cervus canadensis)* by some of the "boys" but evidently not by the hunters. When they reached Fort Snelling, Keating wrote, "Game will be judged to be very scarce where two parties, traveling by land and by water, can kill but two or three dozen of birds upon a distance upwards of two hundred miles. . . . Buffaloes, of the largest size, were formerly found here; but since the establishment of the garrison at Fort St. Anthony they have all been destroyed or have removed farther west."

Leaving the Fort to travel up the St. Peter (now the Minnesota) River, they started with high hopes. But after about 240 km (150 mi) of travel, Keating wrote, "The reports which we had previously received of abundance of game, had not been confirmed; we had, on the contrary, found none at all, and our stores were wasting away." Prairie wolves barked at night and appeared "to be abundant on these prairies."

Just beyond where the Cottonwood River enters the Minnesota River they came upon a few bison, or buffalo *(Bison bison),* and a little farther they met some Indians who invited them to feast upon a recently killed bison. But at the Columbia Fur Company's post at Lake Traverse the inventory of game seen since Prairie du Chien comprised elk *(Cervus canadensis)* and coyotes *(Canis latrans);* also seen were empty houses of muskrat *(Ondatra zibethica).* An enumeration of species taken by the Fur Company along the Red River (of the North) between Lake Traverse and the Canadian border included bison *(Bison bison),* bear *(Ursus* sp.), beaver *(Castor canadensis),* marten *(Martes americana),* otter *(Lutra canadensis),* fisher *(Martes pennanti),* elk *(Cervus canadensis),* wolf *(Canis lupus),* mink *(Mustela vison),* muskrat *(Ondatra zibethica),* lynx *(Lynx canadensis),* rabbit *(Sylvilagus),* wolverine *(Gulo luscus),* moose *(Alces americana),* and fox *(Vulpes vulpes).*

On July 25, the day following their departure from Lake Traverse, they feasted on bison *(Bison bison)* and "for the first time they saw an abundance of game before them." But this lasted only 5 days, for on July 30 near the Wild Rice River they shot their last one and Calhoun (1825) wrote in his diary, "Civilization in its steady march destroys the larger gregarious animals."

Near Pembina (North Dakota) they saw beaver, and Thomas Say shot a "line-tailed Squirrel." The fur records at Pembina included ermine *(Mustela erminea),* grizzly bear *(Ursus horribilis),* and hare *(Lepus).* At Pembina, Calhoun (the astronomer) took observations, and an oak post was driven into the ground to mark the boundary. Beltrami left to continue his search for the source of the Mississippi, but Major Long's group continued down the Red River (or north). Say recorded the pouched rat (pocket gopher, *Geomys bursarius),* the flying squirrel *(Glaucomys sabrinus),* and the Hudson Bay squirrel (fox squirrel, *Sciurus niger).*

At Saline River (in Canada), Say recorded an antelope *(Antilocapra americana)* which he saw while observing insects. On August 11 they arrived at a settlement of Scots called the Selkirk Settlement. It was here that they saw a young bison *(Bison bison)* bull "employed at labor." Evidently the Selkirkers used the buffalo to pull plows and carts. From Selkirk they continued down the Red River to Lake Winnipeg, along the shore to the Winnipeg River, and up the Winnipeg River to the Lake of the Woods without recording any mammals. Although they saw none, they were informed that bear *(Ursus* sp.), otter *(Lutra canadensis),* wolverine *(Gulo luscus),* moose *(Alces americana),* squirrel (sp.?), wolf *(Canis lupus),* beaver *(Castor canadensis),* and muskrat *(Ondatra zibethica)* were occasionally seen, and marten *(Martes americana)* and fisher *(Martes pennanti)* were abundant.

On September 13 they reached Fort William, and in the 200 miles from Rainy Lake to Lake Superior they saw no mammals except a cross fox *(Vulpes vulpes).* Officially the expedition was over, for they were now to go home.

Thomas Say's catalog shows a great accomplishment, although only 10 species of mammals were seen.

For those who speak of the "good old days" with its implied abundance of wildlife, *Keating's Narrative* could be a sobering piece of literature.

The Ohio, the lower Mississippi, and the upper Mississippi were all important in the westward expansion of the United States, but along the Missouri St. Louis was the headland from which the nation overlooked the West. On the morning of March 24, 1833, Baron Braunsberg, or Maximilian, Prince of Wied-Neuweid on the Rhine River, arrived at St. Louis from the Ohio River on the steamboat *Paragon*.

Maximilian (1782–1867) had fought against Napoleon, had been a prisoner of war, decorated with the Iron Cross, and had ridden into Paris as a major general at the head of his division. This distinguished military career had been not of choice but necessitated by circumstance. His choice was a career in science, which, after fulfilling his patriotic obligation, he embarked upon. He spent 2 years in Brazil, studying the natives as well as the flora and fauna, and brought back to Wied a large natural history collection. These studies (which include the first description of the vampire bat, *Desmodus*) won him a distinguished reputation. A brief report of the collections, including a list of type specimens of Brazilian mammals which he collected, has recently been published (Dias de Avila-Pires, 1965). In 1832 at 50 years of age, he embarked on a similar venture to the American West. His first stop was New Harmony, Indiana, where he visited Thomas Say, with whom he had long corresponded. He spent the winter there and left for St. Louis in March 1833.

Although new species from west of the Mississippi were often elaborately described, few had been illustrated. Maximilian changed this, for he brought with him a young Swiss artist, aged 23, by the name of Carl Bodmer (1809–1893), who did 427 watercolor drawings of the flora, fauna, ethnologic aspects, and tribal villages.

The party which Maximilian headed set out on the fur-trading steamboat *Yellowstone* and traveled up the Missouri River to its upper reaches in Montana, making the return trip the following year. It was considered the first major expedition to go up the river since the historic Lewis and Clark expedition. Maximilian Wied's (1839) two-volume *Travels in the Interior of North America (Reise in das innere Nord-America, in den Jahren 1832 bis 1834)* was accompanied by an atlas of paintings by Carl Bodmer. The manuscript and paintings, though ranking as one of the greatest accounts of the West in the early nineteenth century, were unknown for nearly a hundred years. In the early 1950s the original documents, diaries, maps, atlases, and engravings were found in the archives of Neuwied castle, Germany. They are now housed at the Joslyn Art Museum at Omaha, Nebraska, along the very route the explorers traveled. Bodmer included in his paintings bison *(Bison bison)*, pronghorned antelope *(Antilocapra americana)*, white-tailed deer *(Odocoileus virginianus)*, and many more species. He spent the rest of his life in Europe, mostly in Paris, and became a successful and widely recognized artist.

Maximilian's description of a bison slaughter is included to give some indication of the kind of information he recorded.

> The Indians are so skillful in this kind of hunting on horseback, that they seldom have to fire several times at a buffalo. They do not put the gun to their shoulder, but extend both arms, and fire in this unusual manner as soon as they are within ten or fifteen paces of the animal. They are incredibly quick in loading; for they put no wadding in the charge, but let the ball (of which they have several in their mouth) run down to the powder, where it sticks, and is immediately discharged. With this rapid mode of firing, these hunters soon make a terrible slaughter of a herd of buffalo.

The expedition collected many specimens; Bailey (1923) wrote:

> Three years after the death of Maximilian in 1867 his zoological collections were purchased by the American Museum of Natural History and brought to New York, where they became a valuable addition to the Museum collections. The specimens consisted of about 4,000 mounted birds, 600 mounted animals [presumably mammals], and 2,000 fishes and reptiles.

John C. Frémont (1813–1890) was another of the giants of Western exploration who added to our knowledge of mammals. He was the son-in-law of the powerful senator from St. Louis, Thomas Hart Benton, the foremost advocate of the westward expansion. The gathering of information about the West was, during the middle years of the nineteenth century, in the hands of the Topographical Survey under the command of Colonel John J. Abert. Since Abert could not carry on intrigue on westward expansion as Senator Benton could, Benton got most of the credit, but Abert was a most potent behind-the-scenes figure. Frémont, the most famous of the topographic engineers, became a national hero in his own lifetime. He went on five expeditions of exploration in the West, the first, in 1838–1839, made with Joseph N. Nicollet. Between 1841 and 1845 Frémont led four expeditions of his own. Although the scientific contributions to mammalogy resulting from Frémont's expeditions were not as great as those of the other expeditions of the time, his name as well as Abert's appears often in mammalian nomenclature (e.g., *Dipodomys ordii fremonti, Sciurus aberti*).

THE FUR TRADERS AND MOUNTAIN MEN

Although the contribution of the fur traders was more economic than scientific, one cannot completely pass them by in a history of mammalogy. Theirs is a colorful part of American history based almost entirely on one species of mammal, the beaver. This history has been thoroughly documented by Chittenden (1954).

Even though the scientific contributions of the fur trade were not great, the records of the fur companies give us the best, and in many cases the only, information on population dynamics of those early years. Elton (1942) and Elton and Nicholson (1942) based some of their ideas of cycles on the information gleaned from the records of the Hudson's Bay Company.

The fur trade began in the mid-sixteenth century as the Frenchmen along the Gulf of St. Lawrence and the Atlantic Coast began bartering with the Indians. As the fur-bearing animals along the coast became depleted, the trappers and traders moved west, establishing posts and routes along the way to the west end of Lake Superior. Charles II granted some Britishers the exclusive rights to the use of Hudson Bay in 1670. For the next hundred years such names as Pierre Esprit Radisson, Sieur des Grosseilliers, Sieur DuLuth, and Sieur de la Verendrye were associated primarily with the fur trade. When the French forces were defeated at Quebec in 1759 and in Montreal in 1760, control of Canada passed into the hands of the British. The time of the small operator was coming to an end.

Lewis and Clark brought back tales of mountains teeming with beaver and a ready-made transportation system, the Missouri, to get there. Manuel Lisa, a Spanish trader, took a group of 42 "unruly trappers" up the Missouri and Yellowstone to the mouth of the Bighorn, where they built a timbered blockhouse. They spent the winter of 1807–1808 trapping, hunting, and roaming among the Indian tribes of the Northern Rockies.

Lisa came back to St. Louis with one conviction, that only large companies would be successful in the fur trade. Thus the Missouri Fur Company was organized in 1809. Other fur companies were also organized at this time; some survived, others did not. The most successful was that of John Jacob Astor, formed in 1808 but expanded in 1810 with the birth of the Pacific Fur Company. There would be not only fur-trading posts through the Rockies to the Pacific Coast, but also a headquarters at Columbia to handle the China trade!

In 1822 General William H. Ashley, with 43 men, established a trading post on the Yellowstone. Many of these men, such as Robert Campbell, Jim Bridger, William Sublette, Thomas Fitzpatrick, and Moses Harris, became famous in Rocky Mountain history and, together with Colter and others, are now known as the "mountain men."

But the high, prosperous days of the fur trade were coming to a close. By 1832 Hudson's Bay officials considered the region to the Rockies a fur desert. The production of silk had become feasible; people returning from the Far East were replacing their beaver hats with ones of silk, and the sheep of fashion in Paris, London, and New York followed. An extremely colorful era had all but ended.

But where official explorers had failed, fur traders had succeeded. They were probably too busy and too uneducated to seek fame by reporting their own explorations, but these adventurers penetrated every corner of the West between 1800 and 1850, paving the way for the settlers.

THREE ARCTIC EXPLORERS

Using the names of explorers in mammalian as well as in other taxonomy was a very common practice in the nineteenth century; a few such names will be mentioned briefly here. Certainly the excitement and drama of not only the western exploration in North America but also of polar Pacific exploration (i.e., searching for an arctic passage from the Atlantic to the Pacific) must have occupied the "headlines" of the day, when wars were not going on somewhere. Franklin's, Parry's, and Richardson's ground squirrels bear the names of a trio of British polar explorers. Sir John Franklin (1786–1847) explored the arctic from Hudson Bay to the mouth of the Coppermine River, then east along the coast and back to York Factory, in 1819–1822. The surgeon-naturalist of the expedition was Sir John Richardson (1787–1865), who was born in Dumfries, Scotland. The narrative of this expedition was published in 1823. In 1826 both Franklin and Richardson were again exploring the arctic coastline.

The third member of this trio of polar Pacific explorers was Sir William Edward Parry (1790–1855), who was also probing the arctic, making four trips between 1819 and 1829. All were searching for the "Northwest Passage." In the appendix to Parry's *Journal of a Second Voyage for the Discovery of a Northwest Passage . . . ,* which was published in 1824, Richardson described the mammals and birds. He also wrote much of the scientific information for Franklin's narrative of his second expedition. But Richardson is best known for his *Fauna Borealii Americana,* a government publication in which he described the quadrupeds and fishes; Swainson, the birds; and William Kirby, the insects. Ironically, Richardson's last journey to the arctic, in 1848, was in search of any remains that might give a clue concerning the ill-fated Franklin expedition, the earlier disappearance of whose members remains a complete mystery to this day.

THE BEGINNINGS OF MAMMALOGY AS A SEPARATE SCIENCE

During the frenetic activity of the Western expansion in the first half of the nineteenth century, mammalogy was evolving in the more settled Eastern part of the United States to occupy its own separate niche in science.

Bewick's *General History of Quadrupeds,* published in 1804, is probably the first American work on mammalogy. It is generally thought that George Ord (whose name does not appear in the volume) contributed the information on mammals.

In 1825 a friend of John James Audubon, Richard Harlan, published the first installment of *Fauna Americana* devoted exclusively to mammals. Sir John Richardson's *Fauna Borealii Americana* contained a volume on mammals, published in England in 1829. The following year John D. Godman's three-volume treatise on mammals, *American Natural History, or Mastology,* appeared, and in 1842 DeKay's *Zoology of New-York,* Part I, *Mammalia,* dealing only with mammals, was published (Fig. 1–1).

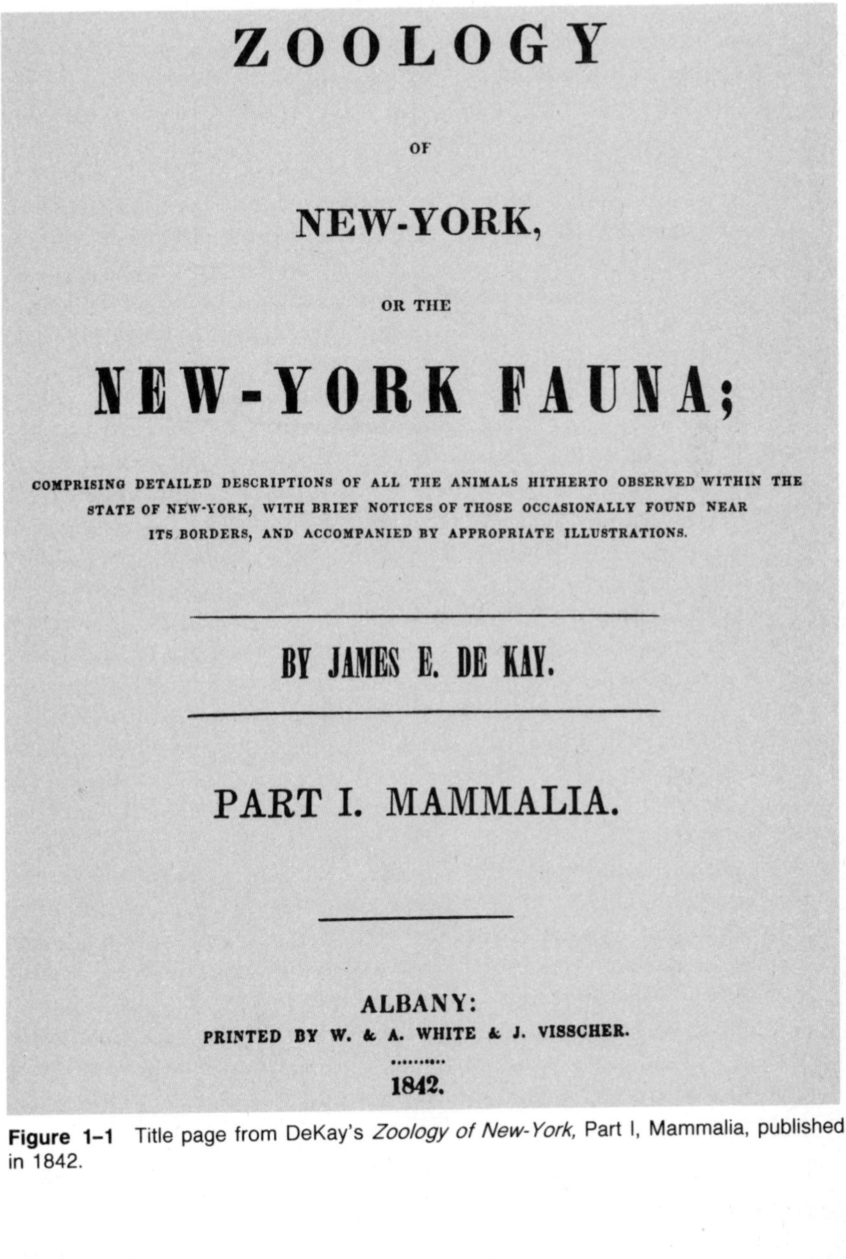

ZOOLOGY

OF

NEW-YORK,

OR THE

NEW-YORK FAUNA;

COMPRISING DETAILED DESCRIPTIONS OF ALL THE ANIMALS HITHERTO OBSERVED WITHIN THE STATE OF NEW-YORK, WITH BRIEF NOTICES OF THOSE OCCASIONALLY FOUND NEAR ITS BORDERS, AND ACCOMPANIED BY APPROPRIATE ILLUSTRATIONS.

BY JAMES E. DE KAY.

PART I. MAMMALIA.

ALBANY:
PRINTED BY W. & A. WHITE & J. VISSCHER.
....
1842.

Figure 1–1 Title page from DeKay's *Zoology of New-York,* Part I, Mammalia, published in 1842.

An important landmark in the mammalogic literature of North America was the publication of *The Viviparous Quadrupeds of North America,* from 1845 to 1854, by Audubon and Bachman. Dr. Bachman was a minister and a learned zoologist. Bachman and Audubon became friends in 1831 and began their work on the quadrupeds volumes in 1839. In 1840 Bachman wrote his friend, "The expenses and the profits shall be yours and the boys." Bachman was the father-in-law of Audubon's sons, John and Victor. The senior Audubon died in 1851, before completion of the work, so it fell to Bachman to write the systematic accounts and contribute to the text. The sons were to color the plates and arrange the editing and selling. The introduction describes the difficulties of vertebrate research of the time. But the work was completed and is available in many research libraries.

The elephant folio *Birds of America* of John James Audubon (1780–1851) so dominates our knowledge of him that we forget that he has also contributed much to mammalogy. He could be counted among the naturalists who explored the West, for in 1843 he made a journey up the Missouri River as far as Fort Union. But he spent most of his time in the Eastern part of the United States, especially along the lower Mississippi River (Louisiana) and the tributaries of the Ohio. His early life was a kaleidoscopic pattern of occupations, failures, and successes. Through it all he painted birds. In 1827 he went to Europe, where he took Liverpool, Edinburgh, London, and Paris "by storm," but the great event of the trip was that the Lizars of Edinburgh agreed to launch the *Birds of America,* a project which Havell of London completed.

Audubon's western journey, recorded in *The Missouri River Journals* (M. R. Audubon, 1897; Dover reprint, 1960), is replete with observations, descriptions, and measurements of birds and mammals, especially the larger mammals such as bison *(Bison bison),* elk *(Cervus canadensis),* pronghorn *(Antilocapra americana),* deer (*Odocoileus* sp.), mountain sheep *(Ovis canadensis auduboni),* and wolves. On May 28, 1843, he recorded an observation on the prairie dog *(Cynomys ludovicianus)* made by his friend: "Harris saw one that, after coming out of its hole, gave a long and somewhat whistling note, which he thinks was one of invitation to its neighbors, as several came out in a few moments." Would this be the first recorded observation of what King (1955) called the "all-clear" signal? On August 19, 1843, he wrote: "Wolves howling and bulls roaring like the long continued roll of a hundred drums." This would be the middle of the rutting season for the bison *(Bison bison);* the constant threatening and challenging of the bulls during the rutting season is an impressive sound. The "American bison" is illustrated in *The Viviparous Quadrupeds of North America.*

Audubon died in Audubon Park, now a part of New York City, in the only house he had ever owned. Lucy, his wife, survived him by some years and gave art lessons to George Bird Grinnell (among others), who founded what was eventually to become the National Audubon Society, was also one of the founders of the Boone and Crockett Club, and often contributed to the *Journal of Mammalogy.*

In his last years Audubon had met and influenced the young Spencer Fullerton Baird, the great American naturalist of the late nineteenth century and architect of American vertebrate zoology. On August 19, 1846, the Smithsonian Institution was established with the maintenance of a museum as one of its primary functions. The act establishing the museum specified "all objects of art and of foreign and curious research, and all objects of natural history." It was a most fortuitous circumstance that Baird served as Assistant Secretary from 1850 to 1878. He was an enthusiastic genius who inspired young collectors with his matchless knowledge of vertebrates. He was in a unique position to turn the nation's consolidation of empire to scientific profit.

PACIFIC RAILROAD SURVEYS

During the first half of the nineteenth century the need for a railroad to the Pacific Ocean had become more and more apparent. It was too huge an undertaking for private interests, and so the federal government became deeply involved. President Jefferson had found it useful to rely upon the scientific information of Lewis and Clark to implement his plans for the West. Thus when issues of public importance occurred, such as the location of the Pacific Railway, it became federal policy to seek answers in the objective judgment of science. The federal government needed to furnish protection and transportation to the fallout of settlers which was a by-product of the fur trade, the gold rush, and finally the Homestead Act of 1862.

The Pacific Railroad Surveys bill was passed in March 1853. It ordered the Secretary of War to submit a report to Congress on all practicable railroad routes from the Mississippi to the Pacific—within 10 months! Simultaneously the Army was adding to its already established system many more fortifications to protect settlers at the very ends of transportation and communication in the West. Congress appropriated $150,000 for field surveys to be made by parties under the supervision of the Topographical Corps, whose commandant was Col. John James Abert. The task was to make broad collections of information, including various aspects of natural history. This aspect was to be undertaken by medical officers of the Army. Geologists, astronomers, and cartographers were also included in these groups.

The Pacific Railroad Surveys did not furnish a conclusive report on the "most practicable and economical route for a railroad from the Mississippi to the Pacific Ocean." Rather, it showed that there were several practicable routes, and that the West had more to offer than had previously been assumed by the Washington policy makers. Ironically the first transcontinental railroad did not follow any of the projected routes.

If the surveys were a failure in finding one practicable route to the Pacific,

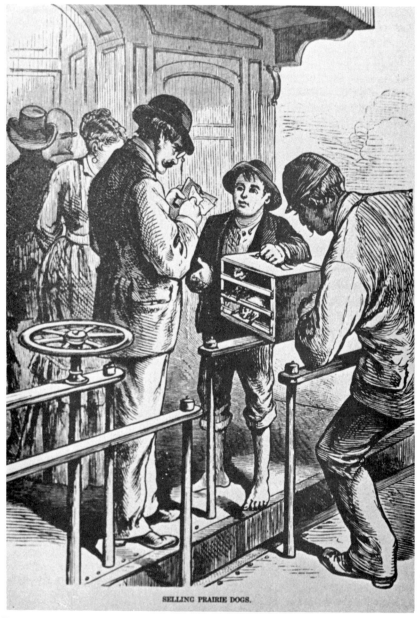

SELLING PRAIRIE DOGS.

Figure 1–2 Passengers on an overland train in 1877 buying prairie dogs, *Cynomys ludovicianus. (From Reinhardt, 1967, courtesy of the American West Publishing Co.)*

the 13-volume scientific report was a monumental achievement. One hundred and six scientifically trained men took part, either as field collectors or as museum classifiers. These museum classifiers did not accompany the field parties but worked in Eastern museums and schools. Baird, who, as mentioned earlier, was Assistant Secretary of the Smithsonian Institution, saw to it that detailed advice on collecting techniques as well as collecting equipment was made available. John Torrey of Princeton controlled the appointment of the botanists.

Separate botanical and geologic reports were included in the record of each expedition, and three of the later volumes were used for a zoologic resumé. There was no general index or inclusive table of contents. The pages were not numbered consecutively. "As a result," wrote Goetzmann (1959), "the Pacific Railroad Reports, though one of the most impressive publications of the time, were a little like the country they were intended to describe: trackless, forbidding, and often nearly incomprehensible."

The three separate resumé volumes, which included four monographs, one each on mammals, birds, reptiles, and fishes, came under Baird's supervision, though he wrote only three of them. The report on fishes was done by Charles Girard. One of the monographs, the monumental *Mammals of North America,* was published in 1859; in it Baird describes 738 species of mammals.

The Pacific Railroad Surveys reports were a summary of all the information gained by the Topographical Corps and its partners up to 1857. They included an immense amount of knowledge; furthermore, enough specimens were brought back to make the Smithsonian one of the world's great museums.

THE EMERGENCE OF MAMMALOGY AS A SEPARATE SCIENCE

In 1879 the United States National Museum, the first offspring from the parent Smithsonian Institution, was organized under Baird's guidance, with Dr. Elliott Coues (1842–1899) as curator of mammals. Baird seems to have been taking care of the mammal collection unofficially, for Coues was the first curator of mammals to be actually appointed. Coues's monograph of the Mustelidae, *Fur-bearing Animals,* published in 1877 is a classic in North American mammalian literature.

Coues was one of that group of nineteenth-century naturalists of the Medical Corps of the United States Army who did such a spectacular job of gathering both information and specimens from the outposts of the Western United States. The primary avocation of these men was ornithology, but their range of interests and competence usually included all the vertebrates. Coues was only 19 years of age when he published his first ornithologic work in the *Proceedings of the Academy of Natural Sciences, Philadelphia* in 1861. Two more articles appeared in the same year, just 2 years before he received his medical degree. It might be noted that the editor of the *Proceedings* was at that time another, older military surgeon and naturalist, Dr. Joseph Leidy. In 1864, when he was not quite 22 years of age, Coues was commissioned an assistant surgeon in the Army and was promptly sent West to New Mexico and Arizona. Captain Curtis wrote of him (in Hume, 1942, p. 57):

> Ornithology was the Doctor's special cult, but he was also prepared to make collections in other branches of natural history. For creeping, crawling, and wiggling things he had brought along a five-gallon keg of alcohol. But the reptilian branch of his researches failed utterly in the early stage of the march, for the soldiers, in unloading and loading the wagon had caught the scent of the preservative fluid, and though it already contained a considerable number of snakes, lizards, horned toads, etc., the stuff, diluted from their canteens, did not prove objectionable to the chronic bibulants. Some of them, however, did look decidedly pale about the gills when the head of the empty keg was smashed in and the pickled contents exposed to view. They had really supposed they had been drinking chemically pure alcohol.

Coues gave a total of 19 years to the military service of his country in New Mexico, Arizona, South Carolina, Maryland, Virginia, Dakota Territory, and Minnesota, as well as serving with the "U.S. Geological and Geographical Survey of Territories under charge of F. V. Hayden, U.S. Geologist." In 1881 he became secretary and naturalist of the Geological and Geographical Survey of the Territories and resigned his commission.

From 1884 to 1891 he supervised the fields of general zoology, biology, and comparative anatomy for the *Century Dictionary.* He furnished over 40,-000 definitions and supervised the preparation of hundreds of illustrations, drawn by E. T. Seton, who was to achieve a prominence of his own in the years to come.

Coues published 15 volumes on the early explorations west of the Mississippi; so thorough was he that he traveled many thousands of miles, making a personal examination of the route of Lewis and Clark and a canoe trip up the Mississippi to determine Pike's route to the headwaters of the river.

He helped organize the American Ornithologists' Union and was its first vice president for 19 years and president from 1893 to 1895. He also started an ornithologic bibliography, which he did not complete. He seemed to regard this as a thankless task and wrote, "Perhaps it is lucky for me that I was forcibly divorced from my bibliographical mania," and in another place, "It takes a sort of an inspired idiot to be a good bibliographer."

His writings were extensive. No complete bibliography exists, but it is believed that he contributed at least 500 titles to ornithology, 30 to mammalogy, 3 to herpetology.

Coues's monograph on mustelids has already been mentioned. His first paper on mammals, published in 1867, was a report on mammals collected in Arizona. He wrote five monographs on Rodentia, which with some reports by Dr. Joel Asaph Allen formed volume 4 of the Hayden Survey Monographs.

Coues was taken sick on a journey to New Mexico and Arizona, and was brought back to Johns Hopkins Hospital, where he died on Christmas Day, 1899. Certainly he stands as a giant among the naturalist-surgeons of the Army Medical Corps and the naturalists of the formative years of the scientific collections in this country.

Space does not allow as detailed a discussion of all the naturalist-surgeons of the United States Army Medical Corps, but a few more must be mentioned, even if only briefly.

One of these, who came later, was Edgar Alexander Mearns (1856–1916), again primarily an ornithologist but one who nevertheless contributed to other fields of vertebrate zoology, including mammalogy. After receiving his medical degree and while waiting for his commission in the Army, he stored his collection of specimens at the American Museum of Natural History and spent a winter there as a temporary curator of ornithology. At that time he established a collection in vertebrate zoology for the use of students. In *The American Museum of Natural History: Its History,* on p. 67, is this statement: "The first material for study collections was given by Dr. E. A. Mearns in 1882."

Mearns began his active duty with 4 years in Arizona, then went to Fort Snelling, Minnesota, in 1888, where he spent 3 years. He was again sent to Fort Snelling in 1903, in between having spent at least some time at the American Museum of Natural History describing new birds and mammals from Arizona. Mammalogists may remember that the type specimen of Mearns's cottontail *(Sylvilagus floridanus mearnsi)* is from Fort Snelling.

From 1892 to 1894 he was naturalist and medical officer of the Mexican –United States International Boundary Commission and sent back 30,000 plant and animal specimens to the United States National Museum! He was stationed at Fort Yellowstone and Fort Myer, had two tours of duty in the Philippines (which resulted in some impairment of his health), and was stationed at Fort Totten, New York.

In 1909 he retired from the Army, only to be "assigned to active duty by his consent" to accompany President Theodore Roosevelt's famous African expedition as collector. In Theodore Roosevelt's *African Game Trails: An Account of the African Wanderings of an American Hunter-Naturalist* (1927), the President often paid tribute to Mearns's collecting ability.

Before Mearns could finish a report of this trip, Childs Frick asked him to go to Africa with him. Two trips were made with Frick to Africa. While working on the African material, Mearns learned that he had diabetes. This was in the preinsulin days, and though he carried on for 2 or 3 years, he finally succumbed on November 1, 1916.

At the time of his death his contributions to the National Herbarium were greater than those of any other man. He also contributed approximately 7000 mammals, 20,000 birds, and 5000 each of reptiles and fishes to the National Museum. Of his 125 titles his most important work is considered to be *The Mammals of the Mexican Boundary of the United States*. The great mass of his Philippine notes has not been edited or published.

As already mentioned, space does not permit the enumeration of all the medicomilitary naturalists who served at far-flung posts, as well as with the Mexican Boundary Survey and the Pacific Railroad Surveys. Besides Baird, Coues, and Mearns, there were many such as C. E. Bendire, J. G. Cooper, B. H. Dutcher, J. F. Hammond, W. A. Hammond, R. W. Shufeldt, G. Suckley, J. Xantus, L. Wood, C. A. Wood, and H. C. Yarrow.

The work produced by these officers was largely inspired by Baird at the Smithsonian Institution. This has been called, by some, the Bairdian period in American natural history. His dreams were of cooperating, not rivaling, bureaus of government science. His wife was the daughter of the inspector general of the Army, and this connection was undoubtedly helpful in obtaining medical officers as naturalists and surgeons for the Western expeditions and railway surveys. One can imagine the grumblings of a recent graduate of a medical school back then who was tempted by the adventurous life at any Army post but whose interests did not include natural history. "Politics! had I wasted my time bird watching rather than studying medicine I might have gotten the job."

MAMMALOGY'S ATTAINMENT OF ACADEMIC STANDING

Mammalogy did not achieve separate status in the academic halls until the twentieth century was well under way. Prior to this, as already noted, various phases of mammalogy were covered in other disciplines, and most studies of mammals were concerned with economic aspects. One of the strongest forces that aided mammalogy to attain separate status was increasing numbers of collections in museums, including museums associated with universities. Those of Harvard and Yale were already established, to be followed by those of California, Michigan, Cornell, and many others. By far the most influential of the early ones was the Museum of Vertebrate Zoology of the University of California at Berkeley, as developed by Joseph Grinnell.

As in so many cases, the success of this institution resulted from the driving force of one man, a man who was primarily an ornithologist, but who was also very successful in training many mammalogists. It is probable that when the eminent paleontologist O. C. Marsh was intervening in Washington in behalf of Chief Red Cloud's people, an energetic small boy by the name of Joseph Grinnell was already a favorite of the chief's. Joseph Grinnell was born February 27, 1877, 40 miles from Fort Sill, Indian Territory, the son of the agency physician. In 1880 the elder Grinnell was an agency physician in the western Dakotas and Nebraska, where Chief Red Cloud's people were gathered, according to H. W. Grinnell (1940), who wrote, "Being the oldest, however, by five years, Joseph relied upon Indian companions for playmates. Undoubtedly his senses were quickened by association with these alert comrades." About 1890 the family moved to Pasadena, California, where they made their permanent home. Joseph Grinnell's college education at Throop Polytechnic Institute was interrupted by a collecting trip to Alaska during 1896. Another trip to Alaska followed his graduation from Throop.

Back in California he enrolled at Leland Stanford University while teaching at Palo Alto High School. At Stanford he became friends with Chester Barlow (then editor of *The Condor*), Walter K. Fisher, and Wilfred H. Osgood. In 1903 his work toward a doctorate at Stanford was temporarily interrupted by typhoid fever. During his recovery he became instructor in biology at Throop Polytechnic Institute. In 1913 he received his doctorate from David Starr Jordan, the renowned ichthyologist, who was then president of Stanford University. Joseph S. Dixon and Walter P. Taylor, each of whom became well-known mammalogists, were his students at Throop.

In 1908 Grinnell became associated with the University of California at Berkeley, where he developed their museum and introduced courses in vertebrate zoology. Graduate students did their thesis work on birds and on mammals.

Joseph Grinnell passed away May 29, 1939. His bibliography consisted of 554 publications, the first in 1893, the last in 1939. Nine birds and four mammals were named after him.

After the introduction of mammalogy courses into college and university curricula after the turn of the century, they increased in number gradually until after World War II, when their number exploded. In part this was simply because of an increasing student population. But a new field, natural resource conservation, started in the midthirties, with the publication of Aldo Leopold's (1933) *Game Management.* This helped to create an even greater demand for mammalogy courses. That the taxonomic, distributional, and natural history information on mammals has received a great deal of attention in the last half century is shown by the literature published. Miller and Kellogg (1955) listed 3622 species and subspecies of mammals in North America. Miller (1912) listed 2100. Seton's *Lives of Game Animals* (1929) provided a great deal of information on carnivores, ungulates, and some rodents. Because Seton apparently recorded every scrap of information he received, the information must be used with discretion. Although not restricted to North America, Max Weber's *Die Säugetiere* (published first in 1904, but the two-volume 1927 edition is the most easily available) described structure and function in mammals and included a worldwide systematic review of the class. The first two volumes, plus a later third on mammals in Pierre Grasse's *Traité de zoologie* (dealing with the entire animal kingdom), appeared in 1955. *Recent Mammals of the World,* by Anderson and Jones (1967), reviews the families. Harold

Anthony's *Field Book of North American Mammals,* published in 1928, was the first popular book on mammals and their identification; it followed the pattern of other identification manuals of the day. In 1952 Burt and Grossenheider published a pocket-sized field guide in the format and series edited by Roger Tory Peterson (2d ed., 1964). In 1954 Palmer's *The Mammal Guide,* containing both identification helps and concise natural history information, appeared. The extensive taxonomic work by Hall and Kelson on *The Mammals of North America* was published in two volumes in 1959, and Walker's ambitious three-volume work on *Mammals of the World* appeared in 1964 (2d ed., 1968). Two of the volumes contain illustrations and a description of each genus of mammals of the world, and the third volume is a very useful bibliography. Both fossil and recent mammals of the world, including geographic and geologic ranges, were listed down to genera in Simpson's (1945) *The Principles of Classification and a Classification of Mammals.* It has become an indispensable reference to mammalogists as well as paleontologists. Accounts of fossil mammals are also found in Osborn's (1910) *The Age of Mammals in Europe, Asia, and North America;* Scott's (1913) *A History of Land Mammals in the Western Hemisphere* (rev. ed., 1937); and Romer's (1966) *Vertebrate Paleontology.* The first mammal "text" was Hamilton's (1939) *American Mammals.* There are numerous regional and state publications as well as species [pronghorn *(Antilocapra americana),* coyote *(Canis latrans),* deer mouse, or whitefooted mouse *(Peromyscus)*] reports or monographs. Studies which fit neither category but represent a new and increasing trend are Darling's (1937) *A Herd of Red Deer;* King's (1955) *Social Behavior, Social Organization, and Population Dynamics in a Black-tailed Prairie Dog Town in the Black Hills of South Dakota;* Schaller's (1963) *The Mountain Gorilla: Ecology and Behavior;* and Mech's (1970) *The Wolf: The Ecology and Behavior of an Endangered Species.*

The wealth of mammalian literature which appeared in the last half century has dwarfed the total of all previously published literature.

MUSEUMS AND COLLECTIONS

Mammalogy has been greatly influenced by museums, which have provided a place for the storage of specimens as well as a place where qualified researchers could work with the specimens. The major museums and many university museums have mammalogists on their staff. Many colleges and universities without museums now have collections housed in the zoology, biology, or wildlife departments. A history of mammalogy would be incomplete without some discussion of museums.

The Smithsonian Institution and, to zoologists especially, the National Museum (a part of the Smithsonian) have become a storehouse of scientific information as a result of the mass of material that has reached their doors. Prior to this, most museums, at least science museums, were more for the display of curiosities than for the storage of scientific treasure.

Credit for being the first museum in the United States goes to Charleston Museum, Charleston, South Carolina, founded in 1773 (Katz and Katz, 1965) and still active. The Charleston Library Society established this museum in the American Colonies two years prior to Paul Revere's ride. At that time (1773) a committee was appointed to collect materials for "promoting a Natural History" of South Carolina.

Scattered throughout this country have been innumerable "museums" of arrowheads, fossils, Indian curios, or antiques, collected by some zealous individual. Soon such a collector's private passions leads to an exhibit of these articles in the living room, garage, or a specially constructed building. Some of these exhibits or collections are small and of little importance; others are carefully done and represent a wealth of material. Some are even profitable, while still others have been assembled with no intent to profit financially but only with a desire to share the findings, good, bad, or indifferent.

Such a collection was that of the celebrated artist-naturalist Charles Wilson Peale of Philadelphia. In 1785 he began the collection of natural history museum specimens illustrating each and all categories in Linnaeus's classification. He turned a wing of his home into an exhibition hall for both his paintings and his specimens, but in 1794 the collection became too big to be contained in his house. Peale was an active member of the American Philosophical Society, and in 1794 they invited him and his family to move to the New Philosophical Hall, where he had both gallery and living space, in return for his services as librarian and curator. But it took him only 8 years to outgrow this space. He then moved to the State House (now Independence Hall), where his exhibits remained for about 20 years.

Since Peale's museum was the "nearest thing to a National Museum then in existence," it was inevitable that his fellow naturalist, Thomas Jefferson, would assign to Peale's care all the material brought back by Lewis and Clark. Few of us probably realized, if we have toured and photographed Independence Hall (the "cradle of democracy"), that at one time it also housed undescribed mammal specimens unique to North America (Fig. 1–3).

Unfortunately Peale's organizational ability was evidently not inherited by his immediate descendants. After his death, funds were borrowed from the United States Bank to build a new museum. Both institutions failed, and the collections were sold to satisfy claims made upon the bank.

In early 1967 the Detroit Institute of Art had a showing of Charles Wilson Peale's art. *Time* (Feb. 24, 1967, pp. 68–69), in reviewing this showing of over 200 of Peale's paintings, wrote:

[He] taught his brothers, sons, daughters and eventually his grandchildren to paint. They in turn taught their children, thus founding the U.S.'s first dynasty of painters.

Just as Charles W. ruled over the family, so the show is dominated by the near-Olympian progenitor who completed more than 1,000 pictures and sired 17 offspring by three successive wives (he died at 85, busy courting a fourth). A man of plain-spun charm, he had fought and wintered at Valley Forge, painted George and Martha Washington, Hamilton, Franklin, Lafayette and many of the other great men of the day in a style renowned for its affable simplicity. Like his lifelong friend Thomas Jefferson, he was an enthusiastic naturalist and inventor, experi-

Figure 1-3 Specimens sent back by Lewis and Clark were housed for a time in Independence Hall, Philadelphia, then called State House, now administered by the National Park Service.

mented with everything from doorbells to apple-peeling machines. In 1786, he opened the nation's first natural-history museum [sic], run by the Peale family and displaying the reassembled bones of a mastodon they had unearthed in Newburger, N.Y., together with 100,000 other stuffed animals and objects.[1]

In New York, one of the first museums (which did not survive) grew out of the cultural ideals of an organization hardly remembered for its cultural ideals, the Society of St. Tammany. Established at the beginning of the nineteenth century, it survived for a while and was sold to P. T. Barnum. It was 10 years later that Barnum bought a part of Peale's Museum. These were the foundations of Barnum's sideshow empire of entertainment.

Peale's Museum in Philadelphia, though a very successful place of amusement, was not a place of service, study, and research. Those who were more seriously "disposed to the study of the laws of the creation" formed the Philadelphia Academy of Natural Sciences in 1812. A decision that religion and politics should not enter the discussions caused the young organization to be labeled atheistic. But the organization grew, and by 1828 the museum was opened to the public on Tuesday and Friday afternoons. By 1862 it was "acknowledged to be the finest on this continent," with 18 departments, which included ichthyology, herpetology, ornithology, and mammalogy. The first official display of Audubon's paintings was held there in 1824, and today Audubon Hall features exhibits of bird biology and natural history.

Harvard College began accumulating "curiosities" about the middle of the eighteenth century. Its early collections of "two complete [human] skeletons of different sexes" and scientific instruments were destroyed by fire in 1765. In 1794 a gift of 50 fossils was the start of Harvard's Mineral Cabinet. But it was Prof. Louis Agassiz who provided the genius which sparked the rapid development of Harvard's Gray Museum of Comparative Zoology, which opened in 1860. Agassiz, educated in Switzerland, came to teach zoology and geology at Harvard and was the most popular public speaker of his time. He had no trouble in obtaining funds, and when he proposed to publish *Natural History of Fishes of the United States* not only money but specimens began to arrive in Cambridge by the "barrelful."

Among his first students was Albert Bickmore, who became a dominating figure in the formative years of New York's American Museum of Natural History. Bickmore and other brilliant students left Agassiz over a disagreement on a matter so basic that it marked a milestone in the history of natural sciences, that of special creation versus evolution.

But when Darwin, in 1859, published his theory of evolution through natural selection, he unleashed a revolutionary idea. The opposing view was quickly marshaled by many leaders, among them Louis Agassiz. His stature lent great weight to the fundamentalist view of the lay public.

[1] Reprinted by permission from *Time, the Weekly Newsmagazine;* Copyright Time Inc.

As Darwinism prevailed among scientists, religion became less and less a motive for the study of natural history. Actually it can be said that at this time the interest of museums changed from the field to the laboratory, or, more picturesquely, from "skin-out" biology to "skin-in" biology.

As a result of Bickmore's disagreement with Agassiz, he left Harvard and traveled extensively in Asia and Siberia. During these travels he carried with him two items: One was a sketch of a plan for a museum in New York, a dream he had evidently started even when a student at Harvard. The other, surprisingly, was a Bible. When he returned from Siberia, he sought the help of William E. Dodge, Jr., a wealthy amateur conchologist, and Theodore Roosevelt (father of President Theodore Roosevelt) as well as Charles Dana (publisher of *The New York Sun*), J. P. Morgan, the City Council, and other men of affluence and authority. Thus began the American Museum of Natural History in 1869. Stevens (1959) described it as "a splendid castle in what was then a wilderness. Only shanties and two or three modest houses were in the area." Hellman (1969) has also written a history of the American Museum. Exhibited within this "castle" was the museum's first major acquisition, a purchase of Prince Maximilian of Weid-Neuweid's collection of mammals, birds, reptiles, and amphibians. The teaching of natural history in museums began to flourish with Bickmore's introduction, in 1880, of an educational program which included yet another pioneering idea, the use of lantern slides. It was also at this museum that exhibits according to habitat group originated.

One cannot escape the feeling that a sense of the dramatic combined with a feeling for showmanship must have been a prime requisite for any staff member in the early days of the American Museum. Such men as Bickmore, Samuel Chubb (whose rearing skeletons of a horse and a man have become the museum's emblem), Carl Akeley with his pioneer movies of Africa, and Roy Chapman Andrews of Gobi Desert fame would certainly seem to support this hypothesis.

J. A. Sheen was made curator of the department of ornithology and mammalogy on May 1, 1885. By 1907 ornithology and mammalogy were divided; Roy Chapman Andrews became an assistant in the mammal section in 1908 and assistant curator in 1911. Other mammalogists early associated with the American Museum were H. E. Anthony, T. Donald Carter, G. H. H. Tate, and John Eric Hill. During its 80 years, the department of mammalogy has had only five department heads.

The beginnings of the National Museum in 1879 have already been briefly noted. It is of course only one branch of the Smithsonian Institution (started in 1846). The Smithsonian has often been called the "nation's attic." In 1826 James Smithson, an Englishman who had never visited the United States, made out his will to include both John Fitall, formerly his servant, and a nephew. Should his nephew die without leaving any heirs the money should be used "to found at Washington, under the name Smithsonian Institution,

an establishment for the increase and diffusion of knowledge among men."
His nephew died within 3 years of Smithson's death and left no heirs. Kellogg
(1946) wrote a brief history of the mammalogists who had served on the staff
of the United States National Museum (Smithsonian Institution). Besides
Baird, whose leadership has already been discussed, there were Harrison
Allen, Elliott Coues, Frederick True, Gerrit S. Miller, Jr., Leonard Stejneger,
Marcus Ward Lyon, Jr., and still others.

The Chicago Academy of Sciences museum, established in 1857, was
known as the "first museum of the West" (Beecher, 1958). Robert Kennicott,
the first director of the Academy, was a protégé of Spencer Fullerton Baird.
Kennicott was leader of the Overland Telegraph Expedition, and it was his
scientific reports on Russian Alaska that spurred federal government officials
to become interested in the purchase of Alaska. To mammalogists he is best
known for his *The Quadrupeds of Illinois, Injurious and Beneficial to the
Farmer* (1857, 1858, 1859). The great Chicago fire of 1871 was the second fire
to destroy an academy site. The museum is now located in Lincoln Park, along
Lake Michigan's shore on Chicago's Near North Side.

Nearly a half century later another museum, the Field Museum of Natural History, was established, also in Chicago. Anthropologic, botanical, geologic, and zoologic exhibits had come to the World's Columbian Exposition
in 1893, from all parts of the world. The Field Museum was organized around
these four divisions of the natural sciences and eventually was housed in a
building at the south end of Grant Park. The museum cost Marshall Field
a little over $9 million. A pair of elephants mounted by Carl Akeley, who spent
12 years at the museum, commands the attention of visitors when they first
enter the museum. The zoologic division started with the exhibits purchased
from Henry A. Ward (founder of Ward's Natural Science Establishment in
Rochester, New York) for the World's Columbian Exposition. Mammalogists
associated in the past with the Field Museum include Girard Elliott (primates),
Wilfred Osgood, (deer mice), and Colin Sanborn (bats).

The gold rush of 1849 began the rush to the West. In 1853 the California
Academy of Sciences, an institution whose officials have been very sympathetic
to mammalogy, was founded in San Francisco. Its first president, Dr. Andrew
Randall, was fatally wounded by a gambler, Joseph Heterington. The academy
held its third annual meeting on the day between Randall's funeral and Heterington's hanging. This was not the only catastrophe to strike this museum.
In 1907 when their schooner *Academy* returned from a collecting trip to the
Galápagos Islands, the explorers found the museum had been leveled by
earthquake and fire. The collection from the Galápagos Islands formed the
nucleus of the new museum, whose home since 1916 has been in Golden Gate
Park.

Periodic reports on collections of recent mammals have been published
by Howell (1923), Doutt, Howell, and Davis (1945), Anderson, Doutt, and

Findley (1963), and the most recent, by Choate and Genoways (1975). These reports reflect the historical development and growth of the collections.

The number of private collections has decreased, especially from 1943 to 1973, and the number of public collections has increased. In the 20-year period between 1943 and 1962 the percentage increase of collections and number of specimens was 20 and 69; in the 10-year period between 1963 and 1973 the figures were 31 and 60 percent. Eighty-four percent of the presently available specimens are in 38 North American collections of 10,000 or more mammals. Each of the following seven institutions contains over 100,000 specimens: National Museum of Natural History and Bird and Mammal Laboratories, 475,000; American Museum of Natural History, 240,000; University of California at Berkeley, 144,000; University of Kansas, 132,000; University of Michigan, 111,800; and Field Museum of Natural History, 110,000

FEDERAL GOVERNMENT AGENCIES

Department of the Interior

U.S. Fish and Wildlife Service   The government agency most closely associated with wildlife, including mammals, is what is now known as the U.S. Fish and Wildlife Service. The first appropriation for what eventually evolved into the U.S. Fish and Wildlife Service was made in 1885 to what was then called the Branch of Economic Ornithology in the Division of Entomology. Dr. Clinton Hart Merriam (1855–1942) became the first director. For a good biography of Merriam see Sterling (1974). Dr. Merriam was born in upstate New York, finished medical school in 1879, and practiced medicine for 6 years in Locust Grove, New York. He was also influential in founding the American Ornithologists' Union in 1883 and was its first secretary. When he became Director of the Branch of Economic Ornithology, he gave up his medical practice and added another medical student, Dr. A. K. Fisher, as assistant ornithologist. In 1886 the branch became the Division of Economic Ornithology and Mammalogy. In 1905 the branch changed its name to the Bureau of Biological Survey; it operated under that name until June 30, 1940, when its name was changed to the U.S. Fish and Wildlife Service. Although that name has remained, there was a reorganization of the U.S. Fish and Wildlife Service in 1956, and it now consists of the Bureau of Commercial Fisheries and the Bureau of Sports Fisheries and Wildlife.

Merriam tended to subdue the economic aspects of the division in favor of other studies, such as geographic distribution and food habits. He also inaugurated the North American Fauna series of monographs and authored the first 11, from 1889 to 1896. Many prominent mammalogists rose to fame under Merriam's guidance. One of these, Vernon Bailey, a farm boy from Sherburne County, Minnesota, not only was one of his outstanding students but also became his brother-in-law. Florence Augusta Merriam became Florence Merriam Bailey and is probably best known for her book *Handbook of Birds of the Western United States* (1924). Osgood (1943) wrote, "Among those with whom [Merriam] was in touch, about 1883, was a farm boy from Elk River, Minnesota, Vernon Bailey by name." This boy's specimens were so well prepared and included so many species thought to be difficult to obtain that Merriam gave him exceptional encouragement. It is related that he once asked Bailey for specimens of shrews, tiny mammals of nocturnal and secretive habits then supposed to be rare. Bailey replied, "How many do you want?" At that time specimens of shrews were mainly something the cat brought in, something that fell in the well, or something found dead and decayed in the road, so Merriam then wrote Bailey, "All you can get." Some time later, Bailey sent him no fewer than 60 shrews, and it is not unlikely that then and there Merriam envisioned the possibilities of a continental campaign of mammal collecting. Others who at one time or another collected specimens under Merriam's direction were E. W. Nelson (1855–1934), later to become chief of the Biological Survey (1916–1927); E. A. Goldman (1873–1946), especially known for his work on the United States–Mexican Border Survey; and W. H. Osgood (1875–1947) of *Peromyscus* (deer mouse) fame. Dr. Osgood's (1909) outstanding work on the taxonomy of *Peromyscus* was published as *North American Fauna,* number 28. Osgood worked for the Biological Survey from 1897 to 1909, when he joined the staff of the Field Museum, where he became director of zoology.

Under Merriam's able leadership as a mentor and as an administrator, great advances were made in an understanding of the fauna of the United States and North America. His own investigations were also of great importance, and he is perhaps best known, or at least most widely known, for his *Merriam's Life Zones,* a work on altitudinal distribution done in the San Francisco Mountains of Arizona. Although there are valid objections to his use of temperature summations, the life zone categories, somewhat revised, are still useful. Merriam is also the recipient of some dubious fame because of his overzealous naming of restricted populations of grizzly bears, 69 species.

Although the economic activities had been somewhat subordinated, a congressional hearing in 1907 brought economic investigations back to prominence. Quite often these investigations took the form of food habit studies in an attempt to determine whether a species was "injurious" or "beneficial." Control methods were instituted, especially for the destruction of "animals injurious to agriculture," which in turn led to the establishment of a Bureau of Predator Control. The activities of this bureau were the subject of great controversy, especially in the 1930s. The bitter debate between naturalists and some government people over use of poison to reduce predators and rodents came to a sharp focus in the early 1930s. The publication of the book *Economic Mammalogy* by Henderson and Craig in 1932 shed some light on the problem, which has not been completely solved yet, although more enlightenment has come with the passage of time, as witness the banishment of the poison 1080 by Congress in 1972. Today the line between "injurious" and "beneficial" animals and "game" and "nongame" species is not so sharply defined. In 1962 a reorganization changed the Bureau of Predator Control into a Division of Wildlife Services, which is concerned not only with control but with methods of control. Much of the earlier controversy centered around the use of poisons and bounty payments. Research is now pointing the way to more sophisticated population control methods such as Balser's (1964), which essentially is the application of birth control methods to coyote populations!

Until the 1930s most state conservation agencies had been concerned with legal protection and enforcement, but during the drought years the nation was shocked by the tragic loss of wildlife and natural resources caused by the drought and resulting dust storms. One of the imaginative solutions proposed by two federal legislators was the Federal Aid in Fish and Wildlife Restoration Act passed in 1937. It is more commonly known as the Pittman-Robertson Act, after the two legislators who proposed it and brought it to realization. This act provided income from an excise tax on small arms and ammunition earmarked for wildlife restoration work and allotted to the states and territories. As a result research work undertaken by state agencies has greatly increased. Though most of it has been directed toward the management of game species, some of it has also been directed at nongame or "subgame" species such as the black bear *(Ursus americanus)* and timber wolf *(Canis lupus)*.

The national wildlife refuges of the U.S. Fish and Wildlife Service, established for the protection of game species, have become increasingly oriented to use as recreational areas. Chincoteague, Blackbeard Island, Big Pine Key, and Fort Niobrara are a part of a far-flung system of federal refuges visited each year by increasing numbers of wildlife photographers, nature enthusiasts, and just plain tourists. Federal refuges total over 300 and include more than 29 million acres.

National Park Service Another federal agency concerned with mammals is the National Park Service, also in the Department of the Interior. The idea of a national park was originally conceived around a campfire in what is now Yellowstone National Park. Here on the night of September 19, 1870, the members of the Washburn-Langford-Doane expedition sat talking around the campfire about the land of northwestern Wyoming through which they had been traveling. Cornelius Hedges, a Montana judge, thought there should be no private ownership of these vast, dense forests, waterfalls, geysers, and abundance of wildlife. On March 1, 1872, President Grant signed a bill establishing Yellowstone National Park.

John Muir's inspiration led to the establishment of Sequoia, Yosemite, and what is now King's Canyon National Park. On August 25, 1916, the National Park Service as we know it now came into being to "conserve the scenery and the natural and historical objects and the wildlife." In 1962 there were 192 units of over 26 million acres in 14 categories (e.g., national monuments, national recreational areas, national historic sites). As far as the mammals are concerned, the national parks serve several functions: They serve as a refuge for vanishing species, they serve as outdoor laboratories for research by both park personnel and cooperative agencies, and they serve to interpret our environment to a great number of people.

Bureau of Outdoor Recreation   More recently (April 2, 1962) the Bureau of Outdoor Recreation was established in the Department of the Interior. Its functions include administering a program of financial assistance grants to states and participating directly in the planning, coordination, and establishment of policies related to recreation and fish and wildlife benefits on public lands.

Department of Agriculture

National Forests   Although not primarily connected with wildlife, the lands of the national forests serve to some extent in the same way as the lands of the national and state parks and the federal and state refuges. Forest reserves were set aside between 1891 and 1905 under the Department of the Interior. In 1905 the Forest Service was established in the Department of Agriculture. There are now 154 national forests and 18 national grasslands. These together with other minor acreages total 186 million acres. In these forests wildlife is managed as a renewable crop.

The parks, wildlife refuges, and national forests give much-needed and sometimes the only protection to bighorn *(Ovis canadensis)*, bison, or buffalo *(Bison bison)*, cougar *(Felis concolor)*, fisher *(Martes pennanti)*, kit fox *(Vulpes macrotis)*, lynx *(Lynx canadensis)*, and timber wolf *(Canis lupus)*. Most of our remaining samples of rare mammals are found in the parks. About a third of the grizzly bears *(Ursus horribilis)* in the United States are in Yellowstone and Glacier parks. A few manatees *(Trichechus manatus)* are protected in the Everglades Park. The bison *(Bison bison)* is one of our highly prized and colorful relics of the recent past, and certainly is not compatible with our present agricultural practices and high human populations. The parks and refuges are the only places where the bison can exist at present.

STATE AGENCIES

Paralleling the federal agencies are state agencies concerned with forests, parks, and wildlife. Originally most of these state agencies were concerned with law enforcement or preservation. Two events of the 1930s dramatically changed all this. The first was the publication of Leopold's (1933) *Game Management,* which led to the establishment of wildlife departments at many universities including the Cooperative Wildlife Units. These research units represent the combined efforts of the Fish and Wildlife Service (which furnishes the unit leader), land-grant colleges and state conservation departments (which furnish housing and other facilities), and the Wildlife Management Institute (which primarily provides funds). The second event was the already mentioned Federal Aid in Wildlife Restoration Act of the United States Congress, in 1937. Much of the research work of state game biologists is relevant to mammalogy. Life histories, aging techniques, habitat requirements, and reproductive data of the larger mammals, including predators, are a few examples of the kind of information now emanating from state agencies. Funds from state game departments have partly or entirely supported the publication of state mammal lists—for instance, those of Kansas and Pennsylvania.

JOURNALS MAINLY DEVOTED TO MAMMALS

The American Society of Mammalogists was founded in Washington, D.C., on April 13, 1919. Its objectives were declared to be:

> the promotion of the interests of mammalogy by holding meetings, issuing a serial or other publication, aiding research, as may be deemed expedient.
>
> One of the principal objects of the society is the publication of the Journal of Mammalogy. It is aimed to make this journal indispensable to all workers in every branch of mammalogy and of value to every person interested in mammals, be he a systematist, paleontologist [sic], anatomist, museum or zoological garden man, sportsman, big game hunter, or just plain naturalist.

Ernest Thompson Seton drew the pronghorn, used on the cover for many years. The report of the organizational meeting was contained in the "Editorial Comment" section of the first *Journal of Mammalogy,* volume 1, number 1, pp. 47–48 (November 1919), and was signed with the initials "N. H.," obviously Ned Hollister, the editor. C. Hart Merriam was the first president. E. W. Nelson and W. H. Osgood were the first vice presidents, H. H. T. Jackson, corresponding secretary, and Walter P. Taylor, treasurer.

The society produces three publications: *The Journal of Mammalogy,* four times a year, a Special Publication series originally called the Monographs, published at irregular intervals, and a new series, Mammalian Species, also published at irregular intervals. The "Literature Review," formerly a part of the *Journal,* has grown until it has become necessary to publish it separately. Indexes for the *Journal* were published in 1945, 1952, and 1961. The society grew from about 443 members in 1920 to 3533 in 1973. Papers presented at annual meetings increased from 20 in 1920 to over 100 in 1973. Storer (1969) has reviewed the history of mammalogy during the life of the society, and Hoffmeister (1969) has reviewed the history of the society.

With the founding of the American Society of Mammalogists and the launching of a journal devoted to mammals, it might be said that the field of mammalogy had reached the status of a separate science. In Germany the *Zeitschrift für Säugetierkunde* was published from 1926 to 1943, and in 1953 the *Säugetierkundliche Mitteilungen* began publication in Germany. In France *Mammalia, morphologie, biologie, systématique des mammifères* was established in 1936. In Poland, *Acta Theriologica,* with the text in English, French, German, and Russian, was started in 1953. In Belgium, the Netherlands, and Luxembourg (Benelux countries) the Vereniging voor Zoogdierkunde en Zoogdierbescherming began publishing the journal *Lutra* in 1959 in Leiden, Holland. The Mammalogical Society of Japan started a *Journal* in 1952. A degree of specialization is seen in two recently established journals, *Primatology,* started in 1957, and *Folia Primatologica,* started in 1963. Although we may assume that mammalogy has reached the status of a specialized field, it is paradoxical that the last *Ulrich's International Periodicals Directory* (1973) has a section devoted to ornithology but none, as yet, devoted to mammalogy.

There are of course many journals other than these that deal with the ecology, natural history, behavior, physiology, and management of mammals. Among the foremost of these are *Animal Behaviour,* started in 1953; the *American Midland Naturalist,* started at the University of Notre Dame in 1909; *Behaviour,* started in 1947 with English and German text; *The Canadian Field Naturalist,* which had its start in 1887; *Ecological Monographs,* started in 1931; *Ecology,* in 1920; *Evolution,* in 1947; *Journal of Animal Ecology* (Britain), in 1932; and *Journal of Wildlife Management,* in 1937.

There are regional journals in North America that are not entirely devoted to mammals but nevertheless contain much information about them. Among the most prominent of these are the *Atlantic Naturalist,* started in 1945 and concerned with the Central Atlantic states; the *Murrelet,* begun in 1920 and dealing primarily with the birds and mammals of the Pacific Northwest; and the *Southwestern Naturalist,* started in 1956.

In addition to the journals there are many series of publications published by museums, government agencies, and research organizations. Walker et al. (1964) has included a selected list of these in volume III of *Mammals of the World.* More detailed lists can be found in *Ulrich's International Periodicals Directory* (1973), and *World List of Scientific Periodicals.*

A REASSESSMENT AND NEW DIRECTIONS

This history has been but a sketch of the circumstances, events, and men that have contributed to forging the science of mammalogy. In a general way the early history of mammalogy repeats the history of other biologic sciences in the *-ology* category. At first some attempt is made at gathering and recording the available knowledge. Next an inventory is made of the kinds of mammals and their relationships—classification. This phase still continues with newer and more sophisticated methods, but now research on mammals has branched out into physiology, ecology, behavior, and innumerable other areas.

Evolutionary studies of the future may not always agree with those of the past. It has been said that the snap trap, first marketed in 1887, made modern mammalogy possible. Traps made it possible to prepare big series of small mammal specimens, each skin with a label as to locality, date taken, size, sex, age, measurements, habitat, and other pertinent data. Descriptions of new species and subspecies were based on these series. A classic example of the results of this kind of work, one which has survived much critical testing, is Osgood's 285-page monograph on the genus *Peromyscus* published in 1909 as *North American Fauna,* number 28. Taxonomic and evolutionary studies of the future, using such modern techniques as electrophoresis, oscillographs, and karyotypes, may not always substantiate earlier works, but the results of some earlier studies survive remarkably well. One must admire the early taxonomists who had only limited techniques at their disposal.

Radiotelemetry and radioactive isotopes are opening whole new areas of study in population dynamics and behavior. The lives of rhinoceroses (several genera), elephants (especially *Loxodonta africana*), polar bears *(Ursus maritimus),* mountain lions, or cougars *(Felis concolor),* and timber wolves *(Canis lupus)* were not very visible to the early field naturalists. New methods and new equipment, not always meant for research, have bared the lives of the larger mammals to scientists using planes, snowmobiles, and radio transmitters and receivers.

The dart gun has been a boon to researchers. Now it is possible to determine age, gather data on temperature, pulse, and blood condition on live animals, even to transmit this information after the individual animal has regained consciousness. Even the underwater love life of seals has been televised!

Interdisciplinary studies are breaking down many barriers. To what discipline can one assign the recent mass of literature on primate behavior? DeVore (1965) wrote:

> In less than a decade, field studies of non-human primates have multiplied at an almost unbelievable rate, and today there are well over 50 individuals from at least nine countries engaged in such studies. Equally remarkable has been the multidisciplinary background of the field investigator including people with training in physical and social anthropology, experimental and comparative psychology and zoology.

There are many exciting avenues in the future for any young person interested in mammalogy.

The Evolution of Mammals

Mammals, as characterized today, have been evolving since the Late Triassic or Early Jurassic geologic period of 180 million years ago. But the evolutionary path leading to mammals started as soon as life on earth began.

THE BEGINNINGS

If life started in the sea, then a giant first step toward mammals would be adaptation to a terrestrial life. This was evidently accomplished by the air-breathing fishes—the lungfishes and the crossopterygian fishes. It is their ability to breathe air which suggests their intermediate position between fishes and land-living vertebrates. The lungfishes living today in Africa, Australia, and South America are not descendants of those considered in the direct line of the progenitors of the amphibians. The direct line toward the early amphibians was probably through the crossopterygian group of fishes, whose single proximal bone in the paired fins is equated with the upper limb bone (humerus in the forelimb and femur in the hind limb) of the land-living vertebrate tetrapod. The next two bones of the fish fin are equated with the radius and ulna of the forelimb and the tibia and fibula of the hind limb. Crossopterygians probably came out on land during the Late Devonian period, a period whose characteristic fauna were fishes.

The advance from fish with gills for respiration to air-breathing fish was a necessary corollary for these early vertebrates. An advanced air-breathing fish soon evolved into an amphibian, which opened up vast new unoccupied areas for these early vertebrates. The Devonian was a time of marked continental sedimentation. It was also a time when plants established themselves on the land, and by Late Devonian times primitive forests provided homes for the ancient insects and other invertebrates. Thus one of the adaptations in mammalian evolution, that of air breathing, began as early as the Devonian period.

Another problem facing animals living out of water was desiccation. Most of the early amphibians never went far from water and returned to it frequently, especially for reproduction, as most modern amphibians do. Fossil evidence indicates that early amphibians retained the scales that had developed in their tough skin. As this outer covering became more impervious, the amphibians became more independent of their aquatic environment.

Without the constant physical support of a water environment, the early amphibian developed a strong backbone and strong limbs, which served not only as a counterforce to gravity but also as a new method of effective locomotion. One problem the amphibians did not solve was the ability to reproduce without moisture.

Thus the basic pattern for a terrestrial life evolved during the Late Devonian period, and during the Carboniferous period, when amphibians were the characteristic fauna.

During the Carboniferous period, provisions for air breathing were evolved further. The external nares were located on the dorsal surface of the skull, rather than more ventrally, and the internal nares became separated from the external nares. Thus there was a well-defined passage leading from the external nares into the throat. This, we shall see later, was an important step toward the development of efficient provision of oxygen, helpful for both warm-bloodedness and great activity. The upper part of the gill arch had also been transformed from a bone which was originally a part of the gill arch to a fish hyomandibular, later becoming a bone to hold the jaws and brain case together, and finally a part of the future ear.

An evolutionary event of profound consequences for the appearance of mammals as well as most other vertebrates was the advent of the amniote egg. The first fossil amniote egg was found in the Early Permian of Texas. This development liberated the terrestrial vertebrates from some of their dependence on water. The egg is fertilized within the animal's body, as opposed to fertilization of free eggs in the water. This amniote egg is sometimes retained within the oviduct of the female until hatched, or the egg can be deposited externally. The egg contains enough nourishment, in the yolk, to nourish the developing young. There are two sacs within the egg. One, the *amnion* filled with liquid (which has been equated with a marine environment by some), contains the embryo, which could not survive without a liquid medium; the other, the *allantois,* receives the waste products of the developing embryo, and for this reason some scientific wags refer to it as a "primitive privy." The amniote egg, along with internal fertilization, which came much later, released animals from still more of their dependence upon an aquatic environment.

The first fossil amniote eggs were found in the red beds of Texas, a formation originally dated as Early Permian. More recent information (fossil plants) may show these beds to be older.

During the Permian an animal appeared which for a long time was held, because of its postcranial skeletal characters, to be a connecting link between amphibians and reptiles. This fossil was found near Seymour, Texas, and bears the likely name of *Seymouria.* However, it is now considered by some paleontologists to be an advanced amphibian with reptilian origins that must lie somewhere within the family of the solenodonsaurids (Carrol, 1969). This family retains a number of primitive features which may be considered as either reptilian or advanced amphibian. As yet, the origin of the reptiles is not precisely known.

THE REPTILES

Of the many subclasses of reptiles, the subclass Synapsida, containing the order Therapsida, held the precursors of modern mammals. The synapsids, which appeared during the Late Carboniferous, are characterized by a single lateral temporal opening, bounded above by the postorbital and squamosal bones, and two coracoid elements in the shoulder girdle. The gap from primitive reptile to primitive mammal was bridged between the end of the Pennsylvanian period and the close of the Triassic.

The synapsids were quadrupedal reptiles with an early trend toward differentiation of teeth into incisors, canines, and cheek teeth. The shoulder girdle contained two coracoid elements, the primitive coracoid, or *procoracoid,* still present in monotremes, and a new bone behind it called the *coracoid.* In general there was an evolutionary trend toward a more actively mobile way of life.

In the Permian the changing land surfaces and diverse and cooler climates, including seasonal changes, provided an opportunity for much experimentation and was the beginning of the Age of Reptiles.

The order Therapsida, of Middle and Late Permian and Triassic age, are known from all continents, but their remains are best known from the karroo sediments of South Africa. Some of the mammallike reptiles approached mammals very closely. Of the three Therapsid reptile groups, the dinocephalians, the dicynodonts, and the theriodonts, the theriodonts developed most rapidly toward the mammals.

The joint between the reptilian lower jaw, which consists of several bones, and the skull is formed by the articular bone of the lower jaw and the quadrate in the skull. Recent (modern) mammals have only one bone in the lower jaw, the *dentary,* and this connects with the skull by way of the squamosal bone. We have already noted that the upper part of the gill arch had earlier moved into position to become a part of the mammalian ear. The articular and quadrate bones of the lower jaw of the reptile also moved into the inner ear to form eventually the malleus (articular) and the incus (quadrate) of the mammal. Remarkable as this may seem, the fossil evidence supports this change.

The differentiation of teeth, the construction of the palate (roof of the mouth), the rerouting of the air around the mouth cavity, the suggestion of a muscular diaphragm (because the rib cage is restricted to the forepart of the

Table 2-1   Geologic Time Divisions

Era	Period	Epoch	Approximate beginning time, millions of years	Remarks
Cenozoic Age of Mammals	Quaternary	Recent	0.006	
		Pleistocene	2	
	Tertiary	Pliocene	10	
		Miocene	23	Appearance of grasslands and associated mammals
		Oligocene	34	
		Eocene	50	
		Paleocene	70	Beginning of great diversification of mammals
Mesozoic Age of Dinosaurs and Reptiles	Cretaceous		135	First eutherian and metatherian mammals
	Jurassic		180	
	Triassic		200	The oldest known fossil mammals from Late Triassic of Europe
Paleozoic Age of Invertebrates	Permian		270	
	Pennsylvanian		320	The first mammallike reptiles, the Synapsida (extinct)
	Mississippian		350	
	Devonian		400	
	Silurian		430	
	Ordovician		490	First vertebrates
	Cambrian		600	
	Precambrian		4500	

axial skeleton), and the possible presence of turbinal bones in the nasal cavity all suggest a much higher metabolic rate than that in the usual reptile.

Cynognathus was a genus typical of theriodonts. It was about the size of a wolf or large dog and was vaguely doglike—hence the name. It had powerful muscles for closing the jaws. The mandible of the dentary was so large as to form almost the entire lower jaw. The teeth were differentiated and highly specialized, and the well-developed canines suggest predatory habits. There was a gap between the canines and the cheek teeth. The cheek teeth were adapted for cutting and crushing the food, an advance from swallowing food whole as reptiles do. A well-developed secondary palate separated the nasal passage from the mouth. The skull articulated to the backbone by a double condyle as in mammals. The articulation of lower jaw to skull was still reptilian, but the articular bone of the lower jaw and the quadrate bone of the skull were getting very small. Whether *Cynognathus* had hair or not and whether it was warm-blooded or not are debatable questions, but these possibilities exist. Hopson and Crompton (1969) believed that these larger and later *Cynognathus* cynodonts were too modified in skull and dentition to be on the direct line to a mammalian group. Nevertheless, they are of great interest because they show a parallel development to a contemporary and poorly represented lineage that was leading to mammals.

Early Triassic cynodonts belonging to the family Galesauridae have been proposed as more direct precursors of mammals. The best known of these is *Thrinaxodon,* whose masseteric fossa extends to the lower border of the dentary, indicating an increased complexity of jaw musculature. In addition the crown pattern of the replacement teeth of *Thrinaxodon* resembles closely that of *Eozostrodon.*

MESOZOIC MAMMALS

At the present time, scientists do not agree on the details of the transition from reptile-mammal to mammal, but newer findings are clarifying the picture. The first groups of animals that fit the mammalian mold were Late Triassic and Early Jurassic European Symmetrodonta and the Morganucodontidae. Pantotheria and Triconodonta appeared in the Late Jurassic deposits of Europe and North America. Triconodonts, symmetrodonts, multituberculates, and unallocated therians are known but are not common in the Early Cretaceous. The multituberculates persisted during the Late Cretaceous when the metatherians and eutherians made their first appearance. The fossil record of these three periods, which together are called the *Mesozoic era,* is incomplete. Nevertheless, it is this record which outlined the origin or origins of mammalian lines from the therapsids and which shows that most of the basic evolution of the class Mammalia occurred during the Mesozoic.

It is widely accepted that the best-documented case for the origin of a higher category of animals is the origin of the class Mammalia from therapsid reptiles. Some paleontologists believed that mammals arose from several lines of therapsids, i.e., that there was a polyphyletic origin for mammals, that the boundary from reptiles to mammals may have been crossed in several groups and at different times. But this view is not shared by all, as we shall see later.

Romer's 1966 classification of monotremes and pre-Cretaceous mammals shown here was little changed from that of Simpson (1945):

Class Mammalia
    Subclass Prototheria
        Order Monotremata
    Subclass uncertain
        Order Triconodonta (extinct)
        Order Docodonta (extinct)
    Subclass Allotheria (extinct)
        Order Multituberculata
    Subclass Theria
      Infraclass Trituberculata (extinct)
        Order Symmetrodonta (extinct)
        Order Pantotheria (extinct)
      Infraclass Metatheria
        Order Marsupialia
      Infraclass Eutheria (the "placental" mammals)

Simpson recognized three subclasses, the Prototheria for the monotremes, the Allotheria for the extinct multituberculates, and the Theria for the marsupials, placentals, and the extinct symmetrodonts and pantotheres. In addition there were two unallocated orders (the subclass is uncertain in Romer): the Triconodonta and the Docodonta. The relationship of the orders Symmetrodonta and Pantotheria to later therians (roughly placental mammals) has been long understood. The relationships of the nontherian orders Monotremata, Multituberculata, Triconodonta, and Docodonta to one another and to therians is not as well understood and has led to separate derivation of each group from therapsid reptiles.

A functional joint between the dentary and squamosal bones is the most widely accepted boundary separating mammals from reptiles. Many paleontologists believe this boundary has been crossed several times, thus giving rise to the belief in polyphyletic origins. The ictidosaurs and maybe the tritylodontids attained this contact between the dentary and the squamosal independently of the ancestors of the main Jurassic as well as later groups of mammals.

Hopson and Crompton (1969) argued for a monophyletic origin of the class Mammalia based, in part, on a broader definition of the class Mammalia. They have expanded the squamosal-dentary contact definition of mammals to include (1) postcanine teeth in which the primary cusp is primitively flanked by anterior and posterior cusps arranged in either a straight line or a triangle and (2) a diphyodont type of tooth replacement (i.e., two sets of teeth, adult teeth replacing milk teeth). Hopson (1970) believed that the occurrence of diphyodonty in several groups of therians and nontherians suggests that this pattern is primitive for the Mammalia. Another feature, the structure of the tooth enamel, which is similar in all the orders of pre-Cretaceous therians and nontherians, but differs from that of therapsids, also suggests a common origin for mammals (Moss, 1969).

Several groups of nontherian mammals possess a specialization of the braincase not present in known cynodont therapsids, from which all mammals were derived (Hopson and Crompton, 1969). In the monotremes, and some of the extinct groups of mammals, the lateral wall of the braincase is formed, in part, by an intramembranous ossification, which is the major element in this region of the skull. In the therian mammals this intramembranous ossification has been replaced by the alisphenoid bone, which is lacking in the monotremes and some other orders of extinct mammals. Some paleontologists feel that the presence of this intramembranous ossification continuously with the periotic indicates affinities within the groups possessing it and separates them from the therians, which do not possess it (Fig. 2–1). Watson (1916) believed that since no similar structure is known in reptiles or in other mammals, it must have been a special feature developed by a common ancestor of the monotremes. The variation in the pattern of ossification that exists between the two monotremes, the platypus *(Ornithorhynchus)* and the echidna *(Tachyglossus),* may indicate that it is a secondary specialization (Hopson, 1970). Watson (1915) felt that the absence of the alisphenoid in the skulls of monotremes showed a wide difference between the monotremes and therians, and that that "alone makes the monotreme skull differ far more greatly from that of any Therian than the members of the latter group do amongst themselves." He pointed out that it was not inconsistent with a close relationship between the monotremes and the therians and was quite possibly a stage in the development of a characteristic mammalian skull. It now became possible to compare the monotreme skull with that of a cynognathid. Many of the remarkable features of the monotreme skull are merely extreme developments of the features found or predicted in the cynognathid skull. Watson (1915) did not regard the monotremes as representative of a different stock from those which give rise to the Theria but believed them to be specialized descendants of very early mammals.

While the assumption is made that the monotremes are primitive mammals, evidence of the lineage is woefully lacking. The fossil record is limited. Prior to the Pleistocene, a fragment from the early Miocene of Australia has been identified as a monotreme, *Ektapodon* (Hopson, 1970). From then to the Pleistocene, fossil material is lacking. Monotreme development has apparently

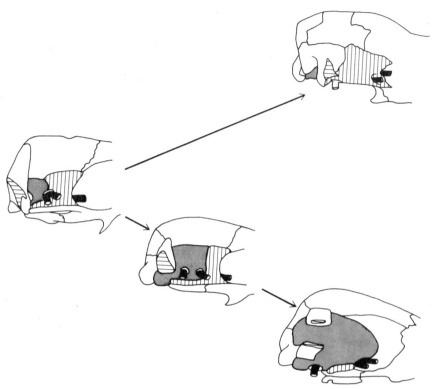

Figure 2-1 Semidiagrammatic representation of the evolution of the braincase in mono-tremes (left) and therians (right) from a cynodont ancestor. Bases of the main trigeminal nerve are indicated as follows: stippled area, periotic; vertical lines, epterygoid-alisphenoid; and horizontal lines, exposed squamosal. *(After Hopson and Crompton, 1969.)*

been separate from that of the other mammals since the Mesozoic. Hopson believed that they have existed as a separate lineage from near the beginning of the Jurassic or from the Late Triassic.

Diphyodonty (replacement of milk teeth by a permanent set of teeth) is considered a basic characteristic of the class Mammalia. Fossil evidence for the origin of this mammalian state of organization eluded scientists until recently. Parrington (1971) has shown fairly conclusively that diphyodonty occurred in *Eozostrodon* specimens examined from the Triassic. *Eozostrodon* is now believed to be an early triconodont, although earlier it was thought to be a docodont. Parrington (1967) and Crompton and Jenkins (1968) independently concluded, from the similarity of the molar teeth, that *Kuehneotherium* and *Eozostrodon* are closely related. Because of these obvious resemblances, the triconodont tooth, as represented by *Eozostrodon,* has been transferred to the therians, through the Kuehneotheriidae. For these reasons Parrington

(1971) and others believed the molar teeth of *Eozostrodon* to be the basic type from which all therian molars have evolved.

On the basis of more recent evidence, reevaluation of older evidence, and a redefinition of the class Mammalia, the consensus has shifted away from the concept of a polyphyletic origin of mammals and toward a monophyletic origin. Hopson and Crompton (1969) present evidence strongly suggesting a monophyletic origin at a very low taxonomic level—family or lower. They have suggested Early Triassic cynodonts from the family Galesuridae as one possibility. Hopson (1970) has restricted the class Mammalia to include the Jurassic and later mammals and the earlier Triassic families Eozostrodontidae and Kuehneotheriidae which are closely related to each other and possess a number of mammalian characters not present in the Ictidosauria. He excludes ictidosaurs, even though one member of the group has a rudimentary contact between dentary and squamosal.

Hopson (1970) has proposed the following scheme of mammalian classification:

Class Mammalia
  Subclass Prototheria
    Infraclass Eotheria[1]
      Order Triconodonta[1]
        Family Morganucodontidae[1]
          (*Morganucodon* is a junior synonym for *Eozostrodon*)
        Family Amphilestidae[1]
        Family Triconodontidae[1]
      Order Docodonta[1]
        Family Docodontidae[1]
    Infraclass Ornithodelphia
      Order Monotremata
        Family Tachyglossidae
        Family Ornithorhynchidae
        Monotremata, *incertae sedis*
        Family Ectopodontidae[1]
    Infraclass Allotheria
      Order Multituberculata[1]
      Suborder Plagiaulacoidea[1]
        Family Plagiaulacidae[1]
      Suborder Ptilodontoidea[1] (family not listed)
      Suborder Taeniolabidoidea[1] (family not listed)
      Multituberculata,[1] *incertae sedis*
        Family Haramiyidae[1]
The rest of the mammals are contained in the subclass Theria.

[1] Extinct

THE CENOZOIC ERA

About 35 orders of fossil and recent mammals are now recognized. Thirty-one of these are assigned to the subclass Theria. The numerous orders of Theria, a group going back to the Jurassic, reflect the great radiation of this group that began in the Late Cretaceous and continued into the Cenozoic, the Age of Mammals.

The order Triconodonta, containing about 10 genera, spanned most of the Mesozoic from the Late Triassic to the Lower Cretaceous. The members of this order were probably carnivorous rather than insectivorous, and some reached the size of a cat.

The herbivorous element of the Mesozoic mammalian fauna was represented by the order Multituberculata, a highly successful side branch of the early mammals, spanning a period of 100 million years from the Late Jurassic of the Mesozoic era to the Eocene of the Cenozoic era. Some of the multituberculates were as large as woodchucks and may have been like them in habits. Brain casts show the olfactory bulbs to have been highly developed, and therefore they must also have had a highly developed sense of smell. The dentition is also specialized (Fig. 2–2). There is a large pair of upper and lower incisors, as in present-day rodents. Smaller lateral incisors persisted in some kinds. There are no canines, only a blank space (diastema) between the incisors and the cheek teeth. Typical of multituberculates are their molars. The lower molars have two parallel rows of cusps, and the upper ones have two, sometimes three, opposing and similar cusp rows. In the lower jaw one or several premolars have been transformed into striated sharp, shearing blades.

The order Symmetrodonta includes some of the oldest mammals, also believed to be related to the most advanced group of mammals, the subclass Theria. The symmetrodonts were predators with three separate cusps in each molar tooth. The three cusps were arranged in a symmetric triangle (hence the name), the base, external above, internal beneath, with the major cusp at the apex of each triangle.

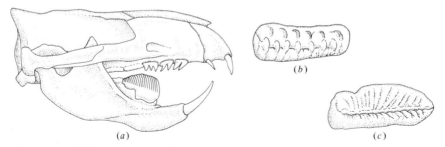

Figure 2–2 Multituberculata skull and teeth. *Ptilodus* skull and occlusal views of upper (upper) and lower cheek teeth. *(After Romer, 1966.)*

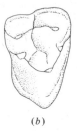

(a)   (b)

Figure 2–3   The order Pantotheria was considered to be the most antecedent to present-day therians. Right upper molars and left lower molar of a Jurassic pantothere. *(After Romer, 1966.)*

The mammals were now inheriting the earth. Extinction of the Mesozoic land reptiles left empty environments which were exploited by the more efficient mammals resulting from the Paleocene experiments.

Members of the order Pantotheria, the most antecedent to the present-day therians, were small, insect-eating mammals, probably most like the shrews of today. The ventral border of the dentary possessed an angular process, as in modern therian mammals. There is a posterior heel on the lower molar of pantotheres which is represented in many therians by the talonid, or the posterior section of the lower molar (Fig. 2–3). The triangular upper molar of the pantotheres is very similar to that of some of the early therians.

The Jurassic and Early Cretaceous were times of experimentation, and the Cretaceous has been called the threshold of the Age of Mammals, which started 65 million years ago. Dinosaurs were still plentiful, including *Tyrannosaurus*. There were smaller carnivourous dinosaurs, too, which probably sometimes fed on the rodentlike multituberculates or some of the primitive hoofed mammals which roamed the forests. There were even some early primates. Birds were by now able to fly, but there also remained a few primitive forms.

But these unusual scenes disappeared. The Late Cretaceous was a time of dying. The dinosaurs and most of the reptiles became extinct, bringing to a close the Age of Reptiles. Many theories have been proposed for this mass extinction, but none is entirely satisfactory.

The greatest radiation of mammals was during the Cenozoic, the Age of Mammals. This was a time of widespread uplift and much erosion, but the erosional deposits were made beyond the margins of dry lands. As a result there are few places where fossil land mammals from the beginning of the Cenozoic, the Early Paleocene, can be found. Those areas are restricted to the region of the Rocky Mountains of North America. Evidently the seed-eating multituberculates were still present, because they continued into the Eocene. The insectivorous forms were mostly marsupials, some of which have not changed much since that time. With them were a few insectivorous placental mammals, classified in the order Insectivora.

Later faunas of the Americas' Paleocene showed the rapid radiation of the placental mammals. The later Paleocene faunas of the World Continent (Africa, Europe, and Asia) show the same expansion. This same expansion also occurred in South America in the Late Paleocene, but the later faunas there developed in isolation for tens of millions of years.

Australia apparently started the Age of Mammals with marsupials and monotremes, and its isolation from other continents (or island continents, according to some) allowed a rapid adaptive radiation of marsupials to fill every niche occupied by placentals elsewhere. There are of course a few native placentals in Australia such as bats and some rodents. Later others were brought by humans.

Among the most primitive of mammals are the monotremes. They reproduce by laying eggs. The skeleton and soft anatomy retain reptilian characters. There are no separate openings for liquid and solid wastes as in mammals, both types of waste going into a common duct, the cloaca, as in reptiles. The evolutionary history of the monotremes is unknown. As already mentioned, there is little fossil evidence prior to the Pleistocene. They represent quite a separate evolutionary line from the rest of the mammals.

In the changing environments many experiments were tried by nature. Some were successful; some were unsuccessful. During the Paleocene shifts in the earth's surface and climate, the dinosaurs were unsuccessful; the mammals as a group were enormously successful, though we know that many "models" of mammals became extinct, even during the Age of Mammals. How many of those now considered successful will eventually become extinct only time will tell. There are those alarmists who strongly feel that the most successful of the mammals, the human being, will soon become extinct if we do not learn to live more successfully within our environment.

Chapter 3

Characteristics of Mammals

It may be both vanity and folly to characterize mammals as the "highest" of animals and human beings as the "highest" of mammals. Biologists especially consider it a heresy to say that any one animal is "higher" than another. Be that as it may, mammals are a group apart from the other animals and human beings are mammals apart from other mammals. If one were to accept the hypothesis that the ultimate in evolution is complete independence of the external environment, then mammals (and perhaps birds) have approached this independence to a greater extent than any other group of animals.

In the previous chapter some of the major evolutionary changes which made the appearance of mammals possible have been discussed. Many of these changes which occurred during the Paleozoic were adaptations for a terrestrial life, such as air breathing, terrestrial locomotion, a body covering to prevent desiccation, the amniote egg, and internal fertilization. The giant evolutionary steps seemed to be from an aquatic to a terrestrial environment.

During the Late Paleozoic some reptiles with mammalian affinities appeared. They in turn gave rise to the mammallike reptiles, the therapsids, which flourished during the Permian and Triassic. Whether any of these were warm-blooded or had hair (two characteristics of recent mammals) is a matter of speculation; these adaptations may have come later.

In the Late Triassic a new level of vertebrate organization uniquely related to later developments in mammalian history emerged. The transitions from reptiles to mammals spanned 100 million years. During that time evolutionary changes occurred which leaned toward a structural organism now called a mammal. The boundary separating reptiles from the fossil mammals includes (1) an articulation between a single lower jawbone (the dentary) and the squamosal of the main skull, (2) postcanine teeth in which the primary cusps are primitively flanked by anterior and posterior cusps, and (3) a limited pattern of tooth replacement, i.e., two sets of teeth, deciduous "milk teeth," replaced by permanent "adult" teeth.

The mammalian characteristics discussed so far are from the fossil record. The fossil record can provide only minimal information on the physiologic, behavioral, and numerous morphologic features, especially of the soft anatomy, which have allowed the mammals to survive through changing environments ranging from millions of years to a daily one of 24 hours. To a great extent that phenomenon—how recent mammals maintain homeostasis—is the subject of this book.

The characteristics discussed in this chapter are those of the recent mammals, and many will be dealt with in greater detail in later chapters. Most of these characteristics serve, in some way, to promote endothermy and to enhance intelligence.

SOFT ANATOMY

The Brain

In mammals the proportion of brain to body size is usually greater than in other animals. The brain is an extension of the spinal cord that has undergone continual development and specialization for higher functions, a process often referred to as *cephalization.* In the embryo three irregular swellings occur at the anterior end of the spinal cord; they are referred to as the forebrain, the midbrain, and the hindbrain. In mammals the brain and spinal cord are wrapped in three protective membranes called *meninges.* The spaces between the three meninges are filled with cerebrospinal fluid. They provide a cushion which protects the nervous tissue from being damaged by the surrounding bone.

In the vertebrate brain the ventral portion of the *medulla,* or hindbrain, became a control center for some autonomic and somatic pathways (heartbeat and respiration), as well as a connecting link between the more anterior parts of the brain. The anterior dorsal part of the hindbrain became the *cerebellum,* which functions as a center for balance, equilibrium, and muscular coordination. The dorsal part of the midbrain evolved into the *optic lobes.* The anterior portion of the forebrain developed as a *cerebrum,* with prominent olfactory lobes. The *corpora quadrigemina,* an elaboration of the midbrain, is found only in mammals. The posterior portion of the forebrain became the *thalamus,* which serves as a relay station for sorting massages, and the *hypothalamus,* which influences the pituitary gland and is involved in what might be called the emotional activities. The hypothalamus may be considered a coordinator between the nervous and chemical control systems.

Most prominent among mammals is the great development of the gray matter, the *cerebral cortex* or *neopallium.* In certain advanced reptiles, a new structure arose on the anterior surface of the cerebrum. This has been called the *neocortex* or *neopallium.* In mammals this has expanded so greatly as to cover most of the forebrain. The old cortex has been pushed to an internal position. As the neocortex continued to expand its dominance over other parts of the brain, it expanded greatly. This greatly convoluted neopallium (cerebral cortex) permits accumulation and utilization of experience to a greater degree than any other center of the brain. In human beings the associative areas constitute by far the largest part of the cortices. It is this characteristic that most markedly distinguishes humans from other mammals. Birds, like mammals, evolved from reptiles, but the reptiles from which birds evolved had not developed a primitive neocortex. These different origins of bird and mammal brains may partially explain behavioral differences, birds being less able than mammals to modify their behavior by learning.

Sense Organs

With some minor variations, which will be discussed briefly, the eyes of mammals are basically similar to those of most amniote vertebrates. In pinnipeds (seals) the eyes are large (except in the walrus), lie in a cushion of fat, are placed well forward, and are of great convexity. The iris accommodates from dark to bright rapidly. Out of water the iris is often a vertical slit; under water it becomes almost spherical. The hooded seal *(Phoca vitulina)* has a fishlike spherical lens, probably useful under water, and a cornea whose inherent astigmatism may be an adaptation for aerial observation. Among the cetaceans (whales) in general, vision does not play as important a part as in other mammals. The keenest vision is found among the porpoises and those dolphins which feed on fish near the surface. At the opposite extreme is the blind river dolphin *(Platinista gangetica),* found in certain muddy waters of India and Pakistan. Their lens is absent, and the retina of the very tiny eye is sensitive only to light and dark and can form no image. Because of adaptations to an aquatic environment, the eyes of whales apparently function less efficiently on the surface, where the images form in front of the retina, i.e., whales become shortsighted. It has been suggested that in some whales aerial images focus on the lower part of the retina, and aquatic images come into focus only on the upper retina. For a discussion of whale vision see Slijper (1962). Much work remains to be done on whale vision.

Among the insectivores are some groups with reduced eyes. In the moles (family Talpidae), most of whose members live underground (i.e., are fossorial), the eyes are small and sometimes hidden under the skin. In the golden moles (family Chrysochloridae) of Southern Africa, the eyes are also hidden under the skin.

Of interest to many people is the "eye-shine" or reflection from eyes of many kinds of mammals (and some birds) picked up in beams of light. This is caused by a reflective structure, the *tapetum lucidum,* in the choroid which improves night vision by reflecting light back to the retina.

The sense of hearing is most highly developed in mammals. Only mammals have an external structure, the *pinna,* which serves to intercept and focus sounds. Pinnae are large in some mammals and missing in others, as in some insectivores, cetaceans, and the "earless" seals. The pinnae of some species can be independently directed, as seen in deer trying to determine the source of a sound.

The tube leading from the pinna to the tympanic membrane, the *external auditory meatus,* is typically long in mammals. The middle ear, containing three ossicles—the *incus, malleus,* and *stapes*—is enclosed by a bony *auditory bulla.* The three bones can amplify the force of the pressure on the eardrum to the oval window of the inner ear as much as 90 times. Two safety devices protect the ear from sudden loud noises: (1) a set of muscles which tightens the eardrum and pulls the stapes away from the inner ear; (2) the eustachian tube, which connects the air-filled middle ear with the mouth cavity and serves as a pressure equalizer.

The inner ear consists of the liquid semicircular canal, which is the body's balancing mechanism, and the coiled cochlea, which contains the "seat of hearing," the organ of Corti. It is this organ which converts mechanical into electrical energy, which in turn is transmitted to the brain by the auditory nerve.

Among mammals variations on this basic pattern are many. One of these variations is found in those mammals (most bats, some whales, a few other kinds) which use acoustic orientation (echolocation) as well as visual orientation. These mammals differ from most others in that the bone enclosing the ear ossicles, the auditory bulla, has no direct connection with other bones of the skull but "floats."

Another variation occurs in mammals living in arid regions. In kangaroo rats the auditory bullae are extremely large and the ear ossicles so constructed as to increase greatly the transformation ratio between the tympanic membrane and the oval window of the cochlea. It is believed that the dry air conducts sound less well than humid air. These variations are dealt with more fully in the section on communication in Chap. 12.

The sense of smell is most highly developed in mammals. The olfactory bulbs and olfactory lobes form a great part of the mammalian brain. The development of the turbinal bones and olfactory surfaces and Jacobson's organ in the nasal cavities enhanced the olfactory senses of mammals. It is generally thought that whales do not have organs of smell or taste, but Kleinenberg, Yablakov, Bel'kovich, and Tarasevich (1964, translated in 1969) have discussed an organ analogous to the olfactory organ of terrestrial mammals. They have assumed that the passing of a school of whales leaves a band of dissolved substances in the waters. The chemoreceptors which could sense these odors cannot be situated in the nasal passages, for these are closed and unused when the whale is submerged. They have found in all toothed whales (suborder Odontoceti) examined, glandular epithelium in the oral cavities, which they suggested could serve as a chemoreceptor. The senses are dealt with in greater detail in Chap. 12.

Hair

One of the typical characteristics of mammals is the presence of hair. Most mammals are covered with hair, but in the whale the only hairs present are vibrissae, a specialized kind of hair. Hair probably developed in therapsid reptiles and has no structural homologue in other vertebrates. The primary function of a coating of hair, the pelage, is for insulation, and it is therefore of great importance in maintaining a high and constant body temperature. Hair is discussed in greater detail in Chap. 6.

Skin Glands

There are a greater number and variety of skin glands in mammals than in any other vertebrate. They serve many functions, from sensing changes in the environment to providing milk for the young. The mammary gland, unique to mammals, is the most important of these glands, and is one characteristic setting the mammals apart from other vertebrates. It is believed to have originated from sweat glands. Scent glands and musk glands are used for communication, either to attract or to dispel. Skin glands are treated in several chapters, including Chaps. 6, 10, and 12.

Circulatory System

A primary requisite for mammals to help maintain their constant and warm body temperatures is an efficient circulatory system. The complete separation of the pulmonary and systemic circulatory systems has come about in mammals. The normal heart rate in mammals can vary from less than 25 per minute in the Asiatic elephant *(Elephas maximus)* to over 1300 per minute in the masked shrew *(Sorex cinereus)*. The heart rate of an individual can vary greatly over a 24-h period. The red blood cells of mammals are concave, and mature cells have no observable nuclei.

Respiratory System

An efficient respiratory system also enhances the efficiency of maintaining a constant, high body temperature in mammals. The lungs are large, but there are no additional air sacs as in birds. The finest division of the lungs comprises the alveoli, which are surrounded by dense capillary beds. It is here that the exchange of gases between the inspired air and the blood occurs. A structure unique to mammals, the muscular diaphragm, increases and decreases the volume of the thoracic cavity during breathing. This muscular diaphragm probably developed as a new device that made possible a great increase in oxygen intake for activity and for the maintenance of an elevated body temperature.

Reproductive System

As mentioned earlier, internal fertilization was a giant step forward in freeing animals from an aquatic environment. In mammals the male copulatory organ, the penis, delivers the sperm to the ova, which remain in the oviducts until fertilized. The embryo develops in the uterus within the fluid-filled amniotic sac. Nourishment from the maternal bloodstream reaches the embryo through the placenta. In mammals (contrary to what occurs in birds) both ovaries are functional in the production of ova, or eggs. However, in some cases both ovaries do not function during each breeding cycle, but alternate, as, for instance, in pinnipeds.

The penis contains erectile tissue. The anterior portion is usually specialized into a *glans penis,* which is enclosed in a sheath of skin, the *prepuce.* The shape of the penis, especially the tip, is often very complicated. In many species the penis contains a bone, the *os penis* or baculum, which has been useful in taxonomic studies. The testes may normally remain in the coelomic cavity at all times, as in elephants and insectivores; they may descend permanently from the coelomic cavity into the scrotum as the individual reaches sexual maturity, as in humans; or they may be withdrawn into the body cavity between reproductive periods, as in rodents. Reproductive organs, reproduction, and associated behavior and events are discussed in Chap. 10.

Digestive System

The food of mammals varies from the very specialized diet of the vampire bat *(Desmodus),* which feeds on blood, to the very cosmopolitan diet of such mammals as bears, which eat nearly anything. Stomachs also vary, from single saclike compartments to the four-chambered stomach of the ruminants, where digestion is aided by microorganisms. A more thorough discussion can be found in Chap. 7.

Muscular System

The bulk of the muscular system of mammals is basically adapted to their various modes of locomotion. Especially well developed in mammals is the dermal musculature, which greatly enhances the maintenance of their endothermy and facilitates communication. Within certain limits the antelope, for instance, can control the angle of the hair-to-skin surface and either conserve or release body heat. It can also erect the hairs on its white rump to serve as a flash signal when danger is suspected. These muscles also control facial expressions which communicate the emotional state of an individual among some mammals, e.g., the timber wolf *(Canis lupus)* or the domestic dog.

THE SKELETON

It appears that a most important morphologic trend in the therapsid-mammalian line was skeletal simplification. Bones in the skull and lower jaw were reduced in size or number. Limbs and limb girdles were simplified. The axial skeleton became more rigid. The mammal skeleton is completely ossified. Skeletal growth is restricted to immature mammals, whereas in most reptiles growth continues at the end of the limb bones throughout life, although the greatest growth occurs early. In mammals, growth during early life occurs at a cartilaginous zone where the articular surface (the epiphysis) and the main shaft of the bone (the *diaphysis*) meet. At maturity this zone ossifies, a fact which is useful in determining the age of an individual. The skeleton supports the body, protects the most important organs, and serves as a point of attachment for muscles.

The Skull

A prominent feature of the skull is the braincase, which is large in mammals. A *sagittal crest* increases the area of attachment of the temporal muscles, which function in closing the jaw. The *zygomatic* arch protects the eye, provides the origin for the masseter muscles, and forms the articulating surface for the lower jawbone (the dentary). The zygomatic arch varies from large in carnivores and some rodents to absent in whales and some insectivores.

A number of *foramina* (openings) penetrate the braincase and allow for the passage of nerves and blood vessels. The incisive foramina in the palate house the olfactory organ (Jacobson's organ). *Turbinal* bones within the nasal cavities support tissues, increase the effectiveness of Jacobson's organ, and produce mucus which filters foreign matter out of the inspired air.

The most remarkable transformation of morphologic structures from one function to another in the history of vertebrate evolution is the change of articulating elements between the jaw and skull in reptiles to auditory elements in the mammals. The articulating bones between the jaw and skull of the reptiles, the *quadrate* and *articular,* have been transformed into the *incus* (quadrate) and *malleus* (articular) of the middle ear. These together with the third of the three middle ear ossicles, the *stapes,* inherited from the reptilian stapes, conduct vibrations from the eardrum to the inner ear.

The *secondary palate* found in mammals is a plate, formed by the premaxillary, maxillary, and palatine bones, lying beneath the original roof of the

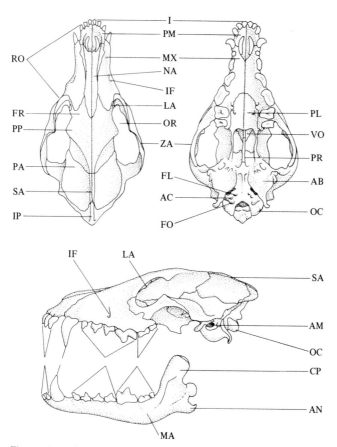

Figure 3-1 Skull of a coyote. AB, auditory (tympanic) bulla; AC, anterior condyloid foramen; AM, external auditory meatus; AN, angle of mandible; C, canine; CP, coronoid process; FL, foramen lacerum posterius; FO, foramen magnum; FR, frontal; I, incisor; IF, infraorbital foramen; IP, interparietal; LA, lacrimal; M, molar; MA, mandible; MX, maxillary; NA, nasal; OC, occipital condyle; OR, orbit; PA, parietal; PL, palatine; PM, premolar; PP, postorbital process; PR, presphenoid; RO, rostrum; SA, sagittal crest; VO, vomer; ZA, zygomatic arch. *(From Gunderson and Beer, 1953.)*

mouth. It forms a passage from the external nares to the internal narial openings toward the back of the mouth. This allows the animal to continue breathing while chewing food and obviously is another adaptation helpful in maintaining a constant body temperature.

The *hyoid apparatus* in the throat region, remnant of the gill arches of the fishes, supports the trachea, the larynx, and the base of the tongue. There is a double occipital condyle, as in reptiles, for the articulation of the skull to the axial skeleton.

The Postcranial Skeleton

The vertebral column, together with the limbs and girdles, makes it possible for mammals to move freely as terrestrial vertebrates. Life in the ocean appears to be a secondary development in mammals. The vertebral column is composed of five differentiated sections of vertebrae, the *cervical, thoracic, lumbar, sacral,* and *caudal.* There are usually seven cervical (neck) vertebrae. Exceptions are found in sloths and manatees. In some groups of mammals—the whales, rodents living in arid regions, and armadillos—some or all of the vertebrae are fused. The first two cervical vertebrae are specialized for articulation of the skull to the axial skeleton. The first cervical vertebra is called the *atlas,* and next, the *axis.*

The *thoracic* vertebrae carry the ribs—normally twelve pairs, but sometimes more, as in the sloths. The *sternum* anchors the ventral ends of most of the ribs to form a rib cage. The *lumbar* vertebrae have no articular facet for the attachment of the rib. The *sacral* vertebrae are fused to form the *os sacrum,* to which the pelvic girdle is attached. The *lumbar* vertebrae are the tail vertebrae and are variable in number.

The *scapula,* or shoulder blade, of mammals has a strong spine running down the middle. Remnants of the *coracoid* and *acromion* of the reptiles remain as processes on the mammalian scapula.

The reptilian interclavicle has disappeared in the pectoral girdle of mammals. A *clavicle* is present in those mammals whose front limbs move in several planes, e.g., squirrels and primates, but is absent where the front limbs move only in one plane, e.g., deer. The elements of the pelvic girdle—the *ilium, ischium,* and *pubis*—are fused to form a solid unit, with the rather enlarged ilium extended forward.

There is a rather standardized pattern of bones in the hand or front foot *(manus)* and in the hind foot *(pes)* of mammals, but variations occur with specialized types of locomotion. General discussions of mammalian skeletal adaptations to locomotion include those of Eaton (1944), Gray (1953), Hildebrand (1952, 1959, 1961, and especially 1962), Howell (1930), Muybridge (1957), and J. M. Smith and Savage (1956). For a discussion on the speed of mammals, Howell (1944) still seems useful.

TEETH: THEIR EVOLUTION AND ADAPTATIONS

Because teeth are of great importance in the study of mammals it seems logical to include them in a discussion of the characteristics of mammals. "Form follows function" is an adage easily demonstrated by teeth. For several reasons, they are extremely important in the study of mammals. Within limits, they reflect the food habits of the mammal. They also reflect genetic relationships. And because they are the hardest tissues in the body, and by the same token the best preserved, they have been very useful in defining the evolution-

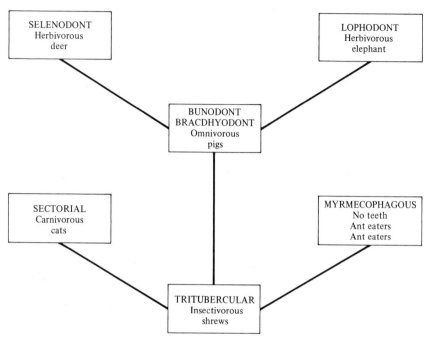

Figure 3-2 Adaptive radiation in teeth. *(From Burt, Mammals of Michigan. Copyright by the University of Michigan Press, 1946.)*

ary history of mammals. Colbert (1961) wrote, "It is very probable that if all the placental mammals, except man were extinct, and represented only by fossil teeth, their basic classification would be essentially the same as the classification now drawn up on the knowledge of the complete anatomy in these mammals." Teeth are not restricted to mammals but certainly have reached their greatest degree of specialization in this group.

Structure of Mammalian Teeth

In the mucous membrane of the mouth calcifications appear which form the teeth, which in mammals occur in a row along each ridge of the jaw. Each tooth is lodged in a socket, or *alveolus,* a condition termed *thecodont.* Teeth are composed of *enamel, dentine, cement,* and *pulp* (Fig. 3–3). The enamel is developed from the epidermis; the dentine, cement, and pulp are developed from mesodermal tissue.

Enamel   The hard part of the tooth is developed by the formation of a cap of *enamel matrix* and *predentine.* These calcify to form the enamel and the dentine. The free surface of the tooth is usually covered with enamel, a dense prismatic material, which is the hardest part of the tooth. There is no enamel on the roots of the teeth. It may in rare instances be either partly or completely missing. It is partly missing on the lingual surface of rodent incisors and on the tusks of the boar and of elephants. It is completely missing on the teeth of armadillos. In these the dentine is surrounded on the outside by cement.

Dentine The dentine is the calcified predentine which makes up most of the crown and root of the tooth. The root of a tooth consists of dentine over which a layer of *cement* is deposited.

Cement Cement is the hard material covering the roots of all mammalian teeth. Crown cement is also present in those mammals whose cheek teeth have to perform punishing chewing functions. These include rabbits, some rodents, perissodactyls, artiodactyls, and elephants, among others. Grazing presents a serious problem and would wear a low-crowned tooth to the roots in short order. The heightening of the dentine-filled bulk of the tooth is not enough, although it occurred in some fossil mammals. Successful forms have a high growth of each cusp, including the enamel, and the whole tooth covered by cement. Wear then takes place through all the resistant layers—enamel, dentine, and cement.

Pulp The roots of teeth are generally open at the base, and nerves and blood vessels enter the pulp of the central area by this lumen, or opening. Eventually these openings may nearly close in an older tooth, but in some teeth the formation of roots is delayed and enamel continues to be added long after the tooth has erupted and come into use. This is true of the incisors of rodents, sometimes of the cheek teeth of rodents and elephants, and of all the teeth

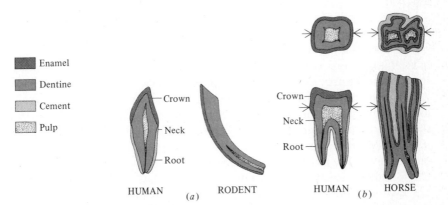

Figure 3–3 Structure of mammalian tooth. Arrows on molar teeth indicate the plane of section, in the figure above or below. *(After Storer, 1943.)*

of rabbits. In the elephant the upper pair of incisors grows continuously from persistent pulps throughout the life of the enamel. The tusks are composed of solid dentine. The molars are very large, and their worn surfaces are marked with transverse ridges. There are a total of six molars on each side, but only one and part of another are functional at once, the more posterior moving forward and taking the place of the anterior as these become worn out and are shed. The walrus has enormously enlarged, permanently growing canines. The narwhal *(Monodon)* exhibits an extreme case of sexual dimorphism in its dentition. In this form the permanently growing (usually left) incisor in the male may reach a length of over 2 m. The right incisor remains hidden in the gum, as do both incisors of the female.

When roots are never formed and the crowns continue to grow, the teeth are said to have *persistent pulps;* the term *rootless* is also used, though this seems illogical. When the tooth ceases to grow, shortly after eruption, and the foramen narrows, the tooth is called a *rooted* tooth (again, in my opinion, an illogical use of the term).

Nomenclature of Teeth

The teeth of mammals are usually differentiated into functional varieties, i.e., the mammals are *heterodont,* but in some species—for instance, dolphins—the teeth are all alike, i.e., they are *homodont.* Among the monotremes (egg-laying mammals), the spiny anteater *(Echidna)* is toothless except for an egg tooth which enables it to escape its egg when hatching. The adult duckbill platypus *(Ornithorhynchus)* has a powerful horny beak which takes over the place of the teeth present in the young animals before they have reached a third of their size. Before that they have multituberculated teeth, four in each upper jaw and three in each lower jaw.

Reduction in number or loss of teeth is found in various other mammals, especially the whalebone whales and the anteaters.

Usually four types of teeth can be distinguished. Most anterior are the *incisors,* single-rooted nipping teeth, usually of moderate size, found in the premaxilla of the upper jaw, as well as in the lower jaw. Next is a single *canine* in each jaw, a single-rooted tooth which in the upper jaw is the most anterior tooth of the maxilla, situated partially on the maxilla or between the premaxilla and maxilla. It is used for defense, grasping, stabbing, and sometimes as a secondary sexual accessory, especially in some artiodactyls. Following this there may be a comparatively long blank space, as in rodents, called the *diastema.* Then there is a series of cheek teeth, usually with a grinding surface on the crown. The more anterior of the cheek are the *premolars.* The posterior cheek teeth are usually larger and have better-developed crowns. These *molars* are also distinguished by the fact that they have no milk predecessors, or conversely no descendants. In the carnivores the cheek teeth tend to be reduced in number and in size, except that in many terrestrial carnivores a single

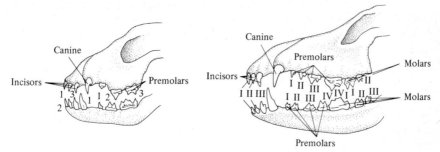

Figure 3–4 Nomenclature of teeth. *(After Storer, Usinger, and Nybakken, 1968.)*

pair on either side have sharp cutting edges for shearing. These slide past each other with a break-and-shearing effect. Such a detition is called *secodont*. In modern carnivores these *carnassials* are always upper premolar 4 and lower molar 1.

A shorthand system has been devised to describe teeth in the complete dentition. The letters I (incisor), C (canine), P (premolar), and M (molar) followed by a number in an upper or lower position will define a single tooth. For instance I^1 refers to the most anterior upper incisor; I_1 refers to the most anterior lower incisor. Usually the small letters are reserved for the lacteal detention. The dental formula 3143/3143 is the primitive placental formula in each half of the upper and lower jaw, which would be 3 incisors, 1 canine, 4 premolars, and 3 molars. This makes 22 teeth which must be multiplied by two to include both the left and right half of the mouth. Thus the total would be 44. For man the formula is $2123/2123 \times 2 = 32$. In ruminants, rodents, and lagomorphs the grinding teeth are separated from the cropping teeth by the *diastema*.

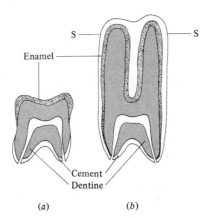

(a)   (b)

Figure 3–5 Development of hypsodont tooth. *(a)* Normal low-crowned tooth. *(b)* Hypsodont tooth, with cusps elevated and the whole covered by cement. As a tooth of this sort wears down to any level, such as that indicated at s–s, it will be seen that no less than nine successive layers of contrasting materials are present across the crown surface. *(After Romer, 1956.)*

The teeth of the early placental mammals were generally *bunodont* (from *bunos,* hill), i.e., they had cone-shaped cusps. Such teeth are also *brachyodont,* i.e., low-crowned with well-developed roots. Pigs and human beings have this crushing kind of teeth. Hypsodont teeth (horses) have high crowns and short roots.

In some mammals the cusps have become connected to form yokes; and such teeth are called lophodont (from *lophos,* yoke) when they are transverse, as in elephants, and selenodont (from *selene,* moon) when they are lengthwise of the tooth, as in deer. The concave side faces lateral in the upper jaw, mediad in the lower jaw. The grinding stroke is outside inward and inside outward, rather than forward backward. This achieves a maximum efficiency in chewing.

Sets of Teeth

Usually mammals have two sets of teeth, the *deciduous* dentition (also called the milk or lacteal dentition) and the *permanent* dentition, which replaces it. Two sets of teeth denote a *diphyodont* type of dentition. When there is only a single set, it is called a *monophyodont* dentition. In the mammals whose dentition is diphyodont, the duration of the lacteal dentition varies greatly. In some of the seals, e.g., the harbor seal and the bearded seal, the lacteal dentition does not reach a functional state, but is largely resorbed before birth. This is also true of the toothed whales. It is not known whether teeth present in the fetus of baleen whales constitute the lacteal or permanent dentition; in either case, they are absorbed before birth. The baleen originates from the epidermal palatal crests. Whether tooth germs ever develop in the members of the family *Myrmecophagidae* is questionable, but there is no doubt that evolutionarily this is a secondary development. In the marsupials only one tooth in each jaw quadrant, the last premolar, is changed, but it is believed that it belongs to the same generation as the tooth it replaces. In a variety of marsupials it replaces not only the last premolar but also the tooth in front of it. The comparison of the tooth generation of the marsupials with those of the placental mammals is still subject to debate. The milk teeth are believed to remain throughout the life of the individual. The teats to which the developing young in the pouch are attached are believed to fill the mouth cavities so completely that the development of the dentition is retarded.

The Evolution of Teeth

The single-cusped haplodont tooth found in some of the mammallike reptiles appears to be the ancestor of the mammalian tooth. To picture the development of simple teeth such as the incisors and canines from a haplodont does not require a great deal of imagination, but how this type of tooth evolved into a complex molar is the subject of several theories and much disagreement (Fig. 3–6).

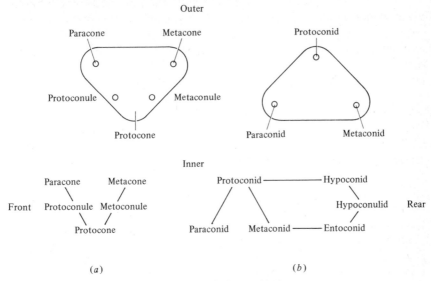

Figure 3–6 The evolution of teeth. *(After Colbert, 1961.)*

The pantotheres are the acceptable ancestors of marsupials and placentals, which include all living mammals but the monotremes. These early mammals had molars which are referred to as tritubercular, tuberculosectorial, or tribosphenic. Their origin can usually be explained by one of two broad theories—*concrescence* and *differentiation*. For a detailed description of these theories the reader is referred to Peyer's (1968) book.

Concrescence Theory   According to this theory a multicuspid mammal molar is formed by the fusion of the germ layers of a number of originally discrete simple conical teeth. In their origin from reptilian teeth there could be two types of fusion: (1) of teeth belonging to the same generation or located along the edges of the jaw in anteroposterior rows; and (2) of a functional tooth with several of the replacement teeth and in a labiolingual direction. The concrescence theory has not been very widely accepted, and has been refuted by paleontologists. Objections are that the earliest elephants had molars that do not correspond with the expectations of this theory; styles, lophs, and cones of present teeth seem never to have existed as separate elements; and if the complication of teeth results from fusion, then species with the most complex teeth should have a greatly reduced number of teeth.

Differentiation, or Cope-Osborn Theory   This theory maintains that even the most complicated molar pattern of mammals has originated from a relatively simple type which is a modification of a single reptilian cone. From a small, simple haplodont tooth a protodont tooth is derived which has small anterior and posterior cusps. The triconodont tooth is derived by forming larger accessory cusps from the insignificant cusps. This triconodont molar constitutes a base from which the varied molar patterns are derived. The accessory cusps change their position so as to form a triangle with the main cusp. The primitive nature of the triconodont molar was established by the American paleontologist Cope (1840–1897), and the nomenclature of the cusps was proposed by his disciple, Henry Fairfield Osborn.

Although the Cope-Osborn theory has been buffeted about greatly since its origin, paleontologists seem to have come back to a general agreement with its major concepts. Parrington (1971) has given a brief summary of recent ideas, reviewed here. He believed that the triconodont tooth, as represented by *Eozostrodon* of the Triassic, is the basic type, from which those of other mammals evolved. Various changes have occurred which have reduced the rate of severe wear to which the teeth were exposed. The first of these, according to Parrington, would be to even the size of the three principal cusps, thus spreading the area of wear at the start of the life of the tooth, and prolonging its life. This change occurred in the typical triconodonts of the Upper Jurassic. The second modification, to reduce wear, was to move the opposing cusps for a better fit. This happened in *Kuehneotherium,* as indicated by fossils found in the Triassic. Here the leading accessory cusp of the upper molars and the posterior accessory cusp of the lower molars had moved away from the principal (largest) cusp of the opposing tooth. At this stage there are two further developments: Triangulation is more fully and evenly developed, and a small "fourth cusp" is reduced. Thus a series of opposed shearing surfaces and a long zigzagged cutting line appears. The retreating leading and trailing accessory cusps continue their movements ahead of other accessory cusps, "allowing for the formation of a full talonid against which the opposing tooth can crush food which has already been cut." In this way the tribosphenic molar evolved in the pantotheres.

Parrington (1971) postulated that this was also the sequence of events in *Docodon* and in the Multituberculata. Some paleontologists do not support the theory of the rotation of the triconodont cusps into a triangle, but rather believe that the accessory cusps were added in a triangular position. Recent findings have also emphasized the numerous resemblances between the dentitions of the *Eozostrodon* and *Kuehneotherium* fossils which occur in the Triassic. These findings add further evidence for a monophyletic origin of mammals, as opposed to a polyphyletic (several times) origin of mammals. The triangular tooth is called *trigon* in the upper jaw, *trigonid* in the lower jaw. Cope and Osborn believed that the tribosphenic upper and molars are opposed reversed triangles. The apex of the *trigon* or triangle points inward in the upper tooth and outward in the lower. The cusp at the apex of the triangle was supposed to represent the original reptilian cone and so was named *protocone* in the upper jaw and *protoconid* in the lower jaw. The other two cusps in the upper molar were named the *paracone* and the *metacone,* and the suffix *-id* was used for their counterparts in the lower jaw—thus paraconid and metaconid.

In addition to the cusps already designated, there are often two intermediate cusps between the main cusps in the upper molar, called *protoconule* and *metaconule.* Three cusps are commonly found around the talonid or posterior basin of the lower molar. The outer one was called the *hypoconid;* the middle and posteriorly placed one was called the *hypoconulid;* and the inner one, the *entoconid.* Sometimes there is a fourth main cusp occupying the posterior inner corner of the upper molar, called the *hypocone.* And lastly there are various crests or ridges called *lophs* (upper) or *lophids* (lower), and small accessory cusps around the edges of the teeth, termed *styles* (upper) and *stylids* (lower), which arise from a basal ring-shaped enlargement of the tooth crown, the *cingulum.*

Within the premolar series, there is usually a transition, from back to front, of a tooth that resembles a molar to one that resembles a canine or incisor. The simple conical teeth of toothed whales and dolphins have originated from more complicated teeth by secondary simplification. Cope's contribution of recognizing the tribosphenic as a basic tooth still stands. Osborn's view as to the origin of the individual cusps has not withstood the test of time, but his terminology has remained.

Adaptive Radiation in Teeth

The teeth of each mammalian order are specialized so that even a few individual teeth will sometimes reveal the identity and food habits of the species to which the animal belongs. The teeth are narrowly conical in moles and bats, which feed on insects; sharp for jabbing and shearing in meat eaters such as the cat and coyote; low and flattened for crushing, as in pigs and human beings; provided with many enamel ridges for grinding vegetation, as in hoofed mammals and rodents. Teeth may be short-crowned as in pigs or high-crowned as in horses. In short they have been adapted for a variety of food habits.

The starting point for all recent placental and possibly marsupial mammals was the insectivore dentition, still retained without much change in shrews, bats, primates, and marsupials. In present-day insectivores the incisors are simple and peglike, usually three in number. The canines are usually the same shape as incisors, sometimes larger. The cheek teeth are low-crowned, tritubercular, and sectorial. The jaw joint is relatively tight, allowing only an up-and-down motion.

The dentition which most closely resembles that of the insectivores is that of the carnivores in the order Carnivora. A baleen whale is carnivorous, but it does not belong to the order Carnivora. There are three incisors in each jaw quadrant; they are small, subequal, and either conical or anteroposteriorly flattened. The four canines are always large and conical for grasping and tearing. The cheek teeth are high-crowned (hypsodont) and sectorial for shearing. They are reduced in number, the ultimate in reduction occurring in the family Felidae, in which the dental pattern is 3/3, 1/1, 2–3/2, 1/1 = 28 or 30. It is in this family that the greatest modifications for a carnivorous diet occur. The teeth are adapted for grasping, holding, and breaking up the food into pieces small enough for swallowing, but they also serve for chewing the food into a pulp. The reduction in size of cheek teeth and the shortening of the rostrum are further modifications for efficiency. In the more omnivorous carnivores the fourth upper premolar and the first lower molar are modified as carnassials. The jaw motion is limited to an up-and-down movement, again in keeping with a nonchewing adaptation.

In the myrmecophagous dentition, secondary modifications include the simple peg as in the teeth of armadillos, or complete loss of teeth, as in South American anteaters of the genus *Myrmecophaga*. In the armadillos the dentition is homodont, and since no visual distinction can be made, the tooth formula is simply 9/9 *(Dasypus)* or 5/4, as in sloths. The enamel layer is absent. The dentine is surrounded by cement, and a hard substance containing vessels fills the pulp cavity. The lumen of the pulp cavity remains wide open, and the teeth grow permanently. The zygomatic arch is weak, and the lower jaw is very simple, for the soft food of armadillos requires little chewing.

Another dental adaptation is to an omnivorous diet, as in pigs and human beings. The incisors are bladelike, with sometimes a reduction in number by the loss of the lateral ones. Canines are not particularly well developed except when they assume secondary sexual functions, as in pigs, some deer, and some primates. The pigs retain the complete dental formula of the placental mammals—3143/3143 = 44—a rare exception in recent mammals. The premolars may be reduced in numbers. The molars are enlarged, and all are bunodont and brachyodont. The jaw joints are loose.

In the herbivorous dentition the incisors tend to be large and modified for cutting or snipping. In some groups, e.g., the Cervidae and Bovidae, there are no upper incisors, the lower incisors pressing against a pad. The canines may become incisiform, reduced, or lost, except in some species where they become large or elongated in the males and have a secondary sexual function, as in the elk and muntjac in the family Cervidae and the dik-dik in the family Bovidae. The cheek teeth develop lophs (in elephants) and crescents (in the deer), with the grooves between filled with cement. As mentioned earlier, the cheek teeth with lophs are called lophodont; those with crescents are selenodont and are believed to have evolved from the bunodont tooth.

The jaw hinges are very loose, moving in all directions; sometimes in chewing they have a rotary motion.

The Cenozoic era has often been called the Age of Mammals. This success can be attributed to the adaptations of the teeth, a success which is reflected by the fossil remains of the hardest of mammal tissues. For this reason much space has been devoted to a discussion of teeth.

Glossary

Because of the volume of new terms the reader has been exposed to in this section, it might be useful to bring them together, so that comparisons or differences will become more obvious.

Brachyodont   Cheek teeth with short crowns, well-developed roots and narrow root canals, as in the molar teeth of human beings.

Bunodont   Tubercles on the crown of molar teeth, as in molars of pigs and some rodents.

Carnassial   Bladelike shearing teeth, found in members of the order Carnivora. In modern carnivores they are the last upper premolars and the first lower molars.

Diphyodont   Two sets of teeth. In mammals this usually refers to having a lacteal (deciduous) set and a permanent set.

Heterodont   Teeth divided into functional varieties, as incisors, canines, premolars, and molars.

Homodont   All teeth alike, as in toothed whales and armadillos.

Hypsodont   Cheek teeth with a high crown and short roots, as in the molars of horses.

Lophodont   Cheek teeth with transverse ridges on grinding surfaces, as in elephants and some rodents.

Monophyodont   Having only one set of teeth in a lifetime.

Myremecophagous   Adapted to the eating of ants and other soft-bodied insects. Comes from a genus of ants, *Myrmecia*. The teeth have degenerated to simple pegs without enamel covering (as in sloths) or are entirely lost (as in spiny anteaters).

Sectorial   Shearing or cutting teeth.

Secodont   A break-shear type of dentition, as in Carnivores.

Selenodont   Longitudinal crescents or "half-moons" on the surfaces of the cheek teeth, as in the artiodactyls.

Thecodont   Teeth set in sockets.

Tritubercular   Teeth with three major cusps.

Locomotor Adaptations

One of the striking characteristics of mammals is their activity. Even the moss-covered back of the South American sloth does not denote a completely sedentary mammal.

D'Arcy Wentworth Thompson, a Scottish scholar and naturalist (1860–1948), is given credit for comparing the outlines of a grazing cow with the structures of the Forth Bridge in Scotland. The legs are the towers of the bridge and the backbone is the arched cantilever system supported by the towers. This whole system serves to carry the animal with all its soft parts to its food, shelter, and other needs. We shall not follow this analogy back to its original development of a tetrapod with a notochord and then a spinal column but will limit this discussion to the adaptations which have occurred since mammals appeared.

General discussions of mammalian skeletal adaptations to locomotion include those of Eaton (1944), Gray (1953), Hildebrand (1952, 1959, 1961, and especially, 1962), Howell (1930), Muybridge (1957), and J. M. Smith and Savage (1956). For a discussion of the speed of mammals see Howell (1944).

In discussing skeletal adaptations for locomotion we can start with the general assumption that ancestral mammals were plantigrade, scampering mammals such as the majority of the small mammals today.

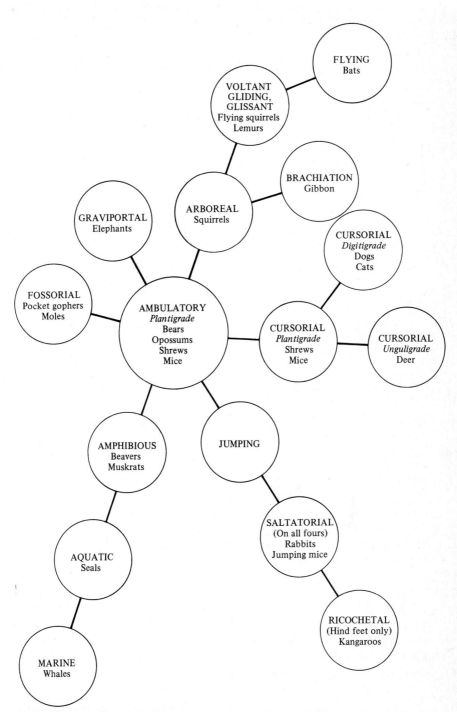

Figure 4–1 Locomotor adaptations. *(From Burt, Mammals of Michigan. Copyright by the University of Michigan Press, 1946.)*

The 12 or 13 (if jumping is divided into saltatorial and ricochetal) types of locomotion will be discussed in detail after being listed:

1. Plantigrade. Feet (soles) and toes on the ground, as in human beings. Opossums, bears *(Ursus),* raccoon *(Procyon lotor),* shrews, mice, and other small mammals serve as examples of this type of locomotion.
2. Cursorial
    a. Digitigrade. Only the toes (sometimes reduced in number) touch the ground. Cats, dogs.
    b. Unguligrade. Only the tips of toes touch the ground. Zebra *(Equus),* deer *(Odocoileus),* pronghorn *(Antilocapra americana),* moose *(Alces americana),* African antelopes.
3. Jumping. Some mammals use both saltatorial and ricochetal jumping, depending on how fast they are moving. Kangaroo rats *(Dipodomys)* and jumping mice *(Zapus)* use both methods.
    a. Saltatorial. All four feet are used. Rabbits and hares are the best examples.
    b. Ricochetal. Only the hind feet are used. The large species of kangaroos are the best example.
4. Amphibious. Muskrat *(Ondatra zibethica),* beaver *(Castor canadensis),* water shrews, otter *(Lutra canadensis),* and nutria *(Myocaster coypus).*
5. Aquatic. The mammals spend the majority but not all of their time in the water. Seals.
6. Marine. These mammals never come on land. Whales.
7. Fossorial. Mammals that are specifically adapted for spending their entire life underground. Pocket gophers, moles.
8. Arboreal. Climbing trees and jumping from limb to limb by use of all four feet. Tree squirrels.
9. Brachiating. Swinging from tree to tree using the front feet (hands?). Primates, such as the gibbon, *Hylobates lar.*
10. Volant. Gliding, glissant. Usually there are some adaptations visible for gliding, such as a dorsoventrally flattened arrangement of hair on the tail and a flap of skin between the front and hind foot on each side. Lemur, flying squirrels, flying phalangers.
11. Flying. Many modifications for flight. Bats.
12. Graviportal. Elephants. Limbs pillarlike. Articulating end flattened.

PLANTIGRADE (AMBULATORY)

This is the central type of locomotion, from which the other types have radiated. Mammals in this category walk on the entire foot and are typically five-toed. The metacarpals (hand bones) and metatarsals (foot bones) are not fused and are longer than the phalanges (toe bones). The ankle and wrist bones permit moderate movement in various planes. In the larger mammals, e.g., the bears *(Ursus)*, the locomotion tends to be more ambulatory; in the smaller mammals, e.g., opossums and shrews *(Didelphis virginianus)*, the locomotion tends toward the running or cursorial types.

In bears the legs are to the side (D. D. Davis, 1949); in shrews and mice the legs tend to be more directly underneath. Many primates, including human beings, use an ambulatory bipedal plantigrade type of locomotion.

The foot structure of bears and that of human beings is somewhat alike, and their footprints are sometimes confused by the uninitiated. Complicating matters still more are the oversized footprints produced by superimposing the hind foot partly over the print made by the adjacent forefoot. It is probably this configuration of the footprints of the Asiatic brown bear *(Ursus arctos isabellinus)* that nourishes the tale of the Yeti, or Abominable Snow Man, of the Tibetans and the romantics.

CURSORIAL (RUNNING)

Cursorial locomotion is restricted to the surface-oriented larger mammals that rely on speed for survival, or for catching prey. This is the type of locomotion found most often in larger mammals, including carnivores, horses, and zebras *(Equus)*, deer *(Odocoileus)*, pronghorn *(Antilocapra americana)*, African antelopes, cattle *(Bos)*, bison *(Bison)*, and giraffe *(Giraffa)*; it reaches its ultimate development in the ungulates living on the plains. Cursorial mammals tend to have an elongated body and neck. The elongated neck can be used to shift the center of gravity forward as the animal gains momentum. A white-tailed deer *(Odocoileus virginianus)*, for instance, when moving at its greatest speed, stretches its neck as far forward as possible. Additional adaptations for cursorial locomotion, which are not skeletal, include the senses of sight, smell, and hearing, which are well developed but variable. In the forest-dwelling white-tailed deer *(Odocoileus virginianus)*, the sense of smell and hearing seem to be keener than that of sight. The external pinnae of cursorial mammals are enlarged and can be independently focused, in the manner of a radar screen. Flash patches, such as the lighter hair on the rump of elk *(Cervus canadensis)*, and pronghorn *(Antilocapra americana)*, are common.

The limbs are lengthened, and there is a tendency toward simplification by fusion or by loss of the metacarpal bones (as in the human palm) and metatarsals (in the human instep) into what are commonly called cannon bones. This was the name applied to them by the scientists (especially paleontologists) of the Western plains, where they often stuck out of the ground at angles remindful of a cannon.

The joint surfaces become tongue-and-groove joints, rather than ball-and-socket joints, allowing the limbs to move mostly in a single plane, parallel to the long axis of the body. Depending on degree of ground contact, cursorial locomotion has been divided into digitigrade (walking on toes, as in most carnivores) and unguligrade [walking on tips of toes, as in bison *(Bison bison)*, deer *(Odocoileus* sp.), and horse *(Equus)*]. Digitigrade locomotion and fore-limb adaptations in such carnivores as the lion *(Panthera leo)*, leopard *(Panthera pardus)*, and cheetah *(Acinonyx jubatus)* have been discussed by Hildebrand (1959, 1961, 1962) and Hopwood (1947). Hildebrand (1961) estimated the maximum speed of the cheetah as a very respectable 56 mi/h. It gallops in a "measuring-worm fashion," adding an extra 15 in. each time the spine is being flexed and again when it is being extended (Fig. 4–2). The synchronous wheeling of a herd of pronghorns *(Antilocapra americana)* in full flight is a beautiful sight, enhanced even more if viewed from a plane. Pronghorns are among the most cursorial of mammals.

One of the early discussions of locomotion in hoofed mammals was that of Gregory's, published in 1912. The lengthening of the limbs, already mentioned, is characteristic of such mammals. The reduction or loss of the clavicle frees the scapula and shoulder joint from a solid connection with the sternum. This reduces the shock to the skeleton when the forefeet strike the ground. Since the scapula is not anchored, the length of the stride can be increased. Weight of the distal parts of the limbs has been reduced by a reduction in number of bones, and therefore in musculature. Most obvious is the reduction of the number of toes—for the horse *(Equus)*, to one. In the horse *(Equus)* are springing ligaments in the hind and front feet (Camp and Smith, 1942)

Figure 4–2 Cursorial locomotion in the horse or cheetah, with maximum flexion and extension of spine and maximum rotation of the scapula. *(After Hildebrand, 1959.)*

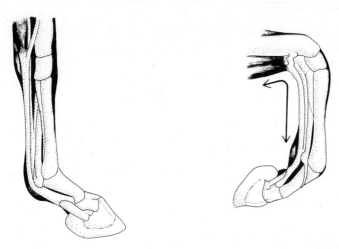

Figure 4–3　The springing ligament of horses (and other ungulates) pulls the hoof up and back as the foot is lifted. *(After Camp and Smith, 1942. Originally published by the University of California Press; reprinted by permission of The Regents of the University of California.)*

(Fig. 4–3). These are also found in other ungulates, such as the pronghorn *(Antilocapra americana)*. They arise on the cannon bone and are attached to the first phalanges of digits III and IV (phalanges 2, 3, and 5 have been lost). The ligament is stretched when the foot is on the ground, and when the foot leaves the ground the phalanges snap backward, thus helping to increase the running speed. This familiar backward flip, so easily observed in running horses, increases speed. It is possible that this ligament is partly or wholly responsible for the clicking noise emanating from the hooves of the Caribou *(Rangifer tarandus)*. This noise is said to be audible for 90 m (100 yd) or more in calm weather. Other explanations include interplay of carpal or sesamoid bones, and the breaking or rejoining of cohesion between various tissue layers (Kelsall, 1968).

JUMPING (SALTATORIAL)

Many morphologic characters of locomotion by jumping are closely similar to those for cursorial locomotion. Some kinds of mammals always move by jumping on all fours; others may at times use all fours but at other times use only the hind feet; and still others primarily use only the hind feet. Though cursorial mammals can elude their predators by speedy running, evasive action (e.g., moving in zigzag patterns) is used by jumping mammals to elude their enemies. Adaptations for travel by jumping vary from slight to great. As already mentioned, many of these adaptations are closely similar to those for running; indeed the mule deer *(Odocoileus hemionus)* of the West does at times move by jumping, yet it has no specialized features for locomotion beyond those of the white-tailed deer *(Odocoileus virginianus)*. Rabbits, hares, and jumping mice have longer hind legs, which are more muscular than the forelegs, and often move by jumping on all fours. There are a few kinds of mammals which sometimes move by jumping on all fours but at other times, when they wish to increase their speed to the maximum, resort to bipedal saltation, which Hatt (1932) called *ricochetal* locomotion. In other words they use both types of locomotion. This is true of kangaroo rats *(Dipodomys)* and many others. Then there are some kinds of mammals whose movements are primarily or only ricochetal, e.g., most species of kangaroos. "Saltatorial" means leaping, jumping, or dancing, and probably suffices to describe all jumping movements; "ricochetal" refers exclusively to bipedal saltation.

Hatt (1932) has discussed ricochetal locomotion in detail. It has evolved independently in several groups, especially among desert or arid-land mammals, e.g., in the African and Eurasian dipodids (Howell, 1932; Bartholomew and Caswell, 1951; Bartholomew and Cary, 1954), some African and Eurasian gerbellines, North American heteromyids, the South African genus *Pedetes,* and the elephant shrews of North Africa (F. D. Evans, 1942). The best-known examples are, of course, the kangaroos of Australia, although they are not all residents of arid lands.

Many morphologic features are associated specifically with saltation. The

Figure 4–4 The kangaroo rat *(Dipodomys)* uses both saltatorial and ricochetal types of locomotion. The long tail, tufted at the end, very long hind legs, and shorter front legs are all adaptations for a jumping type locomotion.

body tends toward stockiness, with the center of mass (gravity) shifted posteriorly, an extreme case being seen in the kangaroos. This posterior shifting of the center of gravity is often aided by an increase in the length of the tail, which frequently ends in a large tuft of hairs. The neck is shortened as a result of the shortening of the cervical vertebrae, but never by their reduction. The tarsi, metatarsi, and toes of the hind limb are elongated, the tarsi may be partly fused (as in the jumping mouse, *Zapus hudsonius*), and sometimes the number of toes is reduced. There is a reduction in the size of the forelimbs, to the point of uselessness for locomotion in some species, which are used only for digging or for manipulation.

Those jumping mammals which live in arid regions with scant vegetation also have greatly inflated bullae and rather delicate bones in the skull. Howell (1932) has described the gross anatomy of the ear of the kangaroo rat *(Dipodomys)*. The distortion in the shape of the skull is caused by the hypertrophy of the mastoid portion of the temporal bone. Webster (1961) has suggested that these modifications are helpful in ricochetal locomotion and in escaping enemies. He wrote, "Sonograms of sounds produced during an owl's flight and a rattlesnake's strike show frequencies within the kangaroo rat's most acute hearing range." Such keen hearing undoubtedly has survival value.

AMPHIBIOUS, AQUATIC, AND MARINE

Adaptations to locomotion and to living in water vary from small to enormous. All are secondary, i.e., they have developed from previously terrestrial mammals. Those mammals with the least modification are categorized as *amphibious;* they include the beaver *(Castor canadensis)*, muskrat *(Ondatra zibethica)*, nutria *(Myocaster coypus)*, otter *(Lutra canadensis)*, and to a lesser extent the mink *(Mustela vison)* among the North American mammals. Usually there is an increase in the thickness and quality of hair in these mammals; in the Temperate Zone it was the pelts from this type of mammal which contributed to the bulk of the trade in the fur industry. Quite often the tail is modified for water locomotion. In the beaver *(Castor canadensis)*, it is flattened dorsoventrally. In the muskrat *(Ondatra zibethica)* and nutria *(Myocaster coypus)* the tail is laterally flattened. The surface area of the feet has been increased in several ways. One of these is webbing in various degrees. In the beaver *(Castor canadensis)* the hind feet are completely webbed; in the muskrat *(Ondatra zibethica)*, they are partly webbed; and in the nutria *(Myocaster coypus)*, the web extends for less than a third of the length of the toes. Another way of increasing the surface area is by the addition of stiff hairs at the lateral edges of the toes and feet. This is the situation, for instance, in the water shrew *(Sorex palustris)* and the muskrat *(Ondatra zibethica)*. In some of these mammals, notably the beaver *(Castor canadensis)*, the lips can close behind the incisors, enabling them to be used under water.

Aquatic mammals are those which spend most but not all of their time in the water. It is usual for them to come to land for reproduction. The seals (order Pinnipedia) come to mind immediately, but the hippopotamus *(Hippopotamus amphibius)* also fits into this group. In the Pinnipedia (as also in the whales, Cetacea, which are marine) the principle mass of the body is anteriorly located, and along with this a certain body rigidity has developed, again most extremely in the cetaceans. The front and hind limbs have been highly modified for swimming, with all the bone elements enclosed in and supporting a paddle or fin.

The seals and walruses (order Pinnipedia) have two types of terrestrial locomotion. In the eared seals (family Otaridae) and the walrus (family Odobenidae) the hind limbs turn forward in a normal four-legged manner for travel on land, but turn backward for swimming in the water, and both forelimbs and hind limbs are used for rapid speeds, but at slow speeds only the forelimbs support the body, as it hitches forward. In swimming, the fore-limbs are pressed against the body and the hind limbs stroke alternately; sometimes both hind limbs are employed simultaneously. The tail of the seals has been severely reduced, nearly eliminated. The trend in the skeleton is toward simplification, the neck is shortened, and altogether the axial skeleton has become rather inflexible. Adaptations, other than skeletal, for an aquatic life include fur as an important insulating material, but a layer of subcutaneous fat also aids in conserving heat. The external pinnae are absent or reduced, and the ear can be closed. The nostrils are modified so that they can be closed. This modification also appears in some marine mammals. The eyes of seals are best adapted for sight below water; they are myopic on land (Walls, 1942). The olfactory bulbs of the brain are poorly developed.

The cetaceans and sirenians represent the greatest adaptations to life in the water, for they never leave it. This adaptation is categorized as *marine* in this discussion. The skulls of these mammals are greatly changed from the skulls of the other mammals. The nostrils are on the dorsal side, with only one external opening. The nostrils can be closed by valves. In the toothed whales (suborder Odontoceti) the skull is asymmetric.

The seven cervical vertebrae are partially fused into one short, rigid neck unit. The posterior appendages are gone, and all that remains of the pelvic girdle are two small bones which are unattached. There is no sacrum. The tail is modified into a horizontal *fluke,* which is the propelling organ.

There are necessarily many other adaptations to a marine life. The ears of whales do not function as ours do but rather in an extraordinary manner. A discussion of ear modifications in whales can be found in the section on communication in Chap. 12.

The cetaceans possess no vocal cords (Haan, 1960), and the sounds they emit probably emanate from a vibrating part of membrane located inside the blowhole. Lawrence and Schevill (1956) have described the anatomy of the blowhole in a delphinid whale. There are no tear ducts, no functional eyelids, and no Meibomian glands in the cetacean's eye.

The body shape is ovoid. The surface of the skin is hairless except for vibrissae or embryonic vibrissae, and underneath the skin is a thick layer of blubber. There are few glands.

FOSSORIAL

Those mammals which spend the majority or all of their time underground and have special adaptations for this way of life are called fossorial. Most of them are small, and of those kinds whose activities are all below the surface, the pocket gophers are the largest. The badger *(Taxidea taxus),* extremely adept at burrowing, and with many fossorial adaptations, still spends much of its time above the surface and might be considered to be semifossorial. In fossorial mammals, such as the pocket gophers, the body is fusiform in shape, quite blunt at the anterior end, with the largest mass and "seat of power" at the pectoral girdle, then tapering toward the posterior end. The profile of the head is triangular and flat; this is especially notable in *Spalax,* a molelike rodent of the eastern Mediterranean area. *Spalax* is unique in using primarily its incisor teeth, rather than its front feet, for burrowing.

In addition to modifications of the skull there are also modifications of the postcranial skeleton. The legs are shortened, with the major modifications in the forelegs and the pectoral girdle, which increases the strength of the legs for digging. In pocket gophers and moles the olecranon process is enlarged, allowing for a larger surface for the attachment of the triceps muscle. The humerus is short and massive, with many tuberosities. The scapula is also massive, with enlarged processes for muscle attachment. In the mole *(Scalopus*

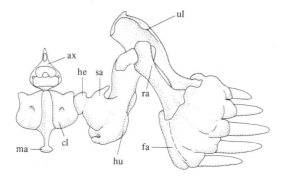

Figure 4–5 Fossorial adaptations for digging as shown in the forelimbs and pectoral girdle of a mole, *Scalopus aquaticus.* The humerus, radius, and ulna are especially robust. ar, articular surface of the clavicle; ax, axis; cl, clavicle; fa, position of the falciform bone; he, head of the humerus; hu, humerus; ma, manubrium of the sternum; ra, radius; sa, secondary articular surface of the humerus; sc, scapula; ul, ulna. *(After Vaughan, 1972.)*

aquaticus), the bones and muscles of the pectoral girdle are greatly modified. The presternum is lengthened so that the pectoral girdle is moved forward. The scapula is greatly lengthened to allow for the forward movement of the girdle (Hisaw, 1923). Sometimes there is even partial fusion of the cervical vertebrae. Kirstin (1929) has given a detailed description of how the clavicle of the mole *(Scalopus, Talpa europaea)* has been adapted for a fossorial life. The clavicle articulates with the humerus. It is a very short, massive bone, in contrast to the normally long, thin clavicle of most mammals. It is carried forward on a very much elongated sternum, causing it to lie much more anterior than the usual clavicle. The humeroclavicular articulation has evolved in response to the great lateral stresses involved in digging. The neck is shortened.

The distal part of the forelimb of fossorial mammals can be modified for digging efficiency in one of two ways. One of these is as a scoop, with extra (sesamoid, heterotropic) bones, and short, rugged claws. In the mole *(Scalopus),* the palmar regions of the forefeet are furnished with a fringe of stiff hairs, some of which are thought to be tactile (Kirstin, 1929). The other way of gaining digging efficiency is found in fossorial mammals such as the pocket gophers and in many of the semifossorial mammals such as the badger *(Taxidea taxus),* whose claws are greatly lengthened. To make up for the wear and tear of digging, some of the front claws grow at a great rate. Howard (1953) measured the growth of the front claws of the western pocket gopher *(Thomomys bottae)* and found the three center claws grew 0.23 mm/day, or over 0.84 cm/year. This was twice as fast as the two outer claws grew!

The fur is short and very dense. The fur of the mole has almost the same appearance whether it is stroked backwards or forwards. The skin is loosely attached to the body, allowing the body to move freely within the skin.

The sense organs of sight and sound are variously reduced. In pocket gophers the eyes are small but functional. In the Bathyergidae (the fossorial mole rats of Africa), the eyes have taken on the functions of detecting air currents and temperature changes (Elloff, 1958). In *Spalax* (the fossorial mole rats of the eastern Mediterranean region), the vestigial eyes are completely covered, as they are in the golden mole *(Chrysochloris),* a fossorial mole of Africa, and the marsupial moles of Australia *(Notoryctes).* Eyes of North American moles are functional (Arlton, 1936). External pinnae are reduced or absent, and auditory bullae, when functional, are small.

The olfactory senses are very important. The sense of touch has been increased by the addition of vibrissae on the nose and of tactile hairs on the tail as well as at the edge of the palmar surfaces on the forelimbs. The tails of fossorial mammals are relatively short and, with their supply of nerves and tactile hairs, are strongly suspected of serving as sensory organs. This would be particularly helpful in navigating backwards in the darkness of their underground world.

GRAVIPORTAL

This type of locomotion is best pictured as a means of locomotion on pillars which are carrying great weight, best illustrated by elephants. These mammals have five digits, distributed in a circle around the edge of each foot, and a pad of elastic tissue under the foot. A common evolutionary trend among very large mammals is the lengthening of the limb bones and the development of short, broad feet. The limbs are straight and pillarlike, the upper part of the limb longer than the lower section. The fibula and radius are almost as large as the ulna and tibia. The bones flatten out on their articulating surfaces. These adaptations are also termed subunguligrade or rectigrade.

ARBOREAL, VOLANT, BRACHIATING, AND AERIAL, OR FLYING

Some mammals, living in forested areas, have become adapted for using the trees. This adaptation has received the name *arboreal.* These individuals are able to climb the trees and use the branches as their highways in search of cover, food, homes, and mates. Several modifications for holding onto tree branches have developed. The tree squirrels and sloths have well-developed nails. Some arboreal primates living in South America have prehensile tails, while the tarsier has adhesive disks on its front toes. Arboreal adaptations may be divided into three categories, illustrated by sloths, squirrels, and primates.

The *sloths* spend the majority of their time hanging upside down in trees. So sedentary are they that moss grows on their fur. The skeletal adaptations of the sloth can most easily be described by saying that the skeleton forms a basket for the animal's soft anatomy, except that the sloth has not learned to view its world in an upside-down manner! Although their necks are very short, they can turn them at least 90°, so their heads remain right side up. In the two-toed sloth *(Choleopus didactylus)* this is accomplished with the usual number of seven cervical vertebrae, but in the three-toed sloth *(Bradypus didactylus)* there are nine cervical vertebrae. The shoulder girdle is very strong, and the clavicle well developed. There is also an increase in the number of ribs over the usual 13 found in most mammals. The bones of the limbs are long and cylindrical, the front limbs being longer than the hind limbs. The joints are of the ball-and-socket type. The radius and ulna are separate. Each digit ends in a long hook-shaped claw. The sacrum is long and solidly fused to the pelvic girdle. The sense organs are poorly developed, and the nostrils and pinnae are small.

The *squirrel* type of adaptation is the one used for climbing and jumping rather than hanging. Among the kinds of mammals adapted for this type of locomotion are the tree squirrels, some opossums, some prosimians, and the tree shrews. In the typical tree-squirrel type the body is elongate and the musculature of the hind legs is more strongly developed than that of the front legs. The claws are always well developed and sharp. The hind legs can be rotated in such a manner that they can be used in climbing up or coming down head first. The sense organs are well developed, as are the eyes and external pinnae. Tactile perception seems important (tree cavities are utilized), and the vibrissae of tree squirrels are longer and more numerous than those of ground squirrels. The vertebral column is unspecialized. The scapula and clavicle are well developed. Although the hind limbs are longer than the forelimbs, all are relatively short, with ball-and-socket joints. When any elongation takes place, it is usually in the digits. The femur is usually both short and slender, and sometimes the tibia and fibula are fused, increasing their strength.

A rather specialized type of arboreal locomotion is *brachiation,* or swinging from branch to branch by use of the forelimbs only. Skeletal modifications for this feat usually involve flexibility, with the greatest emphasis on the forelimbs and the pectoral girdle. The scapula and clavicle are well developed. The forelimbs are greatly lengthened, a modification which involves all the elements except the podials. There is an extreme elongation of the phalanges in some forms, especially notable in the gibbons, and the forelimbs reach the ground. This makes the animals look awkward on the ground when they resort to ambulatory bipedal locomotion. All the joints, except the elbow, are of the modified ball-and-socket type. Sometimes the thumb is reduced to a stump, as in the Pongidae. It would only get in the way in trying to grab a branch. In the primates, sharp, compressed claws are generally replaced by flattened nails, although there are many variations (Clark, 1936). The eyes and ears are well developed, and stereoscopic vision must be very good indeed to allow a gibbon to judge a 13-m (40-ft) jump to another branch with precision.

Volant and *glissant* are interchangeable names applied to a gliding type of locomotion. The flying squirrels (Rodentia), flying phalangers (Marsupialia), and the flying lemur (Dermoptera) have extra surfaces formed by extra flaps of skin and sometimes by a dorsoventral flattening of the fur along the sides and the tail. Although the limbs of sciurids are grossly similar (Peterka, 1936), there is an additional slender cartilage or "oss accessoire" in the wrist joint of *Glaucomys* (Gupta, 1966). This cartilage articulates with the pisiform, the unciform, and the lateral half of the fifth metacarpal. It extends laterally and supports the anterior margin of the gliding membrane. This extra flap of skin, the *patagium,* may extend from front to hind leg on both sides, as in the flying squirrels and phalangers, or it may completely circle the animal, and connect the head, forelimbs, hind limbs, and tail, as in the flying lemur or colugo *(Cyanocephalus).* It is conceivable that by flapping these membranes some type of flight could be achieved in the future. Whether true flight evolved from extreme cursorial adaptations or from the volant type of locomotion is a subject of debate. Savile (1962) and others have argued effectively for the volant type, which in turn developed from simpler arboreal adaptations. Gliding, or parachuting, probably evolved as a means of protection from injury in an accidental fall, and even as a means of escape. According to Savile (1962), gliding (parachuting) means a reduced rate of fall, even at a gradient steeper than 45°. It involves balance but little or no steering. Even red *(Tamiasciurus hudsonicus)* and gray squirrels *(Sciurus carolinensis)* when shaken out of a tree come down in the same position as a flying squirrel, feet spread and "belly down." The first requirement, then, is behavioral. Any small animal that can control its attitude during a fall could become the ancestor of a glider. Gliding has evolved at different times in different groups of animals as well as among different groups in the class Mammalia.

True *flight* exists in only one group of mammals, the order Chiroptera, or bats. The word *chiroptera* literally means "a winged hand." More colorful is the German name for bats, *Fledermaus,* meaning "flying mice." The "winged hands" are the lifting surfaces required for flight. They also serve as a means of propulsion. For greater efficiency in flight the concentration of weight is near the center of gravity. Although both bats and birds use flapping flight (as opposed to gliding flight), their flight styles differ, according to the senses used for orientation. Birds use visual orientation and can recognize objects at considerable distances, thus alleviating the need for instant maneuverability. On the other hand, the majority of bats are nocturnal, pursuing food which moves about. For their orientation they use a very sophisticated means of detecting objects, by receiving echoes of sound signals they have emitted. This system, called *echolocation,* is discussed in Chap. 12. This system usually allows only short-range perception of obstacles or prey. It requires a slow flight and great maneuverability. Toward this objective, many modifications have occurred. What follows is taken from Vaughan's (1959, 1970) interesting reports on his research on chiropteran flight.

The hand and arm have been modified into a wing. The greatly elongated radius (the ulna is reduced and partly fused with the radius) and digits II to IV of the hand support a membrane, the *patagium,* which serves as the wing. The only clawed digit is the short thumb. Muscle strands and elastic fibers further brace the membrane. Specialized wrist and elbow joints help maintain rigidity of the wing during flight by limiting the wing movement to an anteroposterior plane. The clavicle and scapula are also modified. The distal manipulation of the wing is accomplished by a transfer to the pectoralis muscle, which lies closer to the center of gravity, thereby allowing a reduction in the amount of the distal musculature.

Another web, the *uropatagium,* extends between the hind legs and encloses the caudal vertebrae. In addition to supplementing the lifting power of the wings, its increased surface adds to the maneuverability of the bat. During flight the legs are spread wide apart and additional surface is gained by the *calcar,* a small bone that articulates with the calcaneum of the ankle. The calcar serves to spread and reinforce the posterior edge of the uropatagium.

The hind limbs are rotated from the typical terrestrial mammalian position by either 90°, when walking on all fours, or 180° when the individual is suspended upside down.

Classification
and the Kinds of Mammals

CLASSIFICATION

Classification is the naming and organizing of a group of related organisms. A classification serves two purposes, according to Simpson (1945). One is to provide a convenient label by which zoologists (everywhere in the world) and others know what each is writing or talking about. Common names do not serve this purpose, since the name may apply to several kinds of animals. Depending on where you live in North America, a gopher may be a ground squirrel, a snake, or a turtle. The other purpose of a classification is to show affinities or relationships.

The basic unit of classification is the species, of which there are many definitions. Our present idea is that a species is an assemblage of animals which interbreed freely, or can do so when they meet, and do not interbreed with members of other assemblages. This concept does not help much if a classification is built on a small number of dead specimens, but at least it narrows the limits to a definition based on genetics. Although the species is very difficult to define, it is central to taxonomy. Mayr (1942), an ornithologist, called a species "groups of actually or potentially interbreeding natural populations which are reproductively isolated from other such groups." Dobzhansky (1951), a geneticist, called a species "the largest and most inclusive reproductive community of sexual cross fertilizing individuals which share in a common gene pool." Simpson (1945), primarily a vertebrate paleontologist, defined a species thus: "A genetic species is a group of organisms so constituted and so situated in nature that a hereditary character of any one of these organisms may be transmitted to a descendant of any other." Yet another definition is that of Florkin (1966), a biochemist, who wrote:

> The increasingly molecular approach to biology brings us to consider a species as consisting of groups of individuals with more or less similiar combinations of sequence or purine and pyrimidine bases in their macromolecules of DNA, and with a system of operators, controllers and repressors leading to the biosynthesis of similar sequences of amino acids, the integration of which, in one cell, or in a number of variably differentiated cells, leads to similar structural and functional characteristics, adapted to the ecological niche in which the species flourished.

Simpson (1961) has defined an evolutionary species (biospecies) as "a lineage (an ancestral-descendant sequence of populations) evolving separately from others and with its own unitary evolutionary roles and tendencies."

Mayr (1963) has emphasized the three aspects most frequently stressed in the biologic species concept. These are: "(1) species are defined by distinctness rather than difference; (2) Species consist of populations rather than of unconnected individuals; and (3) Species are more unequivocally defined by their relation to non-specific populations ("isolation") than by the relation of conspecific individuals to each other."

The earlier taxonomy, dealing mostly with species descriptions, was based on morphologic characters. The "new systematics" is a synthesis of a variety of approaches, including behavior, cytology, ecology, genetics, geography, morphology, and physiology. A perusal of much recent systematic work, however, shows it to be overwhelmingly morphologic.

The switch from a type concept to a population concept has been gradual and has gone through three stages, according to Mayr, Linsley, and Usinger (1953). They have called the stages: (1) alpha taxonomy (the identification stage), the level at which species are characterized and named; (2) beta taxonomy (arrangement into groups, classifications), the arranging of these species into a natural system of lesser and higher categories; and (3) gamma taxonomy (evolution, species formation), or the analysis of intraspecific variation and the study of evolution. These are not, of course, distinct and separate stages, but all three may be used in the most modern of systematic studies.

MECHANICS OF CLASSIFICATION

The orderly arrangement of species into groups or categories on the basis of their relationships is what is commonly or familiarly called "classification."

Animals are grouped in larger categories and in a systematic framework, or hierarchy.

The hierarchy generally used for most zoologic nomenclature is a sequence of seven levels—kingdom, phylum, class, order, family, genus, and species.

Linnaeus's tenth edition of *Systema Naturae* used only five levels; Simpson (1945) used 21 for mammals, beginning with the kingdom. The addition of *super-* (e.g., superclass, superfamily) above each level, and of *sub-* (suborder, subfamily) and *infra* -(infraclass) below each level would give 34 categories. The use of any number of levels is arbitrary and should be decided upon a criterion of need. For a better understanding of these categories we might take a relatively common mammalian species and place it in this hierarchy. It is, of course, a member of the animal kingdom (kingdom Animalia), so we start in this way:

Kingdom Animalia
  Phylum Vertebrata
    Class Mammalia
      Order Rodentia
        Family Sciuridae
          Genus *Spermophilus*
            Species *Spermophilus tridecemlineatus*

Specific names are made up of two words (binomen). The first is the name of the genus; the second, treated as a Latin noun, is often, but not always, descriptive. *Tridecemlineatus* means "13-lined." Subspecies names, called trinomials (trinomen), are made up of three names, of which the first is treated in the same way as the second word.

The name of the author and the year of first publication (law of priority), discussed later, can be added to the zoologic name. Usually these are omitted except in taxonomic work and listings. Scientific names have usually been italicized, but the author's name is not, and there is no intervening punctuation. In some recent publications names have not been italicized, and this may become more common with newer methods of reproducing written material.

Family names always end in *-idae*—e.g., *Sciuridae.* The superfamily name is formed by adding *-oidea* to the stem; the subfamily name has *-inae* added to the stem; and the tribe has *-ini* added to the stem—thus *Sciurini.*

It is quite obvious that it is not an individual that is classified, although it is the individual to which we apply the handle. But we are describing the individual in terms of its limits of variation, and though we work with what we hope is an adequate sample of the population, it is not the sample but the population that is being described. A population, as already discussed, is usually taken to mean any group of organisms systematically related to one another.

A group of organisms recognized as a formal unit, no matter at what level, is called a *taxon*. For instance, *Bison* (a genus), *Bison bison* (a species), and *Artiodactyla* (an order) are all taxa (the plural form of taxon).

It is now widely recognized that a taxon cannot be characterized by one specimen, the *type* specimen, although one may be named as the name bearer. The modern taxonomist realizes that the type specimen no longer represents the taxonomic sample. Many other kinds of "types" have been proposed, of which only a few are defined here. Other names applied to specimens for mechanical purposes are:

Holotype   A synonym for *type*.
Syntype   Two or more specimens treated as types. Usage common in the past but now discouraged.
Lectotype   A type specimen, picked from the syntypes after the original publication.
Allotype   A paratype of the opposite sex to the *type*.
Neotype   A specimen selected as a type subsequent to the original description, in cases where the original types were destroyed or have become invalid.
Paratype   A specimen other than the type specimen which the author used in his original description.
Topotype   A specimen collected at any time at the locality of the type specimen, but not of the original series.

Zoologic nomenclature is still far from having all its problems solved, but it has come a long way since the first rules were promulgated by a committee appointed by the British Association for the Advancement of Science. Published in 1842, these were known as the Strickland Code. In 1877 the American Association for the Advancement of Science published the "Dall Code," a good essay on zoologic nomenclature but never formally adopted.

It became evident that it would be necessary to formulate international rules if nomenclature was to become an internationally useful tool. The code in use today has been revised many times but was first adopted in 1901 by the First International Congress. The most recent international code published was that of the Fifteenth International Congress, adopted in 1961, with N. R. Stolle as chairman of the editorial committee.

The rules of zoologic nomenclature are contained in the *International Code of Zoological Nomenclature*. The most recent version was dated November 6, 1961, with some changes adopted in the 1964 editions. Some of the more important rules are presented here in a very simplified version. They will point up the complexity of zoologic nomenclature.

The law of priority states that "the valid name of a taxon is the oldest available name applied to it." There are of course exceptions and interpretations. Sometimes after a scientific name has been in use a long time some taxonomist may find an older name in a very obscure publication; the name currently in use must then be changed.

The valid name of the taxon should be a Latin word, a latinized word, or one so constructed that it can be used as a Latin word. The author (authors) of a scientific name is (are) the person (persons) who first publish (es) it. The name of the author does not form part of the name of a taxon, and its citation is optional. The date of a name is the date of publication. It must be reproduced in ink on paper by some method that assures numerous identical copies. Reproduction by microfilm, mention at a scientific meeting, or anonymous publication does not constitute publication. A name must not be based on a fictitious species, as for instance a unicorn or a Martian mouse, but must represent actual specimens.

A very difficult and complicated problem in nomenclature is that of *homonymy.* A homonym is a single name that has been given to two or more different taxa. The earliest such name becomes the senior homonym. The opposite situation, *synonymy,* in which two or more names are applied to the same taxon, also creates confusion. For detailed information the reader can refer to Blackwelder (1967) or Mayr (1969), among others.

THE PHILOSOPHIES OF CLASSIFICATION AND CHARACTERS USED AS TAXONOMIC DATA

The evolutionary (phylogenetic, phyletic) philosophy of classification stems almost from Darwin's time. Each valid taxon should have a common ancestry, but at times convergent evolution may have brought together different populations into a population whose individuals are very similar. Even characters in common do not always indicate a common ancestry. A difficult task for the taxonomist is to determine just what to "split" and what to "lump." This involves the basic problem of distinguishing among various kinds of similarities, and especially of recognizing those which are and are not inherited from a common ancestry. Thus, the phylogenetic taxonomist uses weighted characters, i.e., he gives more importance to one character than to another.

Classification has also embraced the methodology of multivariate statistics and digital computers in a new system generally referred to as numerical systematics. The most recent summary of this system is found in Sneath and Sokal (1973). In taxonomy the key word is "comparison," to find out what is similar or what is different.

Phenetics, or numerical phenetics, is a philosophy based upon the phenotypic similarity between taxa or operational (definable) taxonomic units (OTUs). The method involves the selection of a large number of operational characters, which are, at least initially, assumed to be equal to every other character in importance. The data matrix is then accumulated and processed

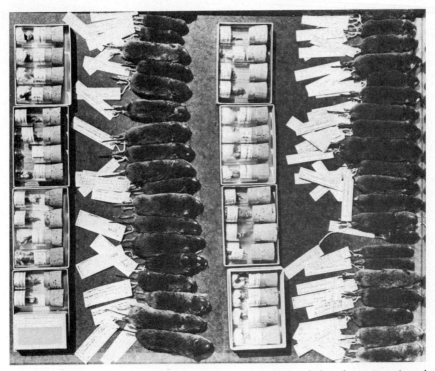

Figure 5–1 Species descriptions are based on a sample population. A great number of specimens must be examined to reflect the characteristics of an entire population.

with the aid of computers and of multivariate statistical techniques. Similarity and distance between OTUs are usually illustrated with a two-dimensional phenogram (Fig. 5–2). The closeness of the terminal points indicates degree of similarity, not phylogeny.

Yet another philosophy of classification is *cladistics (cladism),* or the study of branching sequences in evolution, but various authors have applied different meanings to the term. Phenetic classifications, in the strictest sense, do not employ time sequences (evolutionary history), but Camin and Sokal

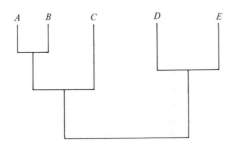

Figure 5–2 Two-dimensional phenogram. Nearness of the terminal points indicates similarity, not phylogeny.

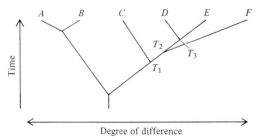

Figure 5-3 Two-dimensional phylogenetic dendrogram of relationships between taxa, showing time sequence.

(1965) discussed a method of adding time sequences to the phenetic system. They attempted to select the fewest number of evolutionary steps that could produce present-day taxa. A cladogram can be constructed using the techniques of Camin and Sokal (1965) which would show branching sequences in various lineages.

The various philosophies of classification overlap, and today statistical methods are used in all schemes; the reader might refer to Hennig (1966), Sokal and Sneath (1973), Camin and Sokal (1965), L. A. Johnson (1968), and Throckmorton (1968) for more discussion of the various philosophies.

Many kinds of characters have been used by taxonomists to show similarities or differences on which to base their taxa. The first and most obvious group would be *morphologic* characters. These were the characters used by the earliest taxonomists, and they are still the ones most often used. *Gross morphologic* characters include external, cranial, and dental measurements. Osgood's (1909b) revision of *Peromyscus* serves as a classic example. A study employing multivariate factor analysis to determine phenetic relationships among various taxa of *Myotis* by Findley (1972) incorporates recent techniques. Functional morphologic features were used by Vaughan (1959) in determining relationships between three genera of bats. Reproductive anatomy has been used by many. Mossman (1953) used the genital system and fetal membranes as criteria for mammalian taxonomy and phylogeny at "higher"

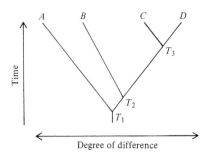

Figure 5-4 Two-dimensional cladogram showing inferred branches in evolutionary history.

levels. The entire glan penis has been used as a taxonomic criterion by Hooper (1959, 1962), Hooper and Musser (1964), and others (Fig. 5–5). The baculum (os penis) is found in many, but not all, mammals. It was first used as a taxonomic criterion in the Old World at the turn of the century (Gilbert, 1892; Tullberg, 1899). American mammalogists who have used this character include Burt (1936, 1960), Chaine (1925), Dearden (1958), and Hamilton (1946), among others. Detailed morphology studies include investigations of chromosomes and spermatozoa. Bender and Chu (1963) wrote, "Of all morphological characteristics, the karyotype or number and form of the chromosomes is the most intimately associated with the genetic makeup of an organism." The most recent review of mammalian karyotypes is that of Hsu and Benirschke (1968). Genoways (1973) utilized sperm morphology as one character in his systematic study of the rodent genus *Liomys*.

Biochemical characteristics—protein electrophoresis and immunodiffusion techniques—are recent additions to the list of characteristics used for taxonomy. Moody (1958) provided serologic evidence to show that the muskoxen *(Ovibos moschatus)* was similar to sheep and goats *(Ovis)* but "markedly dissimilar to domestic cattle and bison." Doolittle and Blomback (1964) studied the amino acid sequences in the blood fibrinopeptides of five artiodactyls. Whey and casein proteins of milk have been studied by Jenness and his associates. Sloan, Jenness, Kenyon, and Regehr (1961) studied the whey and casein proteins of some Artiodactyla, Perissodactyla, Carnivora, Rodentia, Primates, and Marsupialia. They wrote, "In general primitive surviving stocks of any order of mammals have fewer quantitatively important milk proteins than phylogenetically more recent derivatives."

Among *behavioral characteristics,* vocal communication is receiving a great deal of attention as a taxonomic tool. Waring (1970) used call patterns to differentiate three species of prairie dogs.

Attempts to solve the vexing problem of the pandas and their ancestries have used many of the above approaches, including morphology (the first descriptions), serum protein, serum albumin, and karyotype studies. There is still little agreement as to their classification.

Simpson (1960) wrote, "For all its progressive modifications and esoteric intricacies, zoological nomenclature is still far from solving all its problems, or from solving any one of them to the satisfaction of all. Nevertheless it does somehow muddle along well enough to be workable."

In spite of all the changes which plague the nontaxonomist, the objective is to produce a classification which will reflect evolutionary relationships and names which can be universally used and understood. If a scientist does not know the correct name of the organisms with which he works, the work is of no value. Taxonomy is therefore of great importance to all zoologists. As already discussed, taxonomists are evaluating a great variety of information other than morphologic characters. One example of this is a new scientific

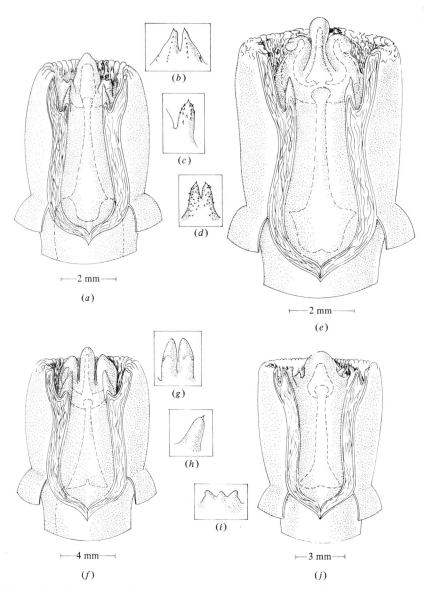

Figure 5–5 The use of glans penis in taxonomy. View of glans (incised midventrally), urethral flap (enlarged ventral aspect), and dorsal papilla (enlarged lateral aspect) in four species: *(a), (b) Thomasomys laniger; (c), (d), (e) Rhipidomys mastacalis; (f), (g), (h) Holochilus brasiliensis; (i), (j) Nyctomys sumichrasti,* laboratory-reared. *(From Hooper and Musser, 1964.)*

series which appeared in 1971, called *Advances in Systematics and Ecology.*

THE KINDS OF MAMMALS

Nineteen orders of living mammals are listed here, but not all mammalogists agree on this arrangement. The orders Marsupialia and Insectivora are sometimes split into two orders. The order Dermoptera is sometimes included as a suborder of Insectivora. The orders Cetacea and Carnivora are sometimes each split into two orders:

Class Mammalia
  Subclass Prototheria
    Infraclass Ornithodelphia
      Order Monotremata
  Subclass Theria
    Infraclass Metatheria
      Order Marsupialia
    Infraclass Eutheria
      Order Insectivora
      Order Dermoptera
      Order Chiroptera
      Order Primates
      Order Edentata
      Order Pholidota
      Order Lagomorpha
      Order Rodentia
      Order Cetacea
      Order Carnivora
      Order Pinnipedia
      Order Tubulidentata
      Order Proboscidea
      Order Hyracoidea
      Order Sirenia
      Order Perissodactyla
      Order Artiodactyla

SUBCLASS PROTOTHERIA

Order Monotremata

These egg-laying mammals, thought to be the sole survivors of the subclass Prototheria, possibly originated from some other stock than that which gave rise to the marsupial-placental line. A coracoid and precoracoid are present, as are the epipubic bones. There is no external pinna on the ear. Teeth are present only in young of the family Ornithorhynchidae. An egg tooth and egg tooth bone are present at hatching, cloaca is present, ureters open in dorsal wall of urogenital passage, penis conducts only sperm, testes are abdominal, oviducts are distinct and enter cloaca separately, there is no uterus or vagina. The single external opening for the reproductive, ruminary, and digestive systems gives this group its original name (i.e., *mono,* "single," *trema,* "hole"). The female is oviparous; mammary glands are without nipples. The egg-laying habits, the structure of the shoulder girdle, and the structure of the female reproductive system are reminiscent of the reptiles. It is generally stated that there is no fossil record of the monotremes prior to the Pleistocene, but Clemens (1970) believed that "the Australian Early Miocene *Ektopodon . . .* is most reasonably interpreted as a monotreme." Found in the Australian region. Families are:

Tachyglassidae, with 2 genera, 5 species
Ornithorhynchidae, with 1 genus, 1 species

SUBCLASS THERIA (MARSUPIALS AND PLACENTAL MAMMALS)

Ears usually with external pinnae. Teeth usually present, a milk set and an adult set (diphyodont), and differentiated as to form and function (heterodont). Generally no cloaca, ureters opening into base of bladder, testes usually in a scrotal sac. Vasa deferentia and bladder opening through a common urethra in penis. Females with two oviducts opening into a distinct and elongated vagina. Viviparous, i.e., young born in a fairly well-developed state.

Infraclass Metatheria, Order Marsupialia

The marsupials are a well-known group of mammals, in which the female usually has a ventral pouch, or marsupium, surrounding the nipples on the abdomen. Epipubic bones are present, and the coracoids are reduced to an apophysis on the scapula. The lower jaws have a different number of incisors than the upper jaws. The scrotum is anterior to the penis, and there is no baculum. The uterus and vagina are double, and the genital and urinary openings of the female are separate. A typical placenta is often lacking. The egg is fertilized internally; after a brief period of development in the uterus, the "larva" leaves and crawls to the marsupium, where it attachs to a mammary nipple and continues to develop until fully formed. The marsupials have diversified to where nearly each kind of placental mammal has an ecologic counterpart among the marsupials. Fossils are first known from the Lower Cretaceous of North America. Except for two families, the Didelphidae and Caenolestidae, all members of this order are found in the Australian region.

Didelphidae (12 genera, 66 species; Neotropical and Nearctic) The only United States opossum, *Didelphus marsupialis,* belongs to this family. The dental formula is 5/4, 1/1, 3/3, 4/4 = 50. The tail is scaly and prehensile. There are up to 22 embryos, which are "born" after a gestation period of 12 days, when they are the size of honeybees. They remain attached to the nipples for 50 to 60 days. Other species are found in Central and South America.

Caenolestidae (3 genera, 7 species) Members of this family, the "rat opossums," are found in western South America, in the Andes, and adjacent coastal areas of south central Chile. The dental formula is 4/3–4, 1/1, 3/3, 4/4 = 46–48. The marsupium is either absent or very small, but epipubic bones are present. The lower incisor is very large and horizontal, and the canine of the male is much larger than that of the female. These small mouse or shrew-like forms are poorly known. The best references are Cabrera (1958) and Osgood (1921).

Dasyuridae (20 genera, 50 species; Australian) The dental formula is 4/3, 1/1, 2–4/2–4, 4/4 = 42–50. Hind limbs are about the same length as front limbs. Dasyuridae are carnivorous or insectivorous, and the pouch may be absent. Their size range is from mouse size *(Sminthopsis)* to the wolf or large dog such as the marsupial wolf *(Thylacinus),* Tasmanian devil *(Sarcophilus),* and tiger cat *(Dasyurus).* A marsupial anteater, *Myrmecobius,* is the ecologic equivalent of anteaters on other continents.

Notoryctidae (1 genus, 2 species) Found in arid parts of northwestern and south central Australia. Fossorial, insectivorous, and marsupial mole, *Notoryctes.* The dental formula is 4/3, 1/1, 2/3, 4/4 = 44. Foreclaws enlarged for digging.

Peramelidae (8 genera, 22 species; bandicoots; Australian) The dental formula is 4–5/3, 1/1, 3/3, 4/4 = 46 or 48. Marsupium opening downward and backward. Incisors, many, small, and specialized; hind foot specializations with the fourth pedal digit the longest, second and third slender and clawed, first rudimentary or absent, fifth digit present and clawed. Chorioallantoic placenta without villi. Terrestrial and fossorial, largely insectivorous.

Phalangeridae (16 genera, 46 species; Australian) This group includes the *phalangers,* cuscuses, and *Phascolarctus,* the koala ("teddy bear"), whose marsupium, unlike those of most of the other members of this family, opens posteriorly. The dental formula is 2–3/1–3, 1/0, 1–3/1–3, 3–4/3–4 = 24–42. These are arboreal, herbivorous marsupials, including *Petarurus,* the flying phalanger, equivalent to the North American flying squirrels. The claws are sharp and the hallux is opposed, as one might expect in a primarily arboreal group.

Phascolomyidae (2 genera, 2 species; Australian)   The dental formula is 1/1, 0/0, 1/1, 4/4 = 24. The wombats are stout marsupials, adapted to a fossorial habitat and herbivorous diet. They are the marsupial equivalent of the placental woodchucks. The marsupium opens posteriorly; there is a pair of mammae. The incisors, 1/1, are large, rodentlike, and separated from the cheek teeth by a wide diastema.

Macropodidae (19 genera, 47 species)   This Australian family contains the wallabies, wallaroos, and kangaroos—or those animals which most typify the marsupials in the public mind. The dental formula is 3/1, 1–0/0, 2/2, 4/4 = 32–34. The incisors are 3/1. During the individual's life, the molars and premolars move forward, by bone deposition posteriorly and absorption anteriorly. The forelegs are greatly reduced, the hind legs greatly enlarged, and the hairy tail is long and thick at the base. All are adaptations for a ricochetal type of locomotion. The forefeet have five digits, but the hind legs have been greatly modified. Digits II and III are slender and united, i.e., syndactylous, digit IV is elongated and bears a long claw, and digit V is similar but smaller; hallux is usually absent. Adults vary in size from the hare wallaby, *Lagorchestes,* about rabbit size, to those which are 6 ft high *(Macropus).* The macropodids are mostly nocturnal, may be terrestrial or arboreal, and are herbivorous. Females usually bear a single young annually.

Infraclass Eutheria: Placental Mammals

The fetus is developed entirely within the body of the female, attached by a placenta to the wall of the uterus, and with growth of a chorion. There is no marsupium, nor are there epipubic bones. The coracoid is only a process on the scapula.

Order Insectivora: Moles and Shrews

Members of this order are generally small, terrestrial forms of ancient origin. Some doubt exists that all families come from a unified origin, and at least one family, the Tupaiidae (tree shrews), has only recently been returned from the order Primates into the order Insectivora (Romer, 1966). The classification of the group presents some difficulties; most of its members are primitive in some respects. This order has sometimes been called the "wastebasket" of the class Mammalia. The feet are usually five-toed with claws. The zygomatic arch is variable, ranging from present to reduced, to absent. The number of teeth is often great; all teeth are rooted, and the molars are tuberculosectorial. A

Figure 5-6 The short-tailed shrew *(Blarina brevicauda)*, a common small mammal.

clavicle is present. The testes are either abdominal or inguinal. There is no scrotum. The tympanic bone is a ring. The eyes are weak. This is a very old mammalian order, probably ancestral to all other placental forms. Fossils date back to the Cretaceous.

Erinaceidae (10 genera, 14 species)   These are the hedgehogs of Africa, Southeast Asia, Eurasia, and Borneo. The dental formula is 2–3/3, 1/1, 3–4/2–4, 3/3 = 36–44. The skull has a slender zygomatic arch that is sometimes incomplete. Postorbital processes are absent, and auditory bullae are incomplete or lacking. The back is covered with spines in Erinaceinae; this is not so in Echinosoricinae. The European hedgehog *Erinaceus* is supposedly somewhat immune to snake venom, bee stings, cyanide, and arsenic. Hedgehogs of Europe hibernate; those of Africa estivate.

Talpidae (about 15 genera, about 22 species; moles and mole shrews; North America, Europe, and Asia)   The dental formula is 2–3/1–2, 1/0–1, 3–4/3–4, 3/3 = 34–42. This is a group of small, fossorial mammals. Zygomatic arches are present. Eyes are small and are sometimes hidden beneath skin; ears are often without pinnae. Forelimbs are usually greatly modified for digging.

Tenrecidae (11 genera, 23 species; tenrecs; Southern Africa)   The dentition is variable. The zygomatic arch is incomplete. The urogenital and anal apertures are included in a cloaca. The morphology of this group is extremely varied, some members looking like hedgehogs *(Tenrec)*, some like shrews, some like moles, and one like a half-grown otter *(Potamogale).*

Chrysochloridae (5 genera, about 11 species)   The golden moles are restricted to Central and Southern Africa. The dental formula is 3/3, 1/1, 3/3, 3/3 = 40. The auditory bullae are present, eyes are hidden, there is no external pinna. The fifth toe of the manus is missing, the central two are armed with large claws for digging.

Solenodontidae (2 genera, 2 species; solenodons)   These are restricted in range to Cuba and Haiti and are becoming rare. The dental formula is 3/3, 1/1, 3/3, 3/3 = 40. Solenodons are among those insectivores which have a poisonous saliva. They look like a large shrew, have small eyes, visible pinnae, and a tail with visible scales.

Soricidae (24 genera, about 290 species; shrews)   This family has a worldwide distribution (except Australia), and in North America the short-tailed shrew, *Blarina brevicauda,* is thought by some to be the most abundant small mammal in the Eastern United States and Canada (Fig. 5–6).

Order Dermoptera: Flying Lemurs or Colugos

The flying lemurs are not lemurs, nor do they fly, but they are superb gliders. The first two lower incisors are transversely widened and are pectinate. A patagium (web), extending from behind the ears, enclosing the limbs, and extending to the tip of the tail, is used to scrape leaves and probably fruits. They may cover 60 m at a jump, and are nearly helpless on the ground. They are nocturnal and rest during the day, hanging either by all fours or by the hind limbs only.

The Dermoptera have been variously grouped with the Chiroptera and with the Insectivora. In this synopsis, Simpson (1945) is followed, and the

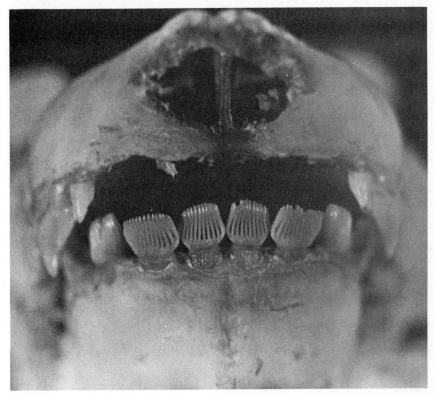

Figure 5–7 The first two lower incisors of the flying lemur *(Cyanocephalus)* are transversely widened and pectinate.

Dermoptera are left as a separate order. Their fossil history is poorly known. Fragmentary specimens from the late Paleocene and lower Eocene of North America have been tentatively assigned to the order Dermoptera.

     Cynocephalidae (1 genus, 2 species) The species are *Cynocephalus variegatus* and *Cynocephalus volans.* The dental formula is 2/3, 1/1, 2/2, 3/3= 34. Characters are the same as for the order.

Order Chiroptera

Bats have wings in common but are highly diversified physiologically and anatomically. Locomotion by other than flight is rare, although some bats, e.g., the vampire bat *(Desmodus rotundus),* are surprisingly agile on the surface of animals or on the ground. Some bats are able to hover momentarily and feed on nectar. The forelimbs have digits II to V greatly elongated; the ulna and fibula are usually rudimentary. In addition to the wing membranes (patagium), there is an interfemoral membrane between the hind limbs and tail. The sternum usually has a keel. There is also generally a cartilaginous calcar on the inner side of the ankle joint to help support the interfemoral membrane. The adult skull shows no evidence of sutures, and the number of teeth varies from 20 to 38. Reproduction is extremely variable, from breeding all year, to ovulation and copulation in spring, copulation and fertilization in the fall followed by delayed implantation, or copulation in the fall followed by delayed implantation, or copulation in the fall with sperm retained in the uterus until fertilization in the spring. The earliest known bat fossil, *Icaronycteris index,* is from the early Eocene from Wyoming.

Suborder Megachiroptera

In this suborder the second digit is rather independent and bears a claw. These are the "flying foxes" of the tropical and subtropical regions of the Old World. They have large eyes, and ears with a tragus. Except for members of the genus *Rousettus,* they apparently do not use echolocation.

Pteropidae (38 genera, 154 species; flying foxes, fruit-eating bats) These are found in the tropical and subtropical regions of the Old World. The dental formula is 1–2/0–2, 1/1, 3/3, 1–2/2–3 = 24–34. In size they range from small nectar-feeding bats to *Pteropus vampyrus* with a wingspan of 5½ ft.

Suborder Microchiroptera

The second digit of the forelimb does not end in a claw. A tragus is usually present. In this group of bats echolocation has reached its greatest development. It is used by these nocturnal mammals for locating their food of flying insects or moving fish, and for avoiding obstacles. Not all bats in this group are insectivorous; some are carnivorous, frugivorous, nectivorous, piscivorous, or sanguinivorous. All the bats of the New World belong to this group.

Rhinopomatidae (1 genus, 3 species) The mouse-tailed bats get the name from their very long tail which extends far beyond a narrow interfemoral membrane. They are found in Northern Africa, Southern Asia, and Sumatra. The dental formula is 1/2, 1/1, 1/2, 3/3 = 28. The large ears are united at the base, across the forehead. There is a ridgelike dermal outgrowth down the center of the muzzle, and the nostrils appear as transverse valvular slits.

Emballonuridae (12 genera, 44 species) These are sac-winged or sheath-winged bats, found in southern Mexico, northern half of South America, most of Africa, and Southern Asia down through Australia. The dental formula is variable. Several species have sacs along the anterior edge of the wing. The distal portion of the tail is free and appears to lie on the surface of the interfemoral membrane.

Noctilionidae (1 genus, 2 species; mastiff bats, or bulldog bats)   These are found in tropical America from southern Mexico to northern Argentina. The dental formula is 2/1, 1/1, 1/2, 3/3 = 28. The bulldog bat, *Noctilio liporinus,* captures fish by utilizing the long, sharp claws of the extremely large and specialized hind feet as gaffs. Echolocation is probably used to locate fish, as the fish break the surface. The other species is primarily insectivorous. Members of this family are restricted to Central America and the northern half of South America.

Nycteridae (1 genus, 13 species; hispid bats, or hollow-faced bats) These bats are found in Africa, Arabia, Israel, Madagascar, and the Malay Peninsula to Borneo and Java. The dental formula is 2/3, 1/1, 1/2, 3/3 = 32. These bats have large ears, connected at the bases with a band of flesh. The long tail is enclosed to the tip by the membrane, and ends in a T-shaped cartilage.

Megadermatidae (4 genera, 5 species; false vampires, or yellow-winged bats)   The dental formula is 0/2, 1/1, 1–2/2, 3/3 = 26–28. Southern and Central Africa, Southeastern Asia, Malayan region to the Philippines and Australia. The large ears are joined halfway up, along the inner edges; i.e., they have a bifid tragus. These bats have a well-developed leaf on the nose. The tail is so short that it seems nonexistent. They do not feed on blood.

Rhinolophidae (11 genera, about 131 species; Old World leaf-nosed bats)   These bats are found in Africa, Southern Europe, and Southern Asia to Japan and Australia. The dental formula is 1/2, 1/1, 1–2/2–3, 3/3 = 28–32.

Phyllostomatidae (47 genera, about 119 species; American leaf-nosed bats)   These are found in Northern Argentina, North to Southwestern United States, and the West Indies. This is the most diverse family of bats. The dentition is variable. A simple nose leaf generally is present, although sometimes it is rudimentary. Tail if present may be longer than, equal to, or shorter than interfemoral membrane. One group (Phyllostomatinae) has very large ears. A wide variety of foods is utilized by the bats in this family. Many are insectivorous, many are frugivorous, picking off fruit while hovering, and others feed on nectar and pollen (especially the Glossophaginae and Phyllonycterinae).

Mormoopidae (3 genera, 8 species; mormoops, leaf-chinned bats; tropical America)   The dental formula is 2/2, 1/1, 2/3, 3/3 = 34. These are small bats whose snout and chin always have cutaneous flaps or ridges. The ears have ventral extensions that curve beneath the eyes. The short tail protrudes from the dorsal surface of the uropatagium. The reduced coronoid process of the dentary allows the jaws to open widely. These are among the most abundant bats of the New World tropical forests. This family is not separated in Simpson's (1945) system, but the work of Vaughan and Bateman (1970) and of J. D. Smith (1972) is a strong argument for a separate family.

Desmodontidae   (3 genera, 3 species; vampire bats)   These bats are restricted to the subtropical New World. The dental formula is 1/2, 1/1, 2/3, 0/0 = 20. They have no tail and a short interfemoral membrane. The most widely distributed species, *Desmodus rotundus,* is a medium-sized bat about 85 mm in length. Vampires have many highly specialized adaptations for a diet of blood, their sole food. They feed entirely on the blood of mammals, making a painless incision with razor-sharp incisors and ingesting the blood through grooves on the underside and edge of the tongue. The premolars are narrow, with sharp-edged longitudinal crowns; the molars are rudimentary or absent. The saliva contains an anticoagulant. They are able to walk with body elevated, "on all fours." Vampires, like other bats, transmit rabies and, in addition, weaken animals by taking blood and by causing secondary infections.

Natalidae   (1 genus, 4 species; long-legged bats, funnel-eared bats)   The dental formula is 2/3, 1/1, 3/3, 3/3 = 38. These bats of the subtropical New World are poorly known, but they are probably all insectivorous, and received their names from their large funnel-shaped ears, with a more or less triangular tragus. There is a glandular structure (natalid organ) on the forehead of adult males.

Furipteridae   (2 genera, 2 species)   These are smoky bats, so named because of their gray pelage, and are found in northern South America. The dental formula is 2/3, 1/1, 2/3, 3/3 = 36. Little is known of the natural history of these two species of small bats. They also have funnel-shaped ears and a grayish fur with a reddish tint.

Thyropteridae   (1 genus, 2 species)   The dental formula is 2/3, 1/1, 3/3, 3/3 = 38. These bats are known from southern Mexico as far south as Brazil and Peru. They also have funnel-shaped ears with tragi. The base of the thumbs and soles of feet have circular sucking disks. These also occur in one other bat, the sucker-footed bat *(Myzopoda aurita)* of Madagascar, and in two genera of hyraxes (Hyracoidea). They are small (less than 75 mm) and insectivorous.

Myzopodidae (1 genus, 1 species; golden bat, or sucker-footed bat) Nothing is known of the biology of this insectivorous bat that is found only in Madagascar. The dental formula is 2/3, 1/1, 3/3, 3/3 = 38. This bat has large ears and suction cups on the thumbs and bottoms of the feet.

Vespertilionidae (35 to 38 genera, about 287 species; vespertilionid bats) This is the most widespread family of bats. The geographic range, with the exception of some islands, extends from the Arctic Circle (tree line) south as far as land masses. The dental formula varies from 1/2, 1/1, 1/2, 3/3 = 28 to 2/3, 1/1, 3/3, 3/3 = 38. They are usually insectivorous; a few are piscivorous *(Pizonyx)*. The majority of physiologic and ecologic studies which have been done on bats have been done on members of this group. As might be expected in such a large group, there is great variation in their natural history. Some are colonial, some are solitary, some migrate, some do not, some mate in the fall, some in spring, some species segregate into nursing colonies, others remain solitary.

Ears are usually separate and with a tragus; there is a complete interfemoral membrane, with a tail that rarely extends beyond the free edge. Size is small to medium.

Mystacinidae (1 genus, 1 species; New Zealand short-tailed bats) These bats are restricted to forested areas of New Zealand and are on the verge of extinction (Dryer, 1962). The dental formula is 1/1, 1/1, 2/2, 3/3 = 28. They live in hollow logs and do little flying. Although they often catch insects in flight, they also secure insects by moving along branches of trees.

Molossidae (11 genera, 88 species; mastiff bats, free-tailed bats) These bats are widespread in the warmer regions of both the Old and New World, also in Australia. The dental formula varies from 1/1, 1/1, 1/2, 3/3 = 26 to 1/3, 1/1, 2/2, 3/3 = 32. The tail extends conspicuously beyond the free edge of a short but complete interfemoral membrane. All are insectivorous and are primarily colonial. *Tadarida brasiliensis* produces guano beds which are used for fertilizer.

Order Primates

The primates are mostly arboreal mammals. The limbs are plantigrade, with five digits. The digits usually have nails, but some have claws. The thumbs and great toes are often opposable. The clavicles are well developed, and the radius and ulna are never united. The nasal region and olfactory lobes of the brain have been progressively reduced. The braincase is becoming relatively larger, and the facial region relatively smaller. Some of the primitive primates are difficult to distinguish from the insectivores. The family Tupaiidae (tree shrews), sometimes included with the primates, has already been discussed under Insectivores. The long shape of the primitive head has been distorted into a round one by the movement of the eyes toward the front of the skull. There is usually a single young, helpless at birth. The geographic range of the order is essentially subtropical in the Old and New World. Human beings are cosmopolitan. The study of primates has literally exploded in the last decade, producing new journals, symposia, books, and research centers. There is much debate over classification within this order.

Suborder Prosimii

In the Prosimii the face is still somewhat elongate, and the orbit is widely confluent with the temporal fossa. The first upper incisors are separated in the midline. These are primitive primates with restricted distribution in the tropics of the Old World.

Lemuridae (5 genera, 15 species) They are found on Madagascar and the Comoro Islands. The dental formula is 0–2/2, 1/1, 3/3, 3/3 = 32–36. The upper incisors are separated by a wide diastema (except in *Lepilemur,* which has no upper incisors). The eyes are large and closely set. These animals range in size from 30 to 140 cm and have long tails.

Indridae (3 genera, 4 species; woolly lemurs; Madagascar) The dental formula is 0–2/2, 1/1, 3/3, 3/3 = 32–36. The digits of the foot are united by a web of skin that extends to the first phalanx.

Daubentoniidae (1 genus, 2 species; aye-ayes) This most unusual group of primates has a range restricted to northern Madagascar, and is surviving under the most rigid governmental protection. The dental formula is 1/1, 0–1/0, 1/0, 3/3 = 18 or 20. Its skull and dentition are rodentlike, down to large, persistently growing incisors, with a large diastema and no canines. The middle digit of the hand has an unusually long claw. In addition to feeding on the pith of bamboo and sugar cane, they utilize their rodentlike dentition to pull apart pieces of wood; the long claw on the middle digit of the hand is a further help in getting larval and adult wood-boring beetles out of their tunnels.

Lorisidae (6 genera, 11 species; lorises, pottos, galagos; Oriental, Ethiopian) The dental formula is 1–2/2, 1/1, 3/3, 3/3 = 34 to 36. They have a round head, very large eyes, and woolly pelage. Members of the family are completely arboreal.

Tarsiidae (1 genus, 3 species; tarsiers; Indonesia, Philippines) The dental formula is 2/1, 1/1, 3/3, 3/3 = 34. These are small insectivorous monkeys with a round head and long tail.

Suborder Anthropoidea

Primates of this group have no rostrum, and the face appears more like a human face than does that of the Prosimii. There is no space between the upper incisors. The digits have nails instead of claws, except in the marmosets. A bony plate separates the orbit completely from the temporal fossa.

Superfamily Ceboidea

These are the New World monkeys and marmosets. There are 36 teeth in the monkeys and 32 in the marmosets. The nasal septum is broad. The thumb is only partially opposable or not at all. It may also be reduced or absent.

Cebidae (11 genera, 29 species; New World monkeys; Neotropical) The dental formula is 2/2, 1/1, 3/3, 3/3 = 36. The nostrils are well separated by a broad internarial cartilage. The tail varies from short to long and also in prehensility. The body is slender; the long slender limbs have nails on all digits. These monkeys are found from Central America to tropical South America.

Callithricidae (4 genera, 33 species; marmosets; Neotropical) The dental formula is 2/2, 1/1, 3/3, 2–3/2–3 = 32 to 36. Except for the thumb, which has a flat nail, the digits have pointed claws. The thumb is opposable; the great toe is not opposable. The tail is long and nonprehensile. The face is naked.

Superfamily Cercopithecoidea

These are Old World monkeys. Members of this group have a narrow nasal septum; therefore the nostrils are closer together than in the Ceboidea. The thumb is usually opposable, the tail is never prehensile.

Cercopithecidae (11 genera, 60 species; Old World monkeys) The dental formula is 2/2, 1/1, 2/2, 3/3 = 32. The bulla is small, the external auditory meatus is long. The thumb and great toe, when present, are opposable. The tail varies in length but is never prehensile.

Superfamily Hominoidea

Members of this superfamily have a large braincase and definite forehead.

Pongidae (4 genera, 8 species; gibbons, orangutans, chimpanzees, gorillas; Oriental, central Ethiopian) The dental formula is 2/2, 1/1, 2/2, 3/3 = 32. The upper jaw is prognathus. The forelimbs are longer than the hind limbs. The fingers and thumb are long. The thumb is opposable. The appearance is quite human. There is no tail. Pongidae are supposedly unable to swim. The gibbons are found in Southeastern Asia, East Indies, Sumatra, and the Malay

Peninsula. The orangutan occurs in Borneo and Sumatra. The chimpanzees and gorillas are restricted to equatorial Africa.

Hominidae (1 genus, 1 species)   This is the human being. Habitat is cosmopolitan.

Order Edentata: American Anteaters, Sloths, and Armadillos

These mammals are found in South America, and north into North America as far as Nebraska. There are no incisors or canines; the cheek teeth are rudimentary or absent, have no enamel, and grow from persistent pulp. Females in this order have a common urinary and genital duct; the testes lie in an abdominal cavity between the rectum and the bladder. The similarity of the reproductive organs is a major reason for placing anteaters, sloths, and armadillos together in the same group. There is a very large urogenital sinus in the female, and the two horns of the uterus are very short, giving rise to one large cavity for the embryos. Extra zygapophysislike (xenarthrous) articulations brace the lumbar vertebrae. The earliest known fossils are from the Paleocene and early Eocene.

Myrmecophagidae (3 genera, 4 species; South American anteaters) They are found from southern Mexico to northern Argentina. Teeth are absent. The zygomatic arch is incomplete, head and snout are elongate. The tongue is slender, protrusible, and sticky. The forefeet have stout claws for digging open ant and termite nests.

Bradypodidae (2 genera, 6 species; tree sloths)   The tree sloths are found in the forests of tropical America. Cheek teeth present include five maxillary and four or five mandibular. Tree sloths have digits with long, recurved claws for hanging onto tree branches. Cervical vertebrae are variable; usually there are six, seven, or nine, but sometimes eight. There are 18 to 27 thoracolumbar vertebrae. Someone has described the postcranial skeleton as simply a "basket" for holding the sloth's soft anatomy. Sloths spend their time in an upside-down position in trees. They seem to have poor temperature control. Sloths often have moss growing on them.

Dasypodidae (9 genera, 12 species; armadillos; Nebraska to Argentina)   Teeth vary in number from 7/9 to 25/25 and are rootless with persistent growth. The zygomatic arch is complete. A major part of the skin is ossified into horny plates covered with a horny, protective shell, divided by transverse furrows. An unusual feature of at least some armadillos is that of delayed implantation. The zygote (fertilized egg) divides into four parts, giving rise to quadruplets, all of the same sex (i.e., identical quadruplets).

Order Pholidota: Pangolins, or Old World Anteaters

The skull has incomplete zygomatic arches. The mandibles are simple blade-like bones, and there are no teeth in the adult forms. The long tongue is highly protractile. The body is covered with overlapping horny scales that form as armor over the dorsal surface of the neck and body and cover the tail completely. Only a few bones from the Miocene and Oligocene of Europe represent early fossils of this order.

Manidae (1 genus, 8 species; pangolins)   Only one family. Characters same as for order. Southeastern Asia and most of southern half of Africa.

Order Lagomorpha: Hares, Pikas, and Rabbits

Generally the hind limbs are longer than the forelimbs. The tibia and fibula are fused. The fibula articulates with the calcaneum. The sides of the maxilla are fenestrated. The teeth are modified, resembling the dentition in rodents, for an herbivorous diet. There are no canines, and the incisors are ever-growing. The lagomorphs have two upper incisors on each side, rather than one as in the rodents. Representatives of the lagomorphs are found on all continents and larger islands, now that they have been introduced into Australia and New Zealand. The oldest known fossil is from the late Paleocene of Mongolia. All recent leporids seem to be derived from Late Tertiary Palaearctic forms.

Ochotonidae (1 genus, about 14 species; pikas, coneys)   They are found in Asia, Korea, and Japan, and in the mountains of western North America. The dental formula is 2/1, 0/0, 3/2, 2/3 = 26. These are the smallest of rabbits, with the hind limbs only slightly longer than the forelimbs. The auditory bullae are enlarged with spongy material. Each maxilla has a large single fenestra. A zygomatic arch present. The ears are short and rounded, and there is no tail. Pikas are very vocal diurnal mammals living mainly on the talus slopes in mountainous regions, although in Asia some species are found in forests or steppes.

Leporidae (8 genera, 49 species; rabbits and hares)   Worldwide now with introductions into southern South America, Australia, nearby islands, and New Zealand. The dental formula is 2/1, 0/0, 3/2, 3/3 = 28. The hind legs are usually much longer than the front legs, the auditory bullae are small and not filled with spongy material, and each maxilla has more than one fenestra. The hares *(Lepus)* are precocial. There is no nest, merely a temporary depression; the young are fully haired at birth and begin moving about shortly after birth. Hares are present in Europe, Asia, Africa, and North America. The rabbits are generally not precocial. The naked, blind young are born into

Table 5-1   Two Systems of Classification of the Rodents

Anderson (1967), modified version of Simpson (1945)	Wood (1955)
SCIUROMORPHA	Suborder SCIUROMORPHA
SUPERFAMILY APLODONTOIDEA	Superfamily Aplodontoidea
Aplodontidae	Family Aplodontidae
SUPERFAMILY SCIUROIDEA	Superfamily Sciuroidea
Sciuridae	Family Sciuridae
SUPERFAMILY GEOMYOIDEA	?SCIUROMORPHA
Geomyidae	Superfamily Ctenodactyloidea
Heteromyidae	Family Ctenodactylidae
SUPERFAMILY CASTOROIDEA	Suborder THERIDOMYOMORPHA
Castoridae	Superfamily Anomaluroidea
SUPERFAMILY ANOMALUROIDEA	Family Anomaluridae
Anomaluridae	SCIUROMORPHA or
Pedetidae	?THERIDOMYOMORPHA
MYOMORPHA	Family Pedetidae
SUPERFAMILY MUROIDEA	Suborder CASTORIMORPHA
Cricetidae	Superfamily Castoroidea
Spalacidae	Family Castoridae
Rhizomyidae	Suborder MYOMORPHA
Muridae	Superfamily Muroidea
SUPERFAMILY GLIROIDEA	Family Cricetidae
Gliridae	Family Muridae
Platacanthomyidae	?Muroidea
Seleviniidae	Family Spalacidae
SUPERFAMILY DIPODOIDEA	Family Rhizomyidae
Zapodidae	Superfamily Geomyoidea
Dipodidae	Family Heteromyidae
HYSTRICOMORPHA	Family Geomyidae
SUPERFAMILY HYSTRICOIDEA	Superfamily Dipodoidea
Hystricidae	Family Zapodidae
SUPERFAMILY ERETHIZONTOIDEA	Family Dipodidae
Erethizontidae	?MYOMORPHA
SUPERFAMILY CAVIOIDEA	Superfamily Gliroidea
Caviidae	Family Gliridae
Hydrochoeridae	Family Seleviniidae
Dinomyidae	Suborder CAVIOMORPHA
Heptaxodontidae	Superfamily Octodontoidea
Dasyproctidae	Family Octodontidae
SUPERFAMILY CHINCHILLOIDEA	Family Echimyidae
Chinchillidae	Family Ctenomyidae
SUPERFAMILY OCTODONTOIDEA	Family Abrocomidae
Capromyidae	Superfamily Chinchilloidea
Myocastoridae	Family Chinchillidae
Octodontidae*	Family Capromyidae
Abrocomidae	Superfamily Cavioidea
Echimyidae	Family Caviidae
Thryonomyidae	Family Hydrochoeridae
Petromyidae	Family Dinomyidae

Table 5-1   Two Systems of Classification of the Rodents *(Continued)*

Anderson (1967), modified version of Simpson (1945)	Wood (1955)
SUPERFAMILY BATHYERGOIDEA	Family Heptaxodontidae
Bathyergidae	Family Dasyproctidae
SUPERFAMILY CTENODACTYLOIDEA	Family Cuniculidae
Ctenodactylidae	Superfamily Erethizontoidea
	Family Erethizontidae
	Suborder HYSTRICOMORPHA
	Superfamily Hystricoidea
	Family Hystricidae
	Superfamily Thryonomyoidea
	Family Thryonomyidae
	Family Petromyidae
	Suborder BATHYERGOMORPHA
	Superfamily Bathyergoidea
	Family Bathyergidae

* Including the family Ctenomyidae. Anderson considered the Ctenomyidae as a separate family.

a nest, where they remain until the eyes are opened and the body is covered with hair.

Order Rodentia

Rodents are the most common mammals in the world, both in numbers and in species. About one-third of the named mammals in the world are rodents. They vary in size from the delicate pygmy mouse *(Baiomys)* to the pig-sized capybara *(Hydrochoerus)*. The outstanding characteristic of rodents is the adaptation of the incisor teeth for gnawing. The first upper incisor on each side of the upper and lower jaw has become greatly enlarged and chisel-shaped. Each grows from persistent pulp, and the enamel is restricted to the anterior face of the teeth, making the posterior part wear faster and creating a chisel effect. There are never any canines, but there is a bare space, the diastema between the incisors and the premolars. The articular condyle of the mandible is capable of a back-and-forth as well as a sideways motion. Rodents are the "primary converters" in the food chain. The earliest known rodents (Paramyidae) are from the late Paleocene of North America and the Eocene of North America and Eurasia. Although the order Rodentia has been taxonomically sliced in many ways on the basis of various morphologic characteristics (Wood, 1955, 1965; Stehlin and Shaub, 1951; Shaub, 1953; Anderson, 1967), the organization presented here is basically that of Simpson (1945), in which there are three suborders. However, a comparison of Anderson's classification (1967) based on Simpson's (1945) with that of Wood's (1955) is included to

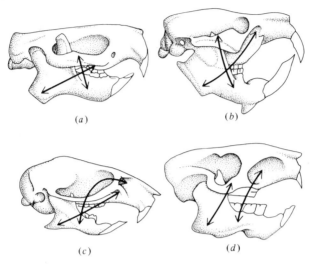

(a)   (b)

(c)   (d)

Figure 5–8 Diagrammatic representations of middle and deep portions of the masseter muscle in various rodent groups: *(a)* aplodontid type, masseter originates from lower edge of zygomatic arch; *(b)* advanced sciuromorph, middle masseter originates from outer side of skull in front of orbit, infraorbital foramen small, not transmitting any muscle; *(c)* myomorph, deep portion of masseter pushes up through orbit and passes through a V-shaped, oval, or round infraorbital foramen; *(d)* hystricomorph, deep portion of masseter enormously developed, with portion passing through a greatly enlarged infraorbital foramen. *(After Romer, 1966.)*

emphasize the lack of uniformity. There is considerable agreement at the superfamily level.

Suborder Sciuromorpha: Squirrellike Rodents

The infraorbital canal is small, and the masseter muscle does not pass through it. The zygomatic arch is slender, and there are always more than three cheek teeth in each half of the jaw.

Aplodontidae (1 genus, 1 species; mountain beaver)   The mountain beaver is restricted to a very narrow range along the west coast of North America from British Columbia into California. The dental formula is 1/1, 0/0, 2/1, 3/3 = 22. The cheek teeth are ever-growing and have a unique crown pattern. The skull is flat and wide. The mountain beaver, about the size of a muskrat, is considered to be the most primitive rodent.

Sciuridae (47 to 51 genera, about 261 species; squirrels—ground, tree, flying—and marmots)   They are worldwide in distribution except for the Australian region, southern South America, Northern Africa, and smaller islands. The dental formula is 1/1, 0/0, 1–2/1, 3/3 = 20–22. A feature of this family that is unique among the rodents is the presence of a postorbital process on the frontal bone. There are four toes on the front feet and five toes on the hind feet, each with a sharp claw. The range of size varies from 1030 mm for the Asiatic and African tree squirrel *Ratufa* to only 70 mm in the African pygmy squirrel.

Geomyidae (8 genera, 40 species; pocket gophers)   These mammals are found from British Columbia and the prairie provinces of Canada south barely into South America, and except for the very southeastern part of the United States, not far east of the Mississippi River. The dental formula is 1/1, 0/0, 1/1, 3/3 = 20. The cheek teeth are ever-growing. The skull is massive and angular with strong-ridged squamosals which in older males tend to form a sagittal crest. There are large, external fur-lined cheek pouches. These animals are highly specialized for a fossorial life, with a short neck, short and dense pelage, small eyes, small ears, short tail with tactile hairs, and forefeet with five digits bearing powerful claws for digging.

Heteromyidae (5 genera, about 75 species; kangaroo rats, kangaroo mice, and pocket mice)   The range of this group is similar to that of the Geomyidae, except that they are not found in the Southeastern United States. The dental formula is 1/1, 0/0, 1/1, 3/3 = 20. These mammals also have external fur-lined cheek pouches, but unlike the pocket gophers are not adapted for a fossorial life. The skull is thin and papery, and the tympanii and mastoid areas are usually highly inflated. In the kangaroo rats and mice the limbs are modified for saltatorial locomotion, with a tail as long as the body or longer. The majority of these animals are physiologically adapted to live in drier regions, and are able to get along with little or no water to drink.

Castoridae (1 genus, 2 species; beavers)   *Castor canadensis* is found in North America, and *Castor fiber* in Eurasia. The dental formula is 1/1, 0/0, 1/1, 3/3 = 20. The skull is massive and has a strong zygomatic arch. These mammals are specialized internally for a diet of bark of trees, which they cut down. Externally they are modified for an aquatic life. The fur is dense, consisting of guard hairs and an exceedingly dense underfur. The hind feet have a web of skin between the digits. The second and third digits of the foot are comblike and used in grooming. The tail is flattened and aids in swimming (though not as a rudder). The nose and short ears have valves which can close under water, and the lips can be closed behind the incisors.

Suborder Myomorpha: Ratlike Rodents

The infraorbital canal is enlarged and usually slitlike. The deep division of the masseter muscle enters the orbit, passes through the infraorbital foramen, and attaches to the side of the rostrum. The suprafacial division of the muscle passes forward in front of the orbit.

Anomaluridae (4 genera, 12 species; scaly-tailed squirrels; central part of Africa)   The dental formula is 1/1, 0/0, 1/1, 3/3 = 20. The upper side of the tail is bushy, but the basal portion of the underside has a double row of raised scales. In all except the genus *Zenkerella,* a membrane is present from the forelimb to the hind limb, which is extended by a cartilage rod attached to the olecranon process.

Pedetidae (1 genus, 2 species; springhaas; Central and Southern Africa) The dental formula is 1/1, 0/0, 1/1, 3/3 = 20. The skull is massive and has greatly inflated mastoids. The zygomatic arch is thickened, and the infraorbital foramen is greatly enlarged. The limbs are adapted for a saltatorial locomotion. The tibia and fibula are fused. The forefeet have five digits. The much-enlarged hind feet have four hooflike claws. The tail equals or is longer than the body.

Cricetidae (97 genera, about 567 species; hamsters, gerbils, voles, lemmings (Fig. 5–9), and New World rats and mice) Worldwide in distribution except for the Austro-Malayan region and some islands. Dental formula is 1/1, 0/0, 0/0, 3/3 = 16. The cheek teeth may be laminate, prismatic, or cuspate, with cusps arranged in two rows (as compared with three in Muridae). The infraorbital canal is generalized, with part of the masseter muscle going through a rounded portion, and the nerve through a narrow, lower portion.

Figure 5–9 The brown lemming *(Lemmus trimucronatus)* is a common and widespread rodent of the Arctic.

The cricetid rodents are primarily terrestrial but may be semiaquatic (musk-rat), fossorial (mole rats), or saltatorial (gerbils). The family is divided into five subfamilies.

Spalacidae (1 genus, 3 species; mole rats)   This family is restricted to the eastern Mediterranean and Southeastern Europe. The dental formula is 1/1, 0/0, 0/0, 3/3 = 16. These fossorial rodents dig with their incisors rather than their front feet. The incisors are even more greatly enlarged than in most rodents. The supraorbital region of the skull slopes forward, and the zygomatic plate is narrowed and turned downward. The fur is short and reversible (as in most fossorial mammals), the eyes are functionless, the pinnae are reduced, and the tail is absent.

Rhizomyidae (3 genera, 18 species; bamboo rats)   These are native to all of Africa and the southern two-thirds of Eurasia. Dentition is 1/1, 0/0, 0/0, 3/3 = 16. Skull is modified for fossorial life. The neural portion of the infraorbital foramen is reduced or obliterated by fusion of the zygomatic plate to the side of the rostrum. Scantily haired tail; ears hidden in pelage.

Muridae (98 genera, about 457 species. Old World rats and mice)   The dental formula is usually 1/1, 0/0, 0/0, 3/3 = 16. The cheek teeth may be laminate or cuspidate, with three rows of cusps, rather than two as in *Criceti-dae.* The size is generally small. Some authors (Ellerman, 1940; Hershkovitz, 1962) include the *Cricetidae* in the *Muridae,* but others (Simpson, 1945; Wood, 1955) separate them. Separating the *Cricetidae* and *Muridae* is usually the interpretation followed in North America. Members of this family have been introduced everywhere man lives, so that they are now worldwide in distribution.

Gliridae (7 genera, 23 species; dormice)   These mammals are present in most of Africa and in England, Europe, Asia Minor, and southwestern Russia. The dental formula is 1/1, 0/0, 1/1, 3/4 = 20. The animals have squirrellike habits and a squirrellike appearance; the tail is long and bushy, and the eyes and ears are large. They hibernate.

Platacanthomyidae (2 genera, 2 species; spiny dormice; Southern China and Southern India)   The dental formula is 1/1, 0/0, 0/0, 3/3 = 16. The cheek teeth have a series of parallel oblique cross ridges. The skull has a series or a single pair of foramina in the palate between the tooth rows. The limbs are adapted for an arboreal life. The thumb is replaced by a pad. *Platacanth-omys* has densely spiny fur. These animals are about the size of house mice.

Seleviniidae (1 genus, 1 species; Selvin's mice, or Betpakdala dor-mice)   The dental formula is 1/1, 0/0, 2/2, 3/3 = 24. These are saltatorial, insectivorous mammals living in arid and semiarid habitats in Kazakhstan, Central Asia. They are stocky rodents, about 150 mm (6 in.) in length. The auditory bullae are greatly inflated.

Zapodidae (4 genera, 11 species; jumping mice, birch mice) They are found across North America and Eurasia. The dental formula is 1/1, 0/0, 0–1/0, 3/3 = 16 to 18. These mammals have inflated bullae and a very large infraorbital foramen. In the jumping mice (subfamily Zapodinae), the hind legs and long tail are adaptations for saltatorial locomotion. Jumping mice hibernate. The birch mice (Sicistinae) of Eurasia are found in birch woods up to an elevation of about 10,000 ft. They also hibernate.

Dipodidae (10 genera, 27 species; jerboas) These African and Asian ecologic counterparts of the North American Heteromidae are found in the Central Asiatic and African deserts. The dental formula is 1/1, 0/0, 0–1/0, 3/3 = 16 to 18. The limbs are modified for saltatorial locomotion. There are 22 or 24 teeth, compared with 16 or 18 in the Zapodidae. The tympanic bullae are expanded, and the infraorbital foramen is greatly enlarged. Digits I and V are reduced or lost in the hind limbs, and digits II, III, and IV are fused. The eyes are large.

Suborder Hystricomorpha

The infraorbital foramen is usually larger than the foramen magnum, and the zygomatic arch is also very large.

Hystricidae (4 genera, 15 species; Old World porcupines) These are found in Africa, Southern Europe, Southern Asia and China, and the southeastern Asiatic islands. The dental formula is 1/1, 0/0, 1/1, 3/3 = 20. The body and tail are covered with spines; in some cases those on the tail are greatly modified as noise makers (*Atherurus* and *Hystrix*). Pneumatic cavities in frontal, lacrimal, and nasoturbinal bones. These are terrestrial or semifossorial animals.

Erethizontidae (4 genera, 8 species; New World porcupines) Members of this family occur all over North America except the southeastern portion of the United States, and their range extends through Central America and into the southern half of South America. The dental formula is 1/1, 0/0, 1/1, 3/3 = 20. The coendous of Central and South America have prehensile tails. These porcupines are arboreal, with short limbs modified for climbing. They have a thick-set, short body covered with long hairs mixed with spines, which increase in number toward the posterior and dorsal part of the body and the tail. These spines, or quills, can be neither "dropped" nor "shot" but can be made loose. Although not usually found great distances from trees, they can exist in a "badlands" habitat of jumbled rock and shrubs.

Caviidae (5 genera, 12 species; guinea pigs, cavies, and Patagonian "hares") The range of this family is restricted to South America. The dental formula is 1/1, 0/0, 1/1, 3/3 = 20. The angular process of the mandible is drawn backward. The body form is robust, and usually the limbs and ears are short, but some are rabbitlike. The guinea pig *(Cavia porcellus),* well-known laboratory animal and pet, belongs to this group.

Hydrochoeridae (1 genus, 2 species; capybara) These animals are found in South America from Uruguay to Panama. The dental formula is 1/1, 0/0, 1/1, 3/3 = 20. The digits are semiwebbed. These mammals wallow in muddy pools and hide in aquatic vegetation, with only their nostrils exposed. They are the largest of rodents, approaching the size of a pig. The young are precocial.

Dinomyidae (1 genus, 1 species; false pacas, or pacaranas; South America) The dental formula is 1/1, 0/0, 1/1, 3/3 = 20. The angular process of the mandible is heavily ridged. The limbs each have four digits. These large (up to 30-lb) rodents are found only in the Andes of South America and are very rare.

Dasyproctidae (4 genera, 11 species; agoutis, pacas; tropical America) The dental formula is 1/1, 0/0, 1/1, 3/3 = 20. The limbs are elongated and unusually modified for cursorial locomotion. The forefeet have four toes with blunt claws that are almost hooflike; thus the digits are best characterized as being subungulate. Pacas have an internal pair of pouches, housed in maxillary concavities, which function as resonating chambers.

Chinchillidae (3 genera, 6 species; chinchillas and vizcachas) With the exception of *Lagostomus,* which is found on the plains of Argentina, these densely furred animals live in the highlands of Argentina, Bolivia, Chile, and Peru. The dental formula is 1/1, 0/0, 1/1, 3/3 = 20. The incisors are relatively narrow, and the cheek teeth are ever-growing. The jugal is usually in contact with the large lacrimal. Hind limbs are long. The female is larger than the male in *Chinchilla.* The body is covered with a very fine, dense pelage; the long tail is also well furred.

Capromyidae (3 genera, 11 species; hutias) These are restricted to the West Indies. The skull is flattened and long, and the paraoccipital processes are usually separate from the bullae. The limbs are short, with five digits on each. This group of mammals is apparently on the way to extinction, unable to cope with human beings and their introduction of dogs and the mongoose *(Herpestes).*

Myocastoridae (1 genus, 1 species; myocastors, coypus) In its native habitat of southern South America the nutria is apparently the ecologic equivalent of the muskrat *(Ondatra zibethica)* in North America. Cheek teeth decrease in size and converge anteriorly. The skull is triangular, with short postorbital processes and long paraoccipital processes. This is a semiaquatic mammal, with limbs modified for aquatic locomotion, and semiwebbed hind feet. Mammae are about midway between the ventral and lateral sides, which allows the young more freedom for nursing even when the female is in the water. They have been introduced into North America to clear aquatic vegetation and as potential furbearers, and have become feral. They are now established in Eurasia and in the southern part of the United States, where they are a detriment. Although they have been liberated in the northern part of the United States (Minnesota, for instance) by disillusioned fur farmers, the climate there is too rigorous for the nutrias to survive.

Octodontidae (5 genera, 8 species; hedge rats, octodonts) These mammals are found from moderate elevations up to 10,000 ft in central South America. The upper cheek teeth have an occluded pattern which is "8"-shaped. The bullae are inflated. The limbs have five digits each, and the thumb bears a nail instead of a claw. The toes on the hind feet have stiff bristles extending beyond claws. Some species *(Spalacopus)* are fossorial.

Ctenomyidae (1 genus, about 26 species; tuco-tucos) These are found in the southern third of South America. The dental formula is 1/1, 0/0, 1/1, 3/3 = 20. The incisors are thickened, and the cheek teeth are simplified. The limbs are short and muscular, the toes have strong claws, longer on front feet than on hind feet. The head is large. The neck is thick, the eyes and ears are small. The tail is short and sparsely covered with hair. These are obviously adaptations for a fossorial life, and the members of this family are very similar in adaptations and habits to the North American pocket gopher, family Geomyidae.

Abrocomidae (1 genus, 2 species; chinchilla rats) These are restricted to the higher elevations in central South America. The dental formula is 1/1, 0/0, 1/1, 3/3 = 20. The rostrum is long and narrow, the bullae are enlarged, the limbs are short with four digits on front limbs, five digits on hind limbs. The hind toes have a fringe of stiff hairs extending beyond the claw. The pelage is dense and fine.

Echimyidae (14 genera, about 43 species; spiny rats) They are found in the tropics of the northern half of South America and extend into Central America. The dental formula is 1/1, 0/0, 1/1, 3/3 = 20. The bullae are enlarged, the frontals are broad. The pelage is coarse and often spiny. These rodents are adapted for a humid life in the tropics and normally consume a great deal of water. Members of this family may be terrestrial, arboreal, or fossorial.

Thryonomyidae (1 genus, 6 species; cane rats; southern half of Africa) The dental formula is 1/1, 0/0, 1/1, 3/3 = 20. The skull is massive, and the incisors are powerful. Each upper incisor has three grooves. The limbs are short, the thumb is absent. These animals are about the size of a muskrat *(Ondatra zibethica)*, swim well, and are found in marshy areas bordering streams and lakes. Cane rats are relished by the natives, who burn the reeds when the water has gone down, thus driving them into the open.

Petromyidae (1 genus, 1 species; dassie rats) These are diurnal mammals living in rocky areas in hills and mountains of southwestern Africa. The dental formula is 1/1, 0/0, 1/1, 3/3 = 20. The skull is flattened, with greatly inflated bullae, and the incisors are weak and ungrooved. The toes have a comb of stiff spines.

Bathyergidae (5 genera, 22 species; mole rats; southern two-thirds of Africa) The incisors are large, and the number of cheek teeth vary greatly, as shown by the dental formula, 1/1, 0/0, 2/2 or 3/3, 0/0 to 3/3 = 12–28. The infraorbital foramen is greatly reduced. The limbs are short, and the tibia and fibula are fused both proximally and distally. The feet are long and strong. The tail is reduced and covered with hair. Eyes do not function for sight, but are supposed to be sensitive to air currents. The pelage may be either short or absent. All these characteristics are adaptations to a fossorial life.

Ctenodactylidae (4 genera, 8 species; gundis; Northern Africa) The dental formula is 1/1, 0/0, 1–2/1–2, 3/3 = 20–24. The flattened skull has a large infraorbital canal and a large lacrimal. There are four digits on each foot. These diurnal rodents live among rocks and ruins in the semiarid regions of North Africa.

Order Cetacea: Whales

The entire life cycle of the cetaceans is spent in a marine (salt water) environment. They do occasionally go up into the lower courses of rivers for brief periods [i.e., belugas *(Delphinapterus)* into the Nelson River, Northwest Territories in spring]. The largest living mammals are whales, and the blue whale *(Balaenoptera)* is usually given the distinction of being the largest living species of mammal. It can attain a length of 30 m (100 ft) and a weight of 118,000 kg (130 tons) although the average length is just under 28 m (79 ft) (Slijper, 1962). The body is fusiform, posterior limbs are absent, pelvic girdle is vestigial, and the forelimbs (flippers) are paddle-shaped and not externally divisible into limb elements. The skull is greatly modified and strongly telescoped. The dentition is varied. The cervical vertebrae are shortened and usually fused. A complex digestive system, in some cases quite similar to that of ruminants, is present in the whales. There is a great reduction or even absence of epidermal glands. The tail (fluke) is flattened. *Retia mirabilia* systems are greatly developed. These are networks of intertwining veins and arteries where heat transfers from outward-traveling arterial blood to inward-traveling venous blood, thus eliminating heat loss to the extremities. The lungs are not divided into lobes. The external narial openings (blowholes) are on top of the skull. The skin is thick and smooth. It is believed that whales diverged early from primitive eutherian stock and adaptation to a marine life is secondary. The progenitors of cetaceans are unknown, but the earliest known fossils are from the Eocene.

Suborder Mysticeti

The dentition is vestigial in embryos and absent in adults. Instead, a cornified epithelial structure, the baleen, is developed along the maxillary dental arch. The skull is nearly symmetric and is greatly arched to accommodate the baleen. The tongue is heavy and muscular.

Balaenidae (2 genera, 3 species; right whales) The baleen plates are long, narrow, and flexible, folding on the floor of the closed mouth. The dorsal fin is absent or falcate. There are no grooves in the throat region. The right whales have been heavily hunted because of their great yield of oil and "whalebone." These large whales—up to 21 m (70 ft) and 90 kg (100 tons)—feed on minute organisms such as crustacea and plankton, which they filter from the water with their baleen plates. The muscular and flexible tongues are an aid in getting this material from the baleen. Found in all oceans, they are now rare but are completely protected and are evidently increasing again.

Eschrichtiidae (1 genus, 1 species; gray whales) Gray whales spend their summer in the North Pacific Ocean. In autumn they migrate southward along the shorelines as far south as Korea and Mexico. The baleen plates are short and few. There is no dorsal fin. There are two to four longitudinal grooves on the throat. These whales, which may reach 15 m (50 ft), also feed on small fish and crustaceans, i.e., "krill."

Balaenopteridae (2 genera, 6 species; finbacked whales, rorquals) They are gregarious and travel in schools. They are found in all oceans and are highly migratory, generally migrating from cold waters in summer to warm waters in the winter. A blue whale has been known to travel 800 km (500 mi) in 88 days. These mammals have a dorsal fin, short baleen, 40 to 100 grooves on the throat. The flippers have four digits. The cervical vertebrae are not fused. These are the largest of whales, the blue whale reaching a length of 30 m (100 ft). These extremely large mammals feed on a variety of small fish, "krill" (a special kind of plankton consisting primarily of small crustaceans, Slijper, 1962), and plankton. These whales are generally bluish dorsally and whitish ventrally.

Suborder Odontoceti

Teeth are present; they may be as few as 0/1 or 1/0 or as many as 260. The skull is asymmetric. It is now believed that all members of this suborder use echolocation, while no members of the suborder Mysticeti are known to echolocate.

Platanistidae   (4 genera, 4 species; river dolphins)   Slijper (1962) has included *Platanista, Inia, Pontoporia,* and *Lipotes* in this family. The teeth vary from 26/26 to 55/55, in parallel rows which are close together in exceedingly long and narrow jaws. The palatines are not united at the midline. The Ganges River dolphin *(Platanista),* or susu, is completely blind. It is found in the Ganges, Indus, and Brahmaputra Rivers of India. The Amazon dolphin *(Inia),* white flag dolphins *(Lipotes)* of China, and Ganges River dolphin *(Platanista)* are freshwater dolphins. The color of the Amazon dolphin ranges from blackish to flesh-colored dorsally and white ventrally.

Ziphiidae   (5 genera, 18 species; beaked whales)   The dental formula varies from 0/1 to 19/27. These mammals are found in all oceans. Dentition is reduced to one or two pairs of functional teeth in the lower jaw. The skull has a slender rostrum—thus the name "beaked whale." There are two to six longitudinal grooves on the throat. These whales are cuttlefish eaters.

Physeteridae   (2 genera, 3 species; sperm whales)   These whales are found in all oceans. Functional teeth are present in the lower jaw only. A facial depression accommodates a greatly developed spermaceti organ. Sperm whales have been recorded at depths of 960 m (3200 ft). They, too, are cuttlefish eaters. The giant sperm whale *(Physeter)* feeds mostly on squid.

Monodontidae   (2 genera, 2 species; narwhals and belugas)   These are in arctic waters, in North America, south to the St. Lawrence River. The cervical vertebrae are not fused. The dentition varies, but there are only two upper teeth on each side. In *Monodon,* one tooth (usually the left) develops into a straight, forward-directed tusk with a spiral groove. The spiral is always sinistral, even when both tusks develop, as sometimes happens. Females usually do not have tusks. The white whales *(Delphinapterus)* are grayish their first 3 years, and become white as adults.

Delphinidae (14 genera, about 32 recent species; ocean dolphins) They are found in all the oceans in the warmer parts of the world, including the mouths of larger rivers. *Tursiops* also inhabits fresh waters in the St. Johns River, Florida. The dentition is extremely variable, teeth numbering anywhere from 4 to 260. The tooth rows diverge posteriorly. Sometimes there is a bulging, globose forehead with a well-developed "melon" which is homologous to the spermaceti organ in the sperm whales. The size varies from 1.5 to 9 m (5 to 30 ft). Dolphins are the most abundant, varied, and, perhaps, gregarious of cetaceans.

Order Carnivora

The carnivores are sometimes divided into two orders, the Carnivora (the terrestrial carnivores) and the Pinnipedia (the aquatic carnivores) (Scheffer, 1958; Grasse, 1955; Ellerman and Morrison-Scott, 1951; Hall and Kelson, 1959). Simpson (1945) combined the suborders Fissipedia (terrestrial carnivores, "toe-footed") and Pinnipedia (aquatic carnivores, "fin-footed") into the order Carnivora. It is well to remember that any carnivore is a meat-eater (i.e., is carnivorous) but is not necessarily a member of the order Carnivora. The incisors are small and pointed; the canines are prominent, pointed, and slightly recurved; the molars tend to decrease. The skull has strong jaws, and the lower jaws have transversely placed condyles that articulate with deep glenoid fossae. The zygomatic arches are strong. The lower jaw can be moved up and down only, not sideways or forward and backward. The primitive dental formula is 3/3, 1/1, 4/4, 3/3 = 44. A pair of teeth on each side (P^4 and M_1), modified for shearing and called *carnassials,* are especially well developed in cats, poorly developed in bears. There are four or five toes, each with sharp, curved claws. Carnivores are worldwide in distribution except on some islands. Fossils are known from the early Paleocene.

The Carnivora are divided into two superfamilies, the Canoidea and Feloidea. During the late Eocene the Feloidea differentiated into the Felidae and the Viverridae, and during the mid-Eocene the Hyamidae arose from the Viverridae. The Canoidea gave rise to the Mustelidae in the late Eocene and the Procyonidae in the late Oligocene. The Ursidae arose from the Canidae during the Miocene.

Canidae (15 genera, 41 species; wolves, foxes, and allies) This group has a worldwide distribution, except for islands. The dingo probably was introduced into Australia. The dental formula is 3/3, 1/1, 4/4, 2 or 3/2 or 3 = 40 to 44. The incisors are unspecialized, the canines are enlarged and pointed, the cheek teeth are tuberculosectorial, and the carnassials (P^4 and M_1) are well developed. The limbs are digitigrade, with five toes on each foot. The claws are nonretractile.

Ursidae (6 genera, 8 species; bears; all continents of the world)   The dentition is modified for an omnivorous diet; 3/3, 1/1, 4/4, 2/3 = 42. The anterior premolars tend to be lost. The second and third premolars are generally absent or reduced in size. The molars are bunodont, the carnassials undeveloped. The limbs are plantigrade and have five digits. The claws are nonretractile. Bears are stocky with a short tail. D. D. Davis (1964) placed the *Ailuropoda* (giant panda) in the Ursidae rather than in the Procyonidae. The complexity of bear taxonomy is exemplified by Merriam's work, which named 87 species of *Ursus*. These have recently been simplified (Rausch, 1953; Couturier, 1954). Further, Simpson (1961) now considers *Thalarctos* to belong in *Ursus*. In northern regions bears become dormant during the winter, but they do not hibernate if one follows the strict definition of hibernation (in mammals) as a lowering of body temperature to near that of the surrounding temperature.

Procyonidae (7 genera, 18 species; raccoons and allies)   Procyonids are American except for the lesser panda *(Ailurus)* of Asia, whose taxonomic affinity to the procyonids is highly suspect. The dentition is 3/3, 1/1, 3 or 4/3 or 4, 2/2 or 3 = 36, 38, or 40. There is a reduction in the number of premolars and molars. The cheek teeth are usually bunodont, and the carnassials are poorly developed. The foot is usually plantigrade. There are five toes on each foot and the claws are semi- or nonretractile. Members of all species can climb, and the gait is bearlike.

Mustelidae (25 genera, 70 species; skunks, weasels, mink, fisher, otters, badgers)   The range of this family is worldwide, except in Australia and Madagascar. The dental formula varies: 3/2 or 3, 1/1, premolars from 2/2 to 4/4, and molars vary from 0/1 to 2/3 = 28 to 36. There is a reduction in the number of posterior molars, the cheek teeth are usually sectorial, and the carnassials are well developed. There are five toes on each foot. The claws are nonretractile. The limbs are either semiplantigrade or digitigrade. The most characteristic feature of this family consists of the well-developed anal musk glands. Members of this family occupy diverse habitats from semiarboreal (fisher) to semiaquatic (otters) to marine (sea otters).

Viverridae (36 genera, 75 species; civets, genets, and mongooses; Southern Europe, Southern Asia, Africa, Malay region, Madagascar, and nearby islands)   They have been introduced in New Zealand, Hawaii, and the West Indies, in all of which they have become serious pests. The dentition is 3/3, 1/1, 3 or 4/3 or 4, 1 or 2/1 or 2 = 32, 36, or 40. The dentition is generally sectorial, the carnassials are well developed, and the middle lower incisor is raised above the level of the other two. The paraoccipital process is in close contact with the auditory bulla, and may at a quick glance be mistaken for it. The legs may be digitigrade to semiplantigrade. There are usually five digits on each foot, though this may vary. The claws are semiretractile.

Hyaenidae (3 genera, 4 species; hyenas) Hyenas are found in Southwestern Asia and Africa. The dentition is 3/3, 1/1, 3 or 4/1, 2 or 3, 1/1 or 2 = 28 or 32. The cheek teeth are adapted for crushing. The carnassials are well developed. The paraoccipital is in close contact with the bulla. The hind limbs are shorter than the forelimbs and are digitigrade, with four or five toes on the forefoot. The claws are nonretractile. Members of this family are specialized as scavengers but are sometimes predacious. *Crocuta* is noted for its laugh and is called the laughing or spotted hyena.

Felidae (4 genera, about 37 species; cats) Their distribution is worldwide except in Antarctica, Australia, Madagascar, and other islands. The dentition is an example of extreme sectorial modification, and is usually given as the best example of adaptation for a carnivorous diet. The canines are large and recurved, the carnassials are large and well developed, and the cheek teeth are reduced. The dental formula is 3/3, 1/1, 2 or 3/2, 1/1 = 28 or 30. There are five toes on the forefeet, four on the hind feet; the claws are sharp, strongly recurved, and retractile (except in the cheetah, *Acinonyx*). The legs are digitigrade. The eyes of cats are larger than those of other carnivores. The senses of smell and hearing are acute. The lynx is usually placed in the genus *Lynx,* but Simpson (1945) and many Europeans place the lynxes in the genus *Felis.* The bobcat *(Lynx rufus)* is among those animals which, like the wolf and the fisher, may feed on porcupines.

Order Pinnipedia: Seals and Walrus

These are the aquatic equivalents of the order Carnivora, and some taxonomists give them subordinal rank within that order. The pinnipeds occur along the coastal regions of the world, as well as in the Caspian Sea and Lake Baikal. They come onto land for the breeding season. The teeth are nearly homodont. In the walrus *(Odobenus rosmarus)* the tusks are the upper canines. The forelimbs and hind limbs are fully webbed, with five digits on each. Hairs are without erector muscles. Blubber is always present. The fat content of the milk is very high. Fossil pinnipeds are first recorded from the Miocene of North America. The Odobenidae probably arose from the Otariidae during the Miocene. The otarids and phocids may have a common ancestor, or each may have arisen from the Carnivora.

Otariidae (6 genera, 12 species; eared seals, sea lions) They are found in marine waters, along coasts and islands, but are absent in the Arctic and Antarctic Oceans. The dental formula is 3/2, 1/1, 4/4, 1 or 2/1 = 34 or 36.

The outer incisors are caniniform. The skull has slightly inflated auditory bullae, and an alisphenoid canal is present. The external ears are small. Hind limbs (flippers) can be placed under the body and are useful on ice and land. There is an external scrotum.

Odobenidae (1 genus, 1 species; walrus)   Walrus *(Odobenus)* are found in the arctic waters of the world, staying close to shores or masses of floating ice. The dental formula is 1 or 2/0, 1/1, 3 or 4/3 or 4, 0/0 = 18 to 24. The walrus is known for its extremely large tusks, which are the upper canines. The anterior portion of the skull is enlarged to hold these tusks. The hind limbs, which can rotate forward and under the body, are useful on ice and land. The baculum is large—up to 64 cm in length (Scheffer, 1958).

Phocidae (13 genera, 18 species; earless seals, hairless seals)   The dentition is variable. The structure of the check teeth is variable. In some species they are simple pegs, but in the crabeater seal *(Lobodon)* the cheek teeth have complex cusps. These seals are found in coastal and pelagic waters of polar, temperate, and subtropical seas. Some occur in freshwater lakes, and some may move into rivers. These pinnipeds have no external pinnae (hence "earless"); the hind limbs (flippers) cannot be placed forward under the body, and therefore are useless on solid surfaces. There is no alisphenoid canal. The teeth of *Lobodon,* the crabeating seal, by forming a sieve, are used for sifting out plankton.

Order Tubulidentata

This order contains only one species, and therefore only one genus and one family.

Orycteropodidae (1 genus, 1 species; aardvarks)   Aardvarks *(Orycteropus afer)* are found in the southern half of Africa. The aardvark looks like a pig that is on a diet. It is an insectivorous mammal, sometimes called the ant-bear. The dental formula is usually 0/0, 0/0, 2/2, 3/3 = 20 (Fig. 5–10). The permanent canines are vestigial and remain in the gum. The permanent cheek teeth are columnar, each tooth consisting of numerous hexagonal prisms of dentine, which surrounds a tubular pulp cavity. The ordinal name is derived from this characteristic. The teeth themselves are without enamel, rootless, and continuously growing. The limbs are plantigrade and show modifications for burrowing. The forefoot does not have a thumb. The other digits are well developed and have strong nails, intermediate between claws and hooves. The hind foot has five toes. The burrowing ability of this mammal stretches credibility. Kingdon (1971) cites a report of three men being lost in an extensive and deep burrow system in Zambia. After 10 days of digging only one body was found. The skull is long and slender with more turbinate bones than any

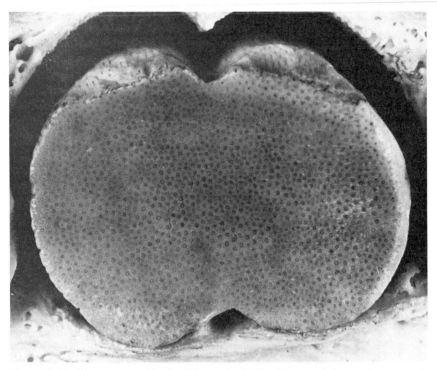

Figure 5-10 The permanent teeth of the aardvark *(Orycteropus afer)* are columnar, each tooth consisting of numerous hexagonal prisms of dentine surrounding a tubular pulp cavity. In the photo this shows best at the left edge of the tooth.

other mammal skull. The zygomatic arch is slender but complete, and the premaxillaries are small. The tongue is long, vermiform, and extensile, another adaptation for a diet of ants and termites. Many of the characteristics mentioned are reminders of other ant-and termite-eating mammals, the Edentata and Pholidata. Fossil aardvarks have been recorded from the Miocene in Africa, and the Pliocene of the Palaearctic. Tentatively assigned to this group are fragments from the Eocene and Oligocene of Europe, and from the lower Eocene in North America.

Order Proboscidea: Elephants

Some of the largest and most spectacular herbivores of the Cenozoic were proboscidians, but during the Pleistocene their diversity was reduced, and today there are only two species. The limbs are immense, pillarlike structures developed to support great body weights (up to 90,000 kg, 6 tons). This modification of limbs is referred to as "graviportal." The dental formula is 1/0, 0/0, 3/3, 3/3 = 26. The upper incisor is ever-growing and is enlarged into a tusk. Either a single molar or parts of two molars function at one time, as a rear molar moves forward to replace the one being worn away. The skull is massive, shortened, and filled with air cells. This increase in surface area, with proportionate decrease in weight, gives more area for muscle attachment, an obvious modification for the long, muscular, flexible proboscis, which has the nostrils at the end. The radius and ulna, and the tibia and fibula are separate. The radius and ulna are permanently crossed in a fixed position of pronation. Elephants are herbivorous. The gestation period is up to 2 years. The fossil record begins in the late Eocene in Egypt. Proboscidians reached North America in the late Miocene, but disappeared during the late Pleistocene.

Elephantidae (2 genera, 2 species) There are two living species of elephants. The Indian elephant *(Elephas indicas)* occurs in India, Burma, part of China, Malay Peninsula, Ceylon, and Sumatra. The Indian elephant is the one seen in circuses and other entertainment features. It has a trunk with a single pointed tip, the head has two rounded bosses, and the ears are small in comparison with those of the African elephant *(Loxodonta africana),* whose home is naturally Africa. The trunk of the African elephant has two pointed tips, it lacks the rounded bosses on the head, is highest at the shoulders and has extremely large ears.

Order Hyracoidea: Coneys

These rabbitlike mammals are structurally unique and ancient ungulates. The forefoot has four toes, but only three are functional; the hind limb has three. The axis of symmetry runs through the third digit. Except for the second digit of the front foot, the digits end in hooves. The soles of the feet are covered with soft elastic pads, which aid in clinging to rock. The soles are kept moist by sudorific glands. The dental formula is 1/2, 0/0, 4/4, 3/3 = 34. The upper incisors are long, triangular in cross section, and continuously growing. The lower incisors are chisel-shaped, with a cutting edge that is tricuspid or becoming unicuspid. The cheek teeth are ungulate-like in shape; the molars are hypsodont or brachydont, and lophodont, resembling the molars of the rhinoceros. The relationship of the hyracoids is puzzling, but they supposedly descended from an early ungulate stock, and are most closely related to elephants. Fossils are known from the lower Oligocene in Egypt.

Procaviidae (3 genera, 11 species; hyraxes or dassies) Hyraxes are found in Southeast Asia and most of Africa. Their characteristics are those of the order. Hyraxes are herbivorous. *Procavia* and *Heterohyrax* are terrestrial; *Dendrohyrax* is arboreal. These are believed to be the "coneys" of the Bible.

Order Sirenia: Manatees, Dugongs, and Sea Cows

Sirenians are probably the mammals that started the mermaid myth. They are fusiform in shape, aquatic, and nearly hairless. The forelimbs are paddlelike and pentadactyl, but the individual digits do not show. There are no hind limbs. The external nares are high on the skull. Nasals are rudimentary or absent. The lower jaw is heavy; the incisors and molars are separated by a wide diastema. The tail is a horizontally flattened fluke, which is spatulate in *Trichechidae* and deeply notched in *Dugongidae*. The pelvis is vestigial (as in cetaceans), the vertebrae are separate, and there is no sacrum. The skeletal elements are massive, the ribs are barrel-shaped. There are thick layers of blubber. The stomach is two-chambered. There are no pinnae; the eyes are small, with well-developed nictitating membranes. The brain is very small and little convoluted. Sirenians and proboscidians probably shared a common ancestor. Sirenian fossils are known from Eocene deposits in Europe, Africa, and the West Indies. Apparently they were far more diverse and widely distributed than they are today.

Dugongidae (1 genus, 1 species) There is only one genus *(Dugong)* unless the Steller's sea cow *(Hydrodamalis)*, exterminated in 1768, is included. Although recent sightings have been reported, they have not been verified, and Geptner (1966) does not consider the reports authentic. The dugongs are found in the tropical portions of the Indian and Pacific Oceans but are becoming extirpated in some areas. The dentition is 1/1, 0/0, 3/3, 3/3 = 28. The molars are replaced by rotation from back to front, as in the elephants. The premaxillae are much enlarged. The jugal is expanded vertically below the orbit, and is in contact with the premaxilla. Dugongs are herbivorous. They are said to be unable to breathe through the mouth.

Trichechidae (1 genus, 3 species; manatees) Members of this family are found along the Atlantic coast of South America, southern North America, and Africa. *Trichechus inunguis* is restricted to the Amazon and Orinoco drainages or interior South America. The premaxillae are relatively small, the jugal not reaching the premaxilla. Functional incisors are absent; the 2/2 of the milk dentition are concealed under a horny plate. The cheek teeth are replaced from the rear.

Order Perissodactyla: Odd-toed Ungulates

This order contains the hooved mammals with an odd number of toes. The weight is usually borne on an axis running through the middle digit, which is the largest. The limbs are unguligrade, with well-developed hooves. The carpals and tarsals are not fused. The astragalus has a deeply grooved, pulley-like articulation proximally and is truncate distally. Thus perissodactyls cannot bend their hind limbs enough to get up "hindfeet first." The skull is elongated, and while there are no antlers or horn cores, there are dermal horns in the rhinoceros, which lie on but are not attached to the nasal bones. The teeth, usually 44, are adapted for an herbivorous diet, with massive and quadrate premolars and molars forming a continuous series. The crown pattern is usually lophodont, but in tapirs it is bunodont. The stomach is rather simple; it may be two-lobed. There is no baculum, gallbladder, or clavicle.

The oldest fossils are known from the Eocene. Fourteen families are known from the Eocene and Oligocene, but by the end of the Oligocene only four remained, and by the end of the Pleistocene all families had disappeared from the Western Hemisphere. The largest land mammal that ever lived was the *Baluchitherium,* a hornless rhinoceros. The shoulder height was 5.5 m.

Equidae   (1 genus, 7 species; horses, zebras, asses)   The dental formula varies (3/3, 0 or 1/0, 3 or 4/3, 3/3 = 36 or 42) and is adapted for an herbivorous diet. The incisors are broad, canines are small and variable, premolars are functional, and molars are hypsodont. The grinding surfaces have exposed enamel, dentine, and cement. Limbs are slender, with extreme adaptation for cursorial locomotion; lateral digits are reduced. The second and fourth digits are represented by metapodials, and only the third digit, terminating in a hoof, is functional. These mammals are generally large. They are gregarious inhabitants of the savannas and open plains. One of the equids, the domestic horse, *Equus caballus (Equus przewalski),* has played an extremely important role in the history of humans. In spite of an extensive literature, the evolution of the horse and the taxonomic affinities of the domestic horse have not been precisely delineated (Simpson, 1961; Romer, 1966). Wild members of the genus are now restricted to Africa, Arabia, Mongolia, and Chinese Turkestan.

Tapiridae   (1 genus, 4 species; tapirs)   They are found in grassy swamps, dense jungles, and forested hill and mountain slopes in tropical areas of the Eastern and Western Hemispheres. In the Eastern Hemisphere this includes the Malay Peninsula, Java, and Sumatra, and in the Western Hemisphere, central and tropical South America. The teeth are more generalized than in other perissodactyls. The dental formula is 3/3, 1/1, 4/4, 3/3 = 44. The incisors are chisel-shaped, the canines are conical and well developed, the upper cheek teeth have simple lophs, and the lowers have transverse protolophs and metalophs. The limbs are short; the forefeet have four evident digits, the hind feet have three. The nasal bones are short and project by themselves, supporting an elongated nose and upper lip to form a short proboscis, with the nostrils near the tip. The skin is very thick.

Rhinocerotidae   (4 genera, 5 species; rhinoceroses)   In Africa they are found south of the Sahara. They are also found in Southern Asia, Sumatra, Java, and Borneo. The dental formula is 0 or 2/0 or 1, 0/0 or 1, 3 or 4/3 or 4, 3/3 = 24, 28, 32, or 34. The incisors and canines are sometimes absent; premolars are molariform. The front limbs have three or four digits, the hind limbs have three. Each digit has a small hoof. The nasal bones stand out freely, and the lip is slightly prehensile. The postorbital process is absent, making the orbital and temporal fossae confluent. These are large, bulky mammals with pillarlike (graviportal) legs. The skin is thick with few hairs. The senses of smell and hearing are well developed, but sight is poorly developed. A solid horn of dermal origin, made up of agglutinated fibers, rests on the median line of the nasals. A second horn sometimes appears behind the first. These huge animals are herbivorous. They may weigh up to 1800 kg (4000 lb). These mammals have been hunted for their meat, hides, bones, but especially for their horns, which have been used for drinking cups and for Oriental medicine. Their supposed medicinal value stems from the idea that in any form the horn serves as an aphrodisiac. The number of rhinoceroses continues to decline.

Order Artiodactyla: Even-toed Ungulates

This order includes all the even-toed hooved mammals. The functional axis of the leg passes between the third and fourth digits, which are usually equal. The second and fifth digits are reduced. The two-toed, or "cloven hoofed," kinds have the two metapodials fused to form a cannon bone. The astragalus has a rolling surface proximally and a pulley surface distally, giving a free ankle motion. Artiodactyls can thus bend their hind legs, bringing them under the body, and get up "hind legs first." The stomach is simple to complex, of the type for ruminating. Twenty families constitute the fossil record from the Eocene. With the reduction in the diversity of the Perissodactyla during the Oligocene and with the diversification of artiodactyls in the Miocene, the recent artiodactyl fauna far overshadows the recent perissodactyl fauna.

Suborder Suiformes

Although the dentition is varied, upper incisors are always present in this suborder. The molars are bunodont. The limbs are short. There are usually four well-developed digits on each foot. There are no horns or antlers.

Suidae (5 genera, 8 species; pigs) The domestic pig, *Sus scrofa*, is worldwide in distribution. Wild species are found in most of Africa and Eurasia, as well as Japan, Taiwan, Ceylon, Madagascar, Ceylon, Philippine Islands, Malaya Region, Sumatra, Java, Borneo, and the Celebes. The dental formula is variable: 2/3, 1/1, 2/2, 3/3 = 34 *(Babyrousa),* to 3/3, 1/1, 4/4, 3/3 = 44 *(Sus).* The middle upper incisors are largest; the lower canines are smaller than the uppers. The canines of both sexes curve outward, forward, and upward. In the male *Babyrousa* the tusklike canines grow continuously, the upper canines curving backward and downward to where they touch the forehead. There is a peculiar prenasal bone at the anterior end of the mesethmoid. Suids wallow in the mud, and grub or root for their food, eating both plant and animal food, including poisonous snakes. The tusks and flat, very sensitive noses help in rooting for food.

Tayassuidae (1 genus, 2 species; peccaries, javelinas) They are found in Southwestern United States south to central Argentina. The dental formula is 2/3, 1/1, 3/3, 3/3 = 38. The upper canine tusks are never recurved, but are directed downward and are smaller than those of the Suidae. They are triangular in cross section. The premolars and molars are bunodont. The stomach approaches that of the ruminants in complexity. The body is covered with bristly hair.

Hippopotamidae (2 genera, 2 species; hippopotamuses; restricted to Africa) The dental formula is 2 or 3/1 or 3, 1/1, 4/4, 3/3 = 38 to 44. The incisors and canines are tusklike and grow continuously. Canines may reach a length of 75 cm (30 in.). The premolars and molars are bunodont. The limbs have four digits, each with a naillike hoof. All four digits support weight. The stomach is three-chambered but nonruminating. Hippopotamuses are gregarious, usually like water, and are expert swimmers. Their nostrils can be closed under water, and the young can nurse in water.

Suborder Tylopoda: Camels

The dental formula is 1/3, 1/1, 2 or 3/2, 3/3 = 32 or 34. The lower incisors are small, procumbent, and spatulate. The molars are selenodont. The metacarpals and metatarsals are fused into a cannon bone. The limbs are digitigrade. The toes have nails protected by heavy ventral pads. The stomach is three-chambered, of the ruminating type. Erythrocytes are ovoid and are without a nucleus when mature, contrary to statements usually found.

Camelidae (2 genera, 4 species; camels, llamas) Native camels are found in the Gobi Desert of Asia and in the mountains of South America from southern Peru, south to Tierra del Fuego. It is thought that the camels from North Africa and Central Asia are not the original stock. Llamas belong to the genus *Lama*.

Suborder Ruminantia

The upper incisors are absent, the upper canines may or may not be present. Cannon bones are present. These are herbivorous mammals. The stomach has three (infraorder Tragulidae) or four (infraorder Pecora) chambers and is of the ruminating type. Horns or antlers may be present.

Tragulidae (2 genera, 4 species; chevrotains, or mouse deer) The chevrotains *(Hyemoschus)* are restricted to rivers and streams of equatorial Africa. The mouse deer *(Tragulus)* are restricted to Southern Asia. They are small (up to 61 cm), secretive animals of the forest. The dental formula is 0/3, 1/1, 3/3, 3/3 = 34. The upper canines are well developed. In the male, they are curved and pointed, and they protrude below the lips. There are no horns or antlers.

Cervidae (16 genera, 37 species; deer and allies) They are worldwide in distribution, with the exception of parts of Africa, the Near East, Australia and New Zealand, and Madagascar and New Guinea (however, they have been introduced into some of these islands). These are all herbivorous mammals. The dentition is 0/3, 0 or 1/1, 3/3, 3/3 = 32 or 34. The molars are brachydont and selenodont. The upper canines may be present (e.g., in *Moschus, Hydropotes,* and sometimes in *Cervus*). The lacrimal and nasal bones are not contiguous. The tarsals and carpals are partially fused, and the principal metapodials are fused into cannon bones. The limbs have the two outer toes reduced ("dewclaws") or absent. Deciduous antlers are present, except in the musk deer *(Moschus)* and water deer *(Hydropotes)*. Only the males have antlers, with the exception of *Rangifer,* the females of which also have antlers.

Giraffidae (2 genera, 2 species; giraffes, okapis) The giraffes live in the savanna country of Africa, and the okapis in the tropical rain forests of Africa. This may explain why the okapi was able to hide from scientists until 1900, when it first became known to them. The dental formula is 0/3, 0/1, 3/3, 3/3 = 32. The molars are brachydont. These animals have horns of bony protuberances of frontal bones covered by unmodified skin. These are two in number and present in both sexes. The extremely large heart, 60 cm (2 ft) long, with 7.5-cm (3-in.) walls and weighing 11 kg (25 lb) and the high blood pressure necessitated by the long neck and tall body of the giraffe have made it necessary for the animal to have mechanisms which prevent hypo- or hypertension. This is accomplished by a "wonder net" of capillaries just below the brain and a series of valves in the jugular vein (Goetz and Budtz-Olsen, 1955; Goetz and Keen, 1957; Goetz, Warren, Gauer, Patterson, Dyle, Keen, and McGregor, 1960). Despite its power, the giraffe is a very sensitive animal and when unduly excited may die from shock. When it is taken alive, it is customary to administer injections of Adrenalin to combat shock.

Antilocapridae (1 genus, 1 species; pronghorn) Pronghorn *(Antilocapra americana)* are restricted to western North America. The pronghorn could well be considered the hallmark of the western North American desert and grasslands (Fig. 5–11). The dentition is 0/2, 0/1, 3/3, 3/3 = 32. The cheek teeth are selenodont. The horns in both sexes (smaller in the female) consist of bony cores with a sheath of fused hairs. The sheaths are shed annually. There are no "dewclaws." The orbits are large, as befits an animal with keen vision. There is an acute sense of smell. These keen senses and a pelage, the hairs of which can be erected or laid down, are all modifications for life on the open prairies, where they browse on grass, buckbrush, and sagebrush.

Bovidae (44 genera, about 110 species; antelopes, bison, buffalo, cattle, gazelles, goats, and sheep) Bovids were native to western North America,

Figure 5–11 The fleet antelope *(Antilocapra americana)* is a common artiodactyl on the Great Plains.

all of Africa, and most of Eurasia. None were found in Central or South America, Madagascar, Australia, New Zealand, or New Guinea, but they have now been introduced into many of these places. These are the hollow-horned ruminants. The dentition is 0/3, 0/1, 3/2 or 3, 3/3 = 30 or 32. The cheek teeth are selenodont. The lacrimal bone and nasal bone are almost contiguous. Lateral digits are reduced or absent. Cannon bones are present. There is one pair of horns, usually in both sexes but larger in males. The horns are composed of a bony core covered with keratin. The stomach is four-chambered and of the ruminating type.

There are five subfamilies: The Bovinae includes the bison and domestic cattle *(Bos taurus)*, among others. The Cephalophina contains the duikers, small (about 60 cm (2 ft) in height) mammals of Africa. The Hippotraginae includes the waterbucks, reedbucks, oryxes, hartebeests, and wildebeests. The Antilopinae contains the gazelles and true antelopes. Within the subfamily Caprinae (the sheep and goats) are four tribes: the Saigini are the chirus and saigas; the Rupicaprini include gorals, serows, mountain goats, and chamois; the Ovibovini are the takins and musk oxen; the Caprini are the goats, ibex, Barbary sheep, domestic sheep, and mountain sheep.

The Skin and Its Derivatives

A mammal's skin envelops its entire body surface, thus separating it from its environment, and serves in many ways to maintain the individual's balance with the environment. Perhaps the two most important selective forces which acted to form the integument in mammals were the prevention of water loss and the retention of heat. But the skin of mammals serves many other functions. It is a protection against mechanical injury, it wards off physical and chemical influences, and it is a defense against the invasion of microorganisms. The skin is also the site of important sensory organs, and the nervous system arises in continuity with the skin ectoderm. This outer seal of mammals is continuously replaced. The external surface, the epithelium, is continuous with the epithelium of the external orifices of the respiratory, digestive, and urogenital systems. Skin is stratified tissue. The surface is composed of *epidermis*. Underneath this is a connective tissue layer called the *dermis,* and immediately under this is a fatty layer, *panniculus adiposus*. The innermost flat sheet of skeletal muscle, called the *panniculus carnosus,* is very common in most mammals but vestigial in human beings.

EPIDERMIS

The epidermis (ectodermal in origin) gives rise to a host of structures such as hairs, glands, claws, hoofs, nails and (together with the dermis) horns and antlers. The epidermis consists of at least three layers of cells, basically the

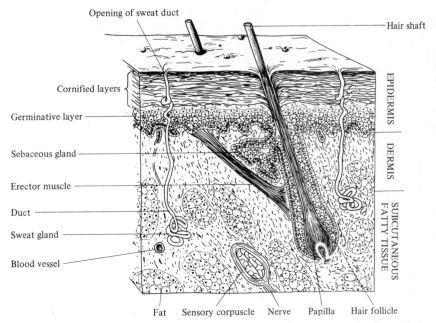

Figure 6–1 Section of skin showing hair follicle, sebaceous glands, sweat glands, and hair follicle. *(From Storer and Usinger, 1955.)*

same but separated in their shape and partly in their function (Fig. 6–1). The outermost layer, the *stratum corneum,* consists of hardened dead cells that are constantly being sloughed off. Active cell division in the deeper layer produces new cells that are pushed outward to take the place of those that are lost. There are no blood vessels in the epidermis, the metabolic needs of which are met by diffusion (H. M. Smith, 1960).

HAIR

The most characteristic epidermal specialization in mammals is hair, which is only secondarily associated with scales, contrary to the situation in birds, whose feathers are derived directly from scales. For a good readable reference on hair, see Searle (1968) or Ryder (1973). Hair, then, is a new structural development in mammals, not a modification of horny scales, as are feathers. Beyond providing the basal papillae, the mesoderm does not participate in the development of hair. Because hair is so characteristic of mammals, earlier names proposed for the class Mammalia were Trichozoa and Pilifera.

The hair owes its remarkable properties to its high content of a sulfur-containing protein, keratin, which is present in both the epidermis and epidermal appendages, such as claws, nails, and horns. These superficial structures, which become impregnated with keratin, arise from a transformation of the

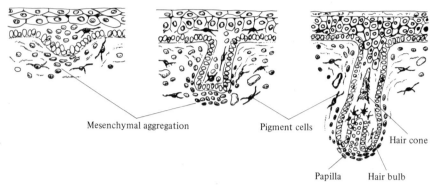

Figure 6-2   Early stages in development of hair follicle and hair, showing migration of pigment cells into the hair bulb. Arrows in I and II show direction of migration. *(After Searle, 1968.)*

cells of the Malpighian layer. Two types of keratin are recognized. Soft keratin is contained in soft structures of the skin, particularly the epidermis, where the keratinized layer is called the *stratum corneum.* Soft keratin has a relatively low sulfur content, is comparatively unreactive, and does not desquamate from the surface. Hard keratin can be found in all appendages, nails, horns, hooves, and hair.

A typical hair includes the projecting shaft and a root which is sunk in a pit in the dermis called the hair follicle (Fig. 6-2). The shaft and the root (except at the base) consist of dead keratinized epidermal cells. Around the root is a sheath which may consist of several layers of dermis and epidermis. Beyond the presence of hair, there is a great variation in the presence or absence of associated sebaceous and sudoriferous glands and the *arrector pili* muscles. When present the sebaceous gland empties its oily lubrication into the follicle. Usually each hair is supplied with a smooth muscle (*arrector pili* muscle), but not always, and sometimes a sweat gland is associated with the hair follicle. In *Bos taurus* each hair follicle is accompanied by an *arrector pili muscle,* a sweat gland, and a sebaceous gland (Carter and Dowling, 1954). At the opposite extreme are the seals. In the southern elephant seal *(Mirounga leonina)* "as in other seals there are no arrectores pilorum muscles," according to Ling (1965). There is a bilobed sebaceous gland on the side of the follicle, and "the acini of the sebaceous glands are enormous, commensurate with their probable function of waterproofing the surface of the skin." Continuing his discussion of the hair follicle of the southern elephant seal, Ling (1965) wrote, "On the other hand, the apocrine sweat glands, each one associated with a pilosebaceous unit are quite rudimentary."

Hair Formation   A hair is formed from the epidermis; its development has been described by Butcher (1951). A hair germ appears as a slightly downward-projecting outgrowth in the deepest part of the epidermis, the *stratum germinativum*. In fetal rats this occurs in 17 to 20 days (Butcher, 1951). Beneath this hair germ there is a condensation of upward-projecting dermal tissue to form the rudiment of a hair papilla. The dermal papillae of hairs and vibrissae are apparently different. Though they may be similar in shape, the dermal papilla of a normal hair is smaller than that of a vibrissa (Cohen, 1965). Just what it is that determines which type of hairs is produced has not been definitely determined. There are indications that both the root sheaths and the dermal papillae help determine hair shape, size, and movement (Cohen, 1965; Straile, 1965).

The hair germ extends and finally surrounds the dermal papilla as a tennis ball would surround a finger pushed into it. The base of the epithelial portion enlarges, and cells arise from this mass of proliferating tissue, the matrix, which eventually gives rise to the cells destined to become cortex and inner sheath, while the matrix cells around the tip of the papilla push outward to become the medulla of the hair cell. The peripheral cells become the surrounding root sheath, which retains its continuity with the stratum germinativum. The inner cells form a core of keratinized material, which eventually becomes the hair.

Anatomy of Hair   It is the cuticle that imparts the character to hair which is useful for identification (Fig. 6–3). Its surface is oily. Cuticle scales completely surround the hair shaft. The minimum number of scales to encircle a hair is two. A cuticular scale is never tubular in form in even the finest hair. There is great variation in scale patterns based primarily on the patterns on the edge of the upper margin of the cuticle.

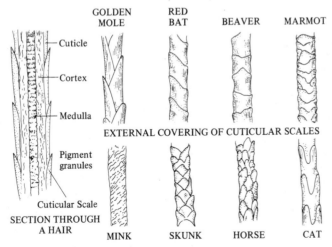

Figure 6–3   Structure of mammalian hair. *(From Storer and Usinger, and Nybakken, 1968.)*

The *cortex* forms the main bulk of the hair; it consists of a column of keratinized cells fused into a hyaline mass with air spaces, called *fusi,* which originally were filled with liquid.

The *medulla* consists of the cornified remnants of epithelial cells connected by a filamentous network and contains the coloring matter which determines the appearance of hair. The medulla is not always present, as already mentioned for sheep. Nason (1948) found no medulla in the hair of bats of eastern North America.

Microscopic examination of hair has revealed black, brown, and yellow pigments (Fitzpatrick, Brunet, and Kukita, 1958). The black-brown pigments are designated as "tyrosine-melanin," and the yellow-red pigments as "pheomelanin." Color of hair depends on the chemical nature, quantity, and distribution of the pigments and is affected by a number of genes—(24 have been recorded in the mouse (Gruneberg, 1952)—some of which are pleiotropic. In many mammals (man is an exception) hair is not uniformly colored, but is barred, with a subterminal band containing pheomelanin, and the base containing melanin. Most rabbits and rodents have this bicolored hair, which has been named the "agouti" pattern.

The finest hairs have no medulla, and the arrangement of the medullary air cells is rather closely correlated with the diameter of hair shaft, rather than with the age of the hair or the species of mammal. The significance of the cuticle, cortex, and medulla in the fur industry has been discussed by Bachrach (1946). Hair in cross section may be round, oval, flattened, and may assume numerous shapes in the same animals.

Kinds of Hair   De Meijere's (1894) classification of hairs was the earliest, but many changes have been made since; the classification most commonly adopted now is Toldt's (1935). There are two major groupings of hair: (1) *hairs with specialized follicles containing erectile tissues* (called "feelers," "whiskers," "tactile hairs," or "vibrissae"), which are said to occur in all mammals except man (Pocock, 1914), and which are of two types, those under voluntary control and those passive hairs which are not under voluntary control; and (2) *hairs with follicles not containing erectile tissue,* includes the remaining types of hair. Sometimes these may have a passive sensory function because the follicles are well supplied with nerves. The remaining types of hair are divided into the coarser guard hairs, overhair or top hair, and fine underhair (Fig. 6–4). Guard hairs may be:

1 Spines, greatly enlarged and modified for defense, such as quills
2 Bristles, which are numerous and provide the protective coating
3 Awn, which are coarse, but with a finer base

Underhair is fine and uniformly soft, and may be divided into fur and vellus, which is the finest and shortest hair, sometimes called "down," "fuzz,"

Figure 6–4 Hair covering of mammals as shown in this prairie dog *(Clynomys ludovicianus)* is divided into a fine underfur, the primary function of which is insulation, and a layer of longer guard hairs which protect the underfur from wear and matting.

or "lanugo." The porcupine *(Erethizon dorsatum)* possesses all the main types of hair. A generalized fur consists of guard hairs (bristles and awns) and an undercoat of fur. In Merino sheep the hairs grow from persistent follicles and hair is not shed.

Functions of Hair   The original function of hair may have been to serve as a tactile organ. The vibrissae around a cat's face (cat's whiskers) are a good example of this type of hair. The hairs on the tail of rodents and fossorial mammals (moles and pocket gophers) are sensory. These tactile hairs are richly supplied with nerves.

Hair color and pattern may serve for concealing [as in the fawn of deer *(Odocoileus* sp.) and in *Peromyscus]* or for warning (as in skunks), as well as for communication, as exemplified by the flash patches of elk *(Cervus canadensis)* and pronghorns *(Antilocapra americana).*

Hair can serve for buoyancy in two ways. The hair of artiodactyls contains many air spaces and is often spoken of as being "hollow." Hair can also trap air and surround the animal with a blanket of air, as seen in beaver *(Castor canadensis),* muskrat *(Ondatra zibethica),* and moles. A mole or water shrew *(Sorex palustris)* under water and in the right lighting almost looks as if it is surrounded by mercury or silver (Lorenz, 1953). Some semiaquatic mammals have nonwettable fur. Johansen (1962) measured the amount of air trapped in the fur of muskrats *(Ondatra zibethica)* living in the Fairbanks, Alaska, area. He found the average amount in 10 muskrats was 176.9 ml or 21.5 percent of their average total dry volume. This air gave the muskrats an average specific weight of 0.790. Air trapped in their fur gives animals a buoyancy which reduces energy expenditure during swimming. Another distinct advantage is provided by the low thermal conductivity of air. A muskrat depleted of its blanket of air was shown by Johansen (1962) to lose heat much faster.

Specialized hairs, such as the quills of porcupines *(Erethizon dorsatum),* can serve for protection. In the African porcupine some quills are modified for rattling.

Hair serves as an aid in locomotion. Muskrats *(Ondatra zibethica)* and the water shrews *(Sorex palustris)* have a fringe of hairs along the outer edge of their feet, and the water shrew supposedly can walk on top of water. The hair on the tail of a flying squirrel *(Glaucomys)* flattened dorsoventrally into a single plane, helps the squirrel in its gliding type of locomotion.

In special situations hair has a sexual function and may show a receptive or aggressive attitude, as in the wolf *(Canis lupus),* or dog *(Canis familiaris).* Hair may also serve to retain odors which may have a sexual or territorial function, as in the shrew, *Blarina brevicauda* (Pearson, 1946), and in the lemur, *Lemur catta* (Petter, 1962).

Certainly the most important function of hair is as a protection from the elements. The mammals with the finest and heaviest coats of hair come from northern regions (Carter, 1965). The addition of 2.5 cm of fine fur permits the environmental temperature to be dropped from 30 to 0°C without altering skin or body temperature (Herrington, 1951). Herrington (1951) presented a good review of the importance of the pilary system in mammals. A genetically controlled condition in which no guard hairs are produced occurs sometimes among mammals. Such individuals are known as Sampson foxes, Sampson mink, etc. During a cycle of warmer-than-average winters the number of Sampson mutations that survive [e.g., among red fox *(Vulpes vulpes)*] causes population increases. Without the protection of guard hairs the underfur becomes matted and worn. When a series of severe winters occurs, these Sampson mutants cannot survive. Quay (1965) has mentioned the tendency to fine hair in the desert species of heteromyids and cricetines. He reported an unusual adaptation to desert conditions for some species of *Dipodomys* and *Perognathus,* in which there was an anteroposterior flattening of the hair shafts forming shinglelike plates of hair affixed to one another at their lateral edges by a lipid film. He suggested that this might help prevent cutaneous evaporative water loss.

PELAGE GROWTH AND CHANGE

The sensory pits of the reptiles were located on the apices of the epidermal scales. The primitive distributional pattern of hair is similar to the distribution of these sensory pits. For the most part this primitive pattern has been lost, but it can still be seen, e.g., in the tail of rodents and on the side of the human hand. Hair follicles most often occur in groups of three with the largest hair in the middle, but the regional variation in density is great (Carter, 1965). The idea of the basic trio as the primitive condition, first expressed by De Meijere (1894), has been accepted as an adequate working hypothesis, but there are many variations.

There seems to be general, though not complete, agreement that the population of hair follicles is completed by the time of birth or shortly thereafter. Wohlbach (1951) wrote, "In the mouse, new follicles do not form in the epidermis, under any circumstances, after the original postnatal formation has been completed." Hayman (1965), in *Hair Growth in Cattle,* wrote, "The total follicle number of an animal is determined by the time it is born."

Individual follicles show alternating phases of activity and inactivity which have been labeled *anagen,* the active growth stage; *catagen,* an intermediate stage; and *telogen,* the resting stage. The development of the pilary system in laboratory rats *(Rattus norvegicus)* and mice *(Mus musculus)* has been described by Butcher (1951). In the laboratory rat, hair germs appear in the 17- to 20-day fetal rats as crowding of cells in the deepest layers of the epidermis. The convexity increases and a slight condensation of connective tissue, the prospective papilla, appears. In a 2-day-old rat the base of the epithelial portion has invaginated to form the bulb of the hair, and the connective tissue papilla protrudes into the bulb. The epithelial cells of the bulb, which represent the future matrix, by active mitosis give rise to cuticle cortex and inner sheath. The matrix cells at the tip of the papilla become the medulla of the hair shaft. The melanophores of black hair follicles are situated in this matrix, giving rise to the medulla. Dendroid processes of the melanophores containing pigment extend into the forming cortical cells. Pigment passes directly into the cortical and medullary cells.

There are two types of follicles in the rat's integument. Large follicles give rise to the longer and coarser hairs; smaller follicles, which are more numerous, give rise to shorter, finer hairs.

Growth of hair continues until the seventeenth day, which terminates the anagen phase. The transformation into a resting stage takes place in 3 or 4 days, which probably corresponds to the catagen stage. The follicle then remains in a resting stage (telogen phase) until the thirty-first or thirty-second day of life. In the rat the second cycle begins on the thirty-second day. The same papilla may function in the growth of another hair, which is probably true for vibrissae, but generally the old papilla ruptures and a new one forms at the base of the follicle. The new hair grows next to the old hair, so that the fine hairs of the first cycle are joined by coarser hairs. The second growing stage in the rat takes about 17 days also, followed by a resting stage. This makes the hair cycle in the rat about 34 days. The growth wave in the rat, as in many other rodents, begins on the venter, spreads dorsally, then anteriorly, and finally posteriorly.

The pattern of hair growth and molt in wild mammals varies greatly and shows some interesting features of adaptation. The hair growth cycle consists of periods of regeneration and increase followed by periods of quiescence and decline. In many mammals this is seasonal; in others it is not (Fig. 6–5).

The classical work of Bissonette and his colleagues (Bissonette, 1935; Bissonette and Wilson, 1939; and Bissonette and Bailey, 1944) on coat changes of ferrets, weasels, and mink has shown that light is an important factor. The change in pelage color was the result of growth of new hair, not of change of color in already present hair. Bissonette and Bailey (1944) felt that temperature change was not a causative factor in these changes, except perhaps indirectly in nature by keeping animals in their warm, dark dens during colder weather, thus changing the individual's time of exposure to light. These authors proposed "that altered daily period of illumination is the factor directly inducing the seasonal molts and changes of coat color of weasels and probably of other birds and mammals studied." Lyman (1942) showed that hares *(Lepus americanus)* turn dark in winter if given long daylight hours and turn white in summer if the daylight period is shortened. He further showed that castrated and thyroidectomized hares undergo the normal cycle, but Bissonette (1935) showed that hypophysectomy abolished the cycle in ferrets. Therefore Bissonette and Bailey (1944) proposed that "this stimulus is received through the eyes and acts through the anterior gland."

More detailed information on how this might occur has recently been offered by Rust and Meyer (1972) and others. They implanted beeswax containing the pineal gland product melatonin, or melanocyte-stimulating hormone (MSH), subcutaneously in male short-tailed weasels *(Mustela erminea)* in the spring. Brown individuals and white individuals undergoing the spring change to brown pelage and reproductive activity, molted, grew a new white coat, and became reproductively quiescent. The pineal gland product MSH appears to initiate changes in the central nervous system and endocrine glands which result in molting, growth of winter pelage, and reproductive dormancy. The hypothalamus apparently inhibits MSH secretion. Light may influence the amount of MSH by a mechanism involving melatonin and an MSH-release-inhibiting factor.

Hadwen (1927) postulated the theory that color changes are primarily for the summer exclusion and winter retention of heat, but it is now generally accepted that changing color makes the individual less vulnerable to predation, as Dice (1947) showed with mice.

Following this general background on hair and hair cycles we shall examine a few kinds of pelage changes of mammals, other than those already mentioned.

Molts of Other Mammals   Bassett and Llewellyn (1948) have described captive mink *(Mustela vison)* as having two seasonal molting patterns very similar to those of the weasels. The adult mink were in their prime between

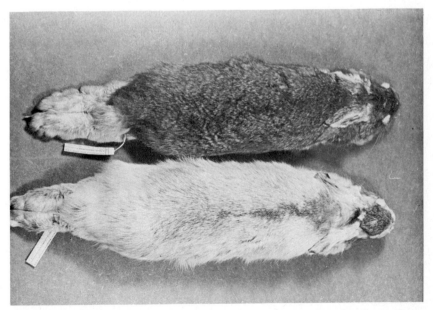

Figure 6–5 Molting occurs over a period of time. These two snowshoe hares *(Lepus americanus),* in different stages of molt, were taken on the same day, November 11.

November 20 and December 1. As the winter progressed there was a gradual decrease in the quality of the pelt. The melanin faded, luster of the guard hairs deteriorated, and the fur showed the effects of wear. The spring molt began when a small amount of new hair appeared just back of the muzzle. The molt wave progressed to include the cheeks, underside of the neck, and shoulders. The front legs, hind legs, and buttocks then began to get new hair, and a streak of new hair extended all along the underside. This spread up the sides, and new hair grew on the tail. By mid-July the winter coat had been replaced by the summer coat. The pelt was prime and the coat was sleek in appearance, but the fur color was light and not so intense, nor was the density so great as in the winter coat.

The summer pelage was retained for only about 3 weeks, from mid-July to early August. The molting of the summer pelage progressed from the tip of the tail forward to the muzzle in the opposite direction of the shedding of the winter coat, and was completed in November. Similar biannual molts have been described for the adults of the snowshoe hare (Severaid, 1945), the cottontail (Negus, 1958), and many rodents.

In contrast to the weasels and mink, which produce two coats of fur a year, the silver fox *(Vulpes vulpes)* produced only one (Bassett and Llewellyn, 1948). This was also the pattern in the coyote *(Canis latrans,* Whiteman, 1940) and other members of the family Canidae. Silver fox pelts, in upper New York, were in their prime by late November or early December. From then on the fur gradually deteriorated from wear and fading until the new molt started in early April. The molt started between the toes and on all four feet and proceeded to spread out until by June 15 the areas of the feet to the carpal and tarsal joints were well shed out. Then there was loosening and replacement of guard hairs and underfur from the lower thigh and the tail. The molt progressed, with the underfur disappearing from the ears, hips, thighs, and also the tail. Following this the abdomen lost its old hair, and the shedding proceeded to the thoracic and lumbar regions. Molting at this stage was characterized by the loss of underfur in large patches.

By the first part of September all old underhair was gone and most of the guard hairs. New fur growth, mostly guard hairs up to 15 mm in length, was plentiful. By October 20 the new fur growth was about three-fourths completed. All fur growth was completed by November 10, but pelt "primeness" was not attained until the skin color became a creamy white. In the latitude of Saratoga Springs, New York, this was after December 1.

In the silver fox the hair was replaced gradually over a period from April until November. The summer coat was thin because it was made up of worn hair, shed hairs, and new fur which was very short, rather than a completely new coat of short hair.

Rodents show a great variety of molt patterns, as would be expected. What causes the pelage to change in such mammals as mice, which are nocturnal and spend much of their time in runways, is unknown. Gottschang (1956) wrote, concerning *Peromyscus leucopus,* "The factors affecting the pelage change are not apparent." He also added, "Weather conditions probably have little or no effect on pelage change. No difference in the onset, progress, or length of time required for the pelage change was noted between spring, summer or fall-born litters."

The pelage and molt of *Peromyscus maniculatus* have been described by Osgood (1909), Collins (1918), and Storer, Evans, and Palmer (1944). The juvenile pelage was dark grayish in appearance, with scattered long hairs. Young *Peromyscus maniculatus* went through a postjuvenile molt when about 6 weeks old. This molt began laterally (between the front and hind legs) and progressed dorsally and then toward the head and tail. The average duration of the postjuvenile molt was 25 days. Osgood (1909) wrote, "In its early stages this adolescent pelage is plainly distinguishable from the adult pelage." The increase of adult hairs changed the appearance from grayish tan to a richer tan. The mice were smaller in the earlier stages of the adolescent pelage and were often designated as subadults. Storer, Evans, and Palmer (1944) wrote, "some individuals were pregnant before they had completed the postjuvenile molt." The confusion in terminology caused by the disconformity between the pelage designations and the breeding condition in white-footed mice (*Peromyscus* spp.) is one of those problems of language which is not easily resolved. Thus comes the paradox of a subadult (so considered using pelage as a criterion) that is breeding (which indicates sexual maturity according to biologic criteria). There are probably two annual molts in the adult.

Ecke and Kinney (1956) have described the molt from juvenile to postjuvenile pelage in *Microtus californicus,* which started at age 23 to 25 days. The molt began in an area directly anterior to the front legs at the lateral line. Within a day or two the ventral area up to and slightly above the lateral lines showed new hair. The molt progressed over the top of the rump to about the middorsal spot on the back. The head area was last to molt. The skin was prime at about 34 days of age. As soon as this molt was completed, the second molt, from postjuvenile to adult pelage, began. This molt pattern progressed in the same way as the first. The adult prime skin appeared when the individual was about 60 days of age. Apparently all adult molts were irregular. Up to the sixtieth day the age of individuals could be determined within an accuracy of 4 days in 88 percent of the cases.

Pinnipeds have quite a different type of pelage, and sequence of pelage changes, than mammals living in other environments. Seals which inhabit the colder regions of the world are highly adapted for an aquatic life, but they haul out on land briefly. A high breeding efficiency, together with a quick annual molt, allows these animals to spend as short a time as possible in a terrestrial habitat where they are out of their normal surroundings. Seals molt once each year; Ling (1965) has described this in much detail for the southern elephant seal, *Mirounga leonina,* of the Southern Hemisphere. In the elephant seal, the *stratum corneum* was a horny layer of continuous sheets, rather than smaller flakes. The *stratum granulosum* was lacking in this species. The *stratum spinosum* was transformed directly to the *stratum lucidum.* The upper part of the hair follicles was funnel-shaped and lined by stratum corneum which was continuous with the surface of the skin down to the constriction of the pilary canal of the hair follicle. Here the stratum corneum seemed to unite with the outer layer of hair. This constriction appeared to be important in the sloughing process of the annual molt. The hairs were lance-shaped and lacked a medulla. There were around 335 follicles per square centimeter, each hair was solitary, and there were no underfur fibers. There were no *arrectores pilorum* muscles in any seals. The hairs overlapped each other and lay close to the body. The highly developed sebaceous glands were a major adaptation to an aquatic habitat. The sweat glands were insignificant; they may be vestigial or they may have a slight function, as in thermoregulation during haul outs.

The elephant seal pups lost their fine, black natal fur soon after birth and were left with a light gray coat similar to that of the adults. The annual molt was unique in that the old hairs were shed along with large sheets of cornified epidermis, during a 4- to 7-week period while the animal was ashore. Molting began in the axillae of the flippers, around the eyes and nostrils, at the base of the tail, and between the hind flippers. Sloughing extended outward over the flippers, to the shoulders and rump and over the face. Newly exposed skin from each side of the pelvic and pectoral regions met and spread over the back. Lastly the skin around the middle of the animal began to break up. Finally the seal had a light gray or blackish gray coat. During the molt, sheets of skin up to 20×15 cm in size were sloughed. There was a loss of weight because of fasting during the molt. The seasonal haul outs put a heavy drain on the blubber. In the southern elephant seal, growth of new hair (which began in July-August) took about 16 weeks. The function of skin and hair in seals appeared to be as a waterproofing at sea and an aid to locomotion on land. Immature seals (up to 4 years of age) molted between November and January, mature cows in January and February, and mature bulls from February to April. In the Southern Hemisphere the breeding season is August to November. In early October no immature seals were found on land. The molting season was November to April. From February to September immature seals and an occasional pregnant cow hauled out for a few weeks and rested. Adult bulls did not come ashore during the winter (February to September) of the Southern Hemisphere.

The sloughing off of sheets of fur described for the southern elephant seal occurs in one completely unrelated and unusual mammal, the Selvin's mouse *(Selevenia betpakdaelensis),* discovered in 1938 in the Betpak-Dala desert in the U.S.S.R. Zwerew (1953) wrote, "Dem Tierchen fallen nicht einzelne Haare aus, sondern die Epidermis schichtet sich Zusammen mit den darauf sitzenden Haarchen ab." That is, the epidermis with the hair growing on it becomes detached, rather than individual hairs falling out. Dense young hairs are already growing where the sheets of epidermis have fallen off. Molting begins between the ears and proceeds along the back and sides. This process takes about a month. In winter the individual hairs attain lengths of 10 mm.

Pelage changes to indicate age of mammals up to maturity, e.g., for the meadow mouse, *Microtus pennsylvanicus* (Ecke and Kinney, 1956), have been used for a number of species, including the cottontail, *Sylvilagus floridanus* (Negus, 1958), and the muskrat, *Ondatra zibethica,* whose pelt primeness during the trapping season was based on sexual maturity (Shanks, 1948).

When no other part of the animal is available, hairs have been used for identification, with varying results. Hausman (1920), Hiner (1938), Mathiak (1938), Williams (1938), and Stains (1962) have provided keys. Nason (1948) examined 18 species of eastern North American bats and found that none of their hairs contained a medulla. As to identification of bats by use of hairs, Nason (1948) concluded, "This study adds support to Cole's findings (1924) that the hair structure of bats is of rather limited taxonomic value."

CLAWS, HOOFS, AND NAILS

Claws, hoofs, and nails (Fig. 6–6) are hard keratinized structures of the epidermis which grow continually outward from a germinative layer near the base. The basic type is the claw, the nail is a broadened modification, and hoofs are

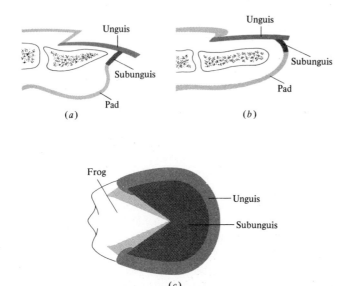

Figure 6–6 Longitudinal section of *(a)* claw of carnivore, *(b)* finger of human, and *(c)* ventral view of horse's hoof.

a development of ungulate mammals. All three have a hard upper unguis and a soft subunguis. In the hoof the whole outer portion, the unguis, is hard. The pad in a horse's hoof is also called the "frog" or the "cuneo." Most primates have nails, some have nails and claws, and primitive primates have claws. In the Hyracoidea (hyraxes), claws and hoofs (or hooflike claws) are found in the same individual. Agoutis and pacas (Central and South American rodents) have hooflike claws. Little work has been done on these structures, and even human nail formation is little understood.

HORNS AND ANTLERS

Horns and antlers are epidermal derivatives, usually including dermal bone, which have roles in reproduction, defense, and offense. Horns are permanent; antlers are deciduous (Fig. 6–7).

Rhinoceros horns are entirely epidermal and were originally thought to be composed of fused hair. They are actually composed of keratinized filaments which differ in structure from hair (Ryder, 1962). Each fiber is separately visible. Rhinoceros horns are never shed, although a new structure is produced if the old one is broken. Indian rhinoceroses *(Rhinoceros)* have one horn, African rhinoceroses *(Diceros)* have two horns. Bovine-type horns are permanent hollow horns which possess a bony dermal core and a hard epidermal covering, which is also permanent. Horns occur in both sexes of bovids, antelope (not the pronghorn), sheep, and goats, but are usually larger in the males. The horn grows in rings around a core from the frontal bone of the skull. These growth rings have been useful in determining age.

The stubby horns of the giraffe *(Giraffa)* and okapi *(Okapi)* remain covered by living skin (velvet) throughout life, and are never shed. In the pronghorn *(Antilocapra americana),* the horns are forked. There is a permanent, horny dermal core which is never shed, as well as a hard epidermal covering that is shed annually.

True antlers possess a bony dermal core and an epidermal "velvet" when growing. When the antler reaches its full growth the velvet is shed, and during the "rut" (breeding season) the antler appears without the velvet. Antlers are deciduous and are possessed usually by the males of the deer family. In caribou *(Rangifer tarandus),* both sexes have antlers, although those of the males are the larger and more branched.

In the white-tailed deer *(Odocoileus virginianus)* antlers begin to grow in April or May, reaching their full growth in August. The velvet is shed in September, and the antlers are shed during midwinter, beginning first with those deer living farthest north. Antlers grow by the addition of new material at the extremities. The bone nearest the antler base is the oldest. The first set of visible antlers is grown during the second spring of the deer's life ("spike buck"). Antler growth can be completed in 14 weeks.

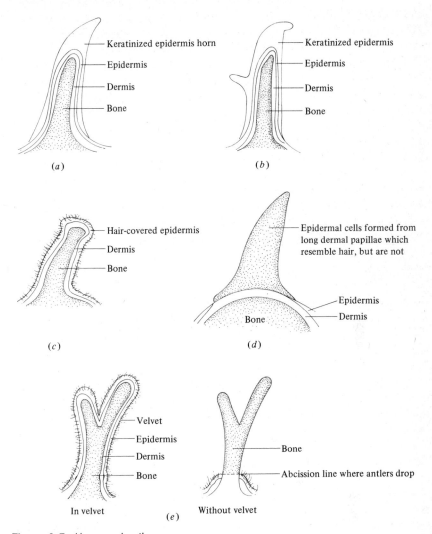

Figure 6-7 Horns and antlers.

Throughout the growth period the antlers are supplied with blood in the arteries of the velvet, and are well innervated by branches of the trigeminal nerve. The dermis of the antler is supplied with sensory endings. It is possible that this protects the developing antler from mechanical injuries caused by bumping them against objects. Wislocki (1956) wrote, "In the largest species of deer [elk *(Cervus canadensis),* caribou *(Rangifer tarandus)*], the rate of growth of the nerve fibers must exceed half an inch a day, establishing a record for the rate of growth of nerves."

The annual growth cycle of the antlers is controlled by a hypothalamic-pituitary-gonad endocrine system. The effects of castration have been reported upon by numerous investigators and have been summarized and added to by Wislocki (1956), who reported that (1) castration in the first 8 months of life results in complete suppression of antler growth, (2) castration after the first set of antlers are in velvet results in failure to shed antlers, which retain the velvet permanently, (3) castration after the antlers have lost their velvet results in immediate shedding, followed by the growth of a new set of antlers which are retained permanently in velvet. The largest antlers are usually worn by deer in the prime of life and with plenty of food. Quite often they are asymmetric. Antlers are poor indicators of age beyond the "spike" stage.

Because antlers are grown at great physiologic cost to the animal, their purpose has been cause for controversy. A pronounced reduction in aggressiveness has been noted in "deantlered elk." In the red deer of Scotland, "hummels," i.e., males that fail to grow antlers, seem to be as successful reproductively as those with antlers, or even more so (Darling, 1937).

EPIDERMAL GLANDS

Epidermal glands play a very prominent role in the lives of mammals. The skin of mammals contains more of these glands, both in number and kind, than the skin of any other vertebrate, but all these glands fit into one of two basic types, which are characterized by the way they secrete their material:

1   *Sebaceous Glands.* These are glands the secretions of which are the product of their own breakdown. This product, called *sebum,* is composed of lipids and cell debris (Montagna, 1956); it served primitively as a lubricating material. Other names include alveolar, mucous, or holocrine glands.

2   *Sweat Glands.* These are glands whose cells secrete a substance. They are also called sudoriferous, tubular, serous, or apocrine glands.

Sebaceous Glands (alveolar, mucous, holocrine)   The sebaceous glands are usually associated with hair follicles and are appendages of the outer root sheath of hair follicles. Sometimes they become quite complex. Some sebaceous glands are free, not associated with hairs, and their ducts open directly into the surface of the skin, but it seems certain that even these developed in association with hair follicles, and that the hair follicles may have dropped out (Montagna, 1956). The sebaceous glands of mammals may therefore be divided into two categories: (1) those of general distribution associated with hair follicles, and (2) those that are highly restricted and modified (but may have a very few hairs). The sebaceous glands of the general body surface are large and active in many desert rodents, particularly in some cricetids and heteromyids, and also in fossorial mammals such as the pocket gopher, *Tho-*

momys talpoides (Quay, 1965). The more active sebaceous secretion of these rodents may help reduce water loss and protect hair and skin surfaces. Captive kangaroo rats (*Dipodomys* sp.) and grasshopper mice *(Onychomys leucogaster)* are among those animals whose pelage is subject to oily matting if sand, sawdust, or similar material is not provided. In the elephant seal of the Southern Hemisphere the sebaceous glands on the side of the follicle are bilobed. The acini are enormous, "commensurate with their probable function of waterproofing the surface of the skin" (Ling, 1965). In human embryos, sebaceous glands associated with hair follicles are functional even before birth. The material encrusting the skin of the newborn baby is secreted in part by the sebaceous glands.

The variety of sebaceous glands not associated with hair follicles (although a few hairs may show) is very great in the mammals. Surprisingly, they are few in human beings and are generally conceded as being present only in the palpebrae (Meibomian glands), the buccal mucosa, the nipples, the prepuce, glans penis, and labia minora.

The short-tailed shrew *(Blarina brevicauda)* has a midventral gland and a pair of side glands. Coues (1896) believed them to be composed entirely of sebaceous tubules. Eadie (1938) as well as Pearson (1946) agreed that the ventral gland was sebaceous, but they thought that the side glands were a mixture of sebaceous and sweat glands. Pearson (1946) suspected them of having a sexual association with territory. Microtine rodents also have side glands.

In the kangaroo rats *(Dipodomys)* there are highly restricted and modified glandular areas called dorsal glands along the middorsal line, in the most exposed area of the arched back. The gross appearance has been described by many authors, but Quay (1953 and 1965) is the most recent to have done work on these glands. The secretion has a sweetish scent and may possibly serve incidentally as a waterproofing agent; more importantly, Quay believes, they serve scent purposes. Glandular activity is subject to seasonal and sexual modifications. Quay (1953) wrote, "It is apparent that the activity of the dorsal gland has taxonomic significance. It is dubious whether we can show as yet any of the phylogenetic trends of the glandular activity within the genus."

Preputial (protometric) glands are common to many mice and especially to the muskrat *(Ondatra zibethica)*, beaver *(Castor canadensis)*, and members of the family Canidae. The glands lie just beneath the skin on either side of the penis, and the ducts open within the prepuce. The secretions are carried by the urine and are believed useful in marking territory. Schenkel (1947) concluded that scent marking serves several functions for wolves *(Canis lupus)*, including "the demarcation of territory, the making of acquaintances, the formation of the pair, and the legitimization of the leader." Kleiman (1966) suggested that scent marking "probably originated as a device for familiarizing and reassuring the animal when he entered a strange environment. . . . Secondary functions have arisen such as the bringing together of the sexes and the maintenance of territory, and these have assumed an important function in the survival of the species."

The anal gland is yet another type of scent gland which has become very well developed among members of the family Mustelidae. Best known of these are the foul-smelling glandular secretions of the striped skunk *(Mephitis mephitis)* and spotted skunk *(Spilogale putorius)*. Although the myth persists that this secretion can cause permanent blindness, it causes only a temporary, though severe, irritation (Hall, 1946). The odor is caused by the organic compound *n*-butyl mercaptan.

Anal scent glands are also prominent in members of the family Viverridae, the mongooses and civets of the Old World. The musk called civet, sometimes used in medicine and commonly used in the perfume industry, is obtained from several genera, such as *Civetticitis, Viverra, Viverricula*. The animals are kept in captivity to obtain their scent. The African civet *Civetticitis civetta* produces up to 4 g a week. In addition to its function as a defense mechanism, the scent of the anal glands must sometimes serve as a territorial pheromone. Hinton and Dunn (1967) reviewed territorial marking among members of the family Viverridae. Similar glands with a less potent odor are found among mice, shrews, and especially in ground squirrels, in which the three white-tipped orifices are especially prominent. When the animals are sufficiently alarmed, the secretions of the glands are forcibly extruded. As in the ground squirrels, the three white-tipped orifices of the anal glands of the woodchuck *(Marmota monax)* lie just within the vent. The musk is released in anger or in fear, and Hamilton (1934) has described its use in communication. In the muskrat *(Ondatra zibethica)* both sexes have special glands situated beneath the ventral skin near the external genitalia. During the breeding season these glands secrete a yellowish musky-smelling substance which is deposited at spots along the travel routes, at the bases of lodges, and other places. The name muskrat obviously comes from this odor. Stevens and Erickson (1942) analyzed the organic materials in the musk as a mixture of cyclopentadecanol, cycloheptadecanol, and other odoriferous ketones.

Meibomian glands lying just within the eyelids serve to lubricate the eye. Quay (1954) has discussed these and their possible use as taxonomic characters. They are believed to be absent in aquatic mammals.

In the artiodactyls are found such glands as the preorbital, tarsal, metatarsal, and interdigital. The interdigital gland, occurring between the digits, as the name implies, has given rise to the myth that deer have an extra set of ears. Although the preorbital gland does not seem to have any great function in American deer, in some African antelope it is so well developed that it can be everted. All these glands probably play some part in communication and in territoriality.

Other examples of these specialized glands are found in mountain goats, which have two behind the horns; the pronghorn, which has ischial glands; and the peccary, which has musk glands over the hips.

Geist (1971) speculated on the evolution of these glands in the mountain goat and the purposes of olfactory marking. "It is highly probable that some olfactory marking systems evolved as intimidation mechanisms evolved. Marking probably evolved in territorial forms, where it would conceivably be most effective. Scent acts as an extension of the animal. Its presence is revealed to conspecifics although they may not see or hear the animal that left the mark."

Sweat Glands (sudoriferous, tubular, apocrine)   Of the glands which function as secretory organs, the sweat glands are best known. In these kinds of glands the material is secreted by the cells and does not consist of the breakdown of cellular material. These glands are also called tubular, serous, or apocrine. The secretions of the sweat glands contain fatty substances, salts, and pigments, but are more watery than the secretions of the sebaceous glands. Sweat of one kind of African antelope is said to be slightly pink, that of another kind slightly blue. Sweat glands serve to rid the body of salt and water and also serve in thermoregulation, although these are not necessarily separate functions. Sweat glands have been characterized as to size and complexity. The larger, more complex ones are sometimes called epicrine or apocrine and usually open into a hair follicle; the smaller ones, which are just simple tubes, are called eccrine and open directly onto the epidermis.

The secretory portion of an apocrine sweat gland is a simple, compactly coiled tube with adjacent loops joined by shunts or sometimes ending in blind sacs. The duct, which is perfectly straight, opens into the hair follicle just above the entrance of the sebaceous duct into the hair follicle. Apocrine sweating is largely a response to adrenergic stresses such as fear, pain, or sex.

Sweat glands occur in many mammals. In human beings, eccrine sweat glands are found all over the body (except on the lips, glans penis, clitoris, and inner surface of the prepuce) and are much more numerous than the larger apocrine sweat glands, which occur in restricted areas. On pigs and horses, apocrine sweat glands are especially numerous and are the only sweat glands found on these animals. Sheep, cattle, and dogs have them on the muzzle. Although desert rodents were once thought to lack sweat glands, Quay (1964) has provided evidence of their presence in restricted areas, viz., the anal region, the middorsal skin, skin of the oral angle and lips, as well as on the palmar and plantar soles. It is commonly believed that moles, sloths, scaly anteaters, sirenians, and cetaceans do not have them. Concerning the Pinnipedia, Ling (1965) described conditions in the southern elephant seal: "On the other hand the apocrine sweat glands, each one associated with a pilosebaceous unit, are quite rudimentary. Only a few tiny coils lie almost hidden within the fatty tissue of the dermis below the follicle. . . . No eccrine sweat glands have been found."

The mammary glands are thought to have evolved from the progenitors of the apocrine sweat glands. In monotremes they are in fact but slightly modified from sweat glands. This evolution is discussed more thoroughly in Chap. 10.

The last group of glands, of ecodermal origin, considered in this chapter are the paired salivary glands of the buccal cavity in mammals. These are the submaxillary, sublingual, parotid, malar, and suborbital glands, which serve, in the absence of a water medium, as aids in moistening and swallowing food. The secretions of these glands are in some cases and under certain conditions poisonous to other individuals. The saliva of a human being when introduced into the circulatory system of another person can be poisonous! In snakes and gila monsters the producers of the poison are the parotid glands.

The short-tailed shrew *(Blarina brevicauda)* has been shown to possess a venom with which to subdue its prey (Pearson, 1942). Probably other shrews also possess this venom. Extracts from the submaxillary glands of *Blarina,* when injected into mice, proved lethal. Pearson (1955) wrote, "There seems little doubt that the venom could kill a human being if injected intravenously." The shrew, of course, has no way of injecting this poison such as a rattlesnake has. The effect on humans when bitten can be extremely painful.

It is a puzzling fact that poison glands of some form are found in all classes of vertebrates but birds. In mammals they are found even in the hind foot of the male duck-billed platypus.

Other miscellaneous epidermal structures which should be mentioned but will not be discussed are the whalebone, or baleen, of whales, palatine rugae on the roof of the mouth, ischial callosities of primates, and the integumentary armor of pangolins.

RECEPTORS IN THE SKIN

The skin also serves as a very important sense organ for the reception of touch, pain, pressure, and temperature, although these receptors are not epidermal in origin.

The presence of the Meissner corpuscle (touch) in the North American opossum led Winkelmann (1965) to remark that the structure was developed many millions of years ago, for the North American opossum of today is entirely comparable to that of the Cretaceous era. Winkelmann (1965) further commented, "The presence of a common nerve end organ in the Metatheria and in the Eutheria seems logical, for they must have come from some early progenitor."

Krause end bulbs (bulbs of Krause) have usually been designated as cold receptors. The heat receptors are termed Ruffini organs. Hair follicles are richly supplied with nerve endings and can also be considered true receptors.

The myth that the raccoon *(Procyon lotor)* washes all its food has some basis of truth. Raccoons do use water, when available, but they will certainly decimate a row of sweet corn without the presence of water. It is entirely possible that the water is used not for washing but to increase tactile sensitivity. The raccoon has in its skin encapsulated end organs, the insulated structure and deeper location of which decrease the sensory capacity. Welker and Seidenstein (1959) showed that the brain of the raccoon has a well-developed sensory cortex with an individual gyrus for each digit of the forepaw and eminence of the palm. A raccoon's sensory capacity seems to be due more to the development of an expanded sensory cortex than to the peripheral sensory system in the paws. The use of water when available may increase the tactile perceptivity in the paw.

Tactile, cold, heat, and pressure receptors, innumerable kinds of glands, and other derivatives of the skin make it more than just an insulating material. The very thin layer of skin and its derivatives which surround a mammal provide a remarkable shield against the environment.

From this brief review it becomes apparent that this thin layer which separates a mammal from its environment is a very complicated organ that functions admirably in helping to maintain homeostasis between the mammal and its environment.

Digestive Systems and Digestion

During their evolution, mammalian digestive systems have become adapted to utilize a wide variety of food materials, although every mammal begins its life on a diet of milk. Digestion is the transformation of food to metabolically useful nutrients that can be absorbed by the body and assimilated into body tissue. In some species the diet is quite generalized, but in others, such as the vampire bat, it has become extremely specialized. Diverse modifications of the digestive tracts of various mammals are striking and applicable to both functional and systematic problems.

Present knowledge does not allow anything but the broadest systemic generalizations, although the relation of gastric structure to food habits can be very striking in some species.

Food habits cannot be used as a systematic criterion, although mammals are sometimes broadly grouped according to their modes of feeding. The order Carnivora is considered to consist of meat-eating (or animal-eating) mammals, but the giant panda, a member of this order, feeds almost exclusively on bamboo shoots.

Adjectives describing either the mode of eating or the kind of food eaten have come into common usage, and it might be pertinent to list them here.

Carnivorous   Eating animal material, or, more specifically, meat- or flesh-eating. Use of the word as an adjective is sometimes confused with its use as a noun, the order name Carnivora. Not all carnivorous mammals belong to the Carnivora, nor are all Carnivora meat eaters. Whales feed on animal material but do not belong to the Carnivora, and bears eat a lot of plant material but do belong to the Carnivora. Animal material is mostly protein and is considered a "high-quality food," because it is converted to energy more efficiently than plant material. Generally, carnivores have a simple stomach and short intestines. The loosely used word "predator" connotes a meat-eating animal. Coyotes, wolves, weasels, mink, bob-cat, lions, and tigers are included in this group.

Durophagous   This specialization is an adaptation for handling hard food, especially hard-shelled animals.

Frugivorous   This term designates a diet of fruit. Many bats, especially the large "flying foxes," are popularly known as fruit bats. The Jamaican fruit bat *(Artibeus jamaicensis)* has become a pest to the orchardists of the Caribbean. Some semi-tropical bats are known to move northward with the ripening of fruit. The flying lemur *(Cynocephalus volans)* of Southeast Asia includes fruit in its diet and has a dentition and palate remarkably adapted for this diet.

Herbivorous   This adjective refers to a diet of plant material, though in practice it is limited to the larger mammals. Mice are rarely referred to as herbivores; the name is usually applied to the ungulates, a term which refers to specific morphologic characters. In plant eaters the intestine is long; the stomach is often complicated by internal folds and may be divided into several functionally different chambers. This specialization reaches its greatest complexity in ruminants, where the digestive process is aided by microfauna which help in the breakdown of cellulose. Timid ruminants can at times snatch and swallow food at intervals without chewing between "bites." Later this food can be digested in safer circumstances. This ruminantlike digestion has also been reported for other mammals, including marsupials, edentates, and primates.

Bacterial fermentation occurs in only a few kinds of nonruminant mammals lacking a compartmentalized stomach for storage. In some cases it occurs in the stomach, in others in the cecum. A bizarre adaptation for extending the time available for bacterial fermentation has been found in pocket gophers, rats, and rabbits. It consists of the reingestion of the expelled pellets, sending them through the digestive tract again. This is called *coprophagy.*

Insectivorous   This word refers to a diet consisting of insects. One might call it a highly specialized carnivorous diet. Many species of bats are insectivorous, and of course a whole order of mammals, the Insectivora, feed primarily on insects.

Nectivorous   This refers to the diet of mammals that feed on nectar and inadvertently perhaps on pollen. One example would be the long-nosed bat *(Leptonycteris nivalis).*

Omnivorous   This describes the diet of mammals which feed on a great variety of food material. Bear (*Ursus* spp.), humans *(Homo),* opossum *(Didelphis virginianus),* and pig *(Sus),* might be good examples.

Sanguinivorous   This refers to a diet of blood, best exemplified among the mammals by the food habits of the vampire bat *(Desmodus rotundus).*

ANATOMY OF THE ALIMENTARY TRACT

A brief consideration of a rather simple (but not the simplest) alimentary tract such as that of the wood rat, *Neotoma cinerea* (Howell, 1926), will serve as a basis for the further consideration of the more complicated digestive systems discussed later (Fig. 7–1).

In the wood rat the stomach is U shaped (Howell, 1926), with a cardiac horn, a pyloric horn, and a fundic region between the two horns. These regions are histologically different, with specific types of glands in each region.

The cardiac glands are made up of columnar cells. The fundic region is made up of mucous cells; chief cells, which secrete pepsin; and parietal cells, which secrete hydrochloric acid. In the fundus mostly mucus is produced.

The small intestine is a slender tube about 600 mm in length. It is divided into two tracts, called by Howell (1926) the duodenum and Meckel's tract. Usually three divisions are named for most mammals: the short duodenum; the long central part, the jejunum; and the remaining part, the ileum. The absorptive surface of the small intestine is increased by fingerlike projections, the villi.

The large intestine, or colon, is quite variable in mammals. A cecum, situated at the junction of the large and small intestine, is usually present in mammals but varies greatly; sometimes there is more than one, sometimes it is large, and in some mammals—civets, monkeys, and some rodents—the distal end is degenerate and is distinguished as the appendix. The cecum is

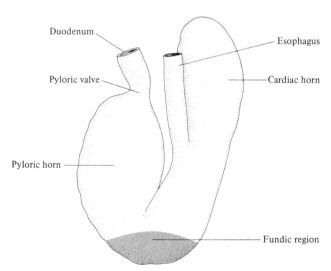

FIGURE 7-1 Stomach of wood rat *(Neotoma)* showing glandular regions. *(From Howell, © 1926, The Williams and Wilkins Co., Baltimore, Md.)*

greatly developed in the wood rat. Digestion of plant material is far more complicated than the digestion of animal material, because of the cellulose in plant material.

The salivary gland, liver, and pancreas are usually discussed as a part of the digestive system but do not concern us here.

COMPARATIVE GASTRIC MORPHOLOGY AND FUNCTION

The simplest of mammalian stomachs include those of the spiny anteater *(Echidna tachyglossus)* and of carnivores. The echidna's food consists of ants and termites; that of carnivores, mostly of protein. Rodent stomachs, although more complex than those of the echidna or of carnivores, are still not as complicated as those of whales or ruminants. The stomach of the cetaceans approaches, in morphologic complexity, the stomach of subruminants and ruminants but lacks their microbial fauna. The most complex digestive systems have been found in the ruminants, grazing marsupials, bradypodid edentates, and some of the leaf-eating monkeys. It is more than likely that more such complex systems will be found. In these ruminant or ruminantlike digestive systems the rumen functions as a fermentation and mixing vat, where, with the aid of microbial fauna, cellulose is digested.

Because of the importance and complexity of ruminant digestion and the effort that has been spent on its research, it will receive a large share of attention in this chapter.

Specialized stomachs for specialized diets which will be emphasized are those of the echidna *(Tachyglossus aculeatus)*, vampire bat *(Desmodus rotundus)*, and the grasshopper mouse *(Onychomys leucogaster)*.

Spiny Anteater (Monotreme)

The echidna will serve as an example of a mammal whose diet consists primarily of termites. There are no glands—cardiac, fundic, or pyloric—in its stomach. The stomach surface is covered only with stratified, cornified epithelium. Therefore the stomach cannot contribute digestive enzymes. The saliva contains amylase and is active in the stomach, where, mixed with the homogenized insects, it degrades glycogen to carbohydrates of low molecular weight. The termites themselves contain amylase, and this also helps in digestion of carbohydrates of high molecular weights (Griffiths, 1968).

Carnivores

The simplest stomachs, as already indicated, are those of carnivores. Most of them are little more than expanded hollow tubes with muscular walls which store and partly mix the food, while the intrinsic glands intermittently add enzymes, mucus, and hydrochloric acid. The dog's stomach is a good example of a carnivore stomach. A detailed description of the dog's digestive system

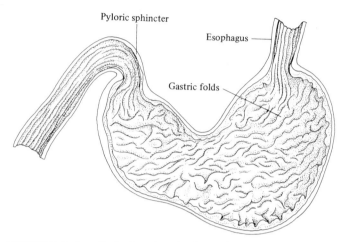

FIGURE 7–2 Longitudinal section of a dog's *(Canis familiaris)* stomach to show the simple carnivore stomach. *(From Miller, Christensen, and Evans, Anatomy of the Dog, 1964.)*

can be found in Miller, Christensen, and Evans (1964) and is the basis for the description given here (Fig. 7–2).

The dog's stomach is shaped somewhat like the letter C, with the greater curvature of about 30 cm long when the stomach is moderately filled. The capacity of the stomach varies; it has been cited as being 100 to 250 ml/kg of body weight (Ellenberger and Baum, 1943).

The stomach of the dog cannot be grossly differentiated but does consist of four parts common to most stomachs (as in the pack rat). The esophagus blends into the cardiac portion. The fundic region is a large outpocketing to the side of the esophageal entrance. The large middle portion of the organ is the body of the stomach. The distal third of the stomach is the pyloric region. The gastric groove is the shortest path the food can take from the cardiac to the pyloric portion along the lesser curvature of the stomach.

The small intestine of the dog is about 3.5 times the length of the body of the dog. The large intestine is short and unspecialized; its important function is the dehydration of the food material. The large intestine is made up of the cecum, colon, rectum, and anal canal. The cecum in the dog exists as a diverticulum of the proximal portion of the colon. In live animals the cecum may be about 5 cm long and 2 cm across the base.

Bats

The greatest variety of specialized food habits within one group of mammals is found among the bats. Most North American bats feed on insects, but a few kinds are specialized for feeding on fruit, nectar (and pollen), fish, or blood. The majority are insectivorous or carnivorous (insects, fish, lizards, birds, and bats) and have typical insectivore teeth, which are modified tuberculosectorial or tribosphenic. Since these are similar to the teeth of other Mesozoic mammals, it is also assumed that it is the primitive dentition in bats. The sharp-edged crescents, reshaped from the conical cusps of the primitive molar, can quickly blend the chitin of insects into a purée (Glass, 1970).

Bats which feed on fruit have molars whose sharp cusps have been dulled and the crowns flattened so that they can crush the fruit. In nectivorous bats the teeth have no function, in fact some of the incisors are absent, and the jaws are elongated and weak. The tongue is long and often there are long papillae at the tip to soak up the nectar. In the digestive systems of those North American bats investigated by Park and Hall (1951) no distinctions were found between the small and large intestine, and there was no cecum. There were, however, differences in the stomach. The stomach of insectivorous (carnivorous) bats, considered the generalized and ancestral type for bats, contains gastric glands of the tubular variety seen in mammals generally.

The most unusual bat stomach is that of the vampire bat *(Desmodus rotundus),* whose diet is primarily blood. The ingestion of blood is facilitated by an anticoagulant. Insects have been found in its stomach contents, but these may have been accidently ingested. A blood diet is unique among mammals, for even in the carnivores the feeding of blood in substantial quantities causes difficulties (McKay, 1949). Not only does the composition of the blood present problems, but its liquid state compounds the difficulty.

No other bat's food habits, stomach anatomy, and digestive physiology have received as much attention as those of the vampire bat. The gross morphologic characteristics have been described by Park and Hall (1951), among others; the histologic characteristics have been described by Rouk and Glass (1970). The dentition of vampire bats has been highly modified for scooping out a shallow depression in the capillary bed beneath the epidermis of mostly larger animals such as cattle and horses, but also of chickens and sheep (Fig. 7–3). The molars have been reduced in number and are bladelike. The upper canines and incisors have been enlarged and compressed into anteroposterior-facing razor-sharp blades. The lower incisors and canines are not quite as extremely modified.

Glass (1970) has described the modification and function of the tongue. The tongue has a pair of grooves at each border which serve as drinking straws. The musculature of the tongue passes the blood through these grooves in peristaltic waves by moving the tongue in and out with great rapidity. In movies of this action, no blood appears on the surface of the tongue. The explanation is that at the tip of the tongue the grooves are deep and near the lower surface, but they diverge toward either side of the tongue, and in the rear of the mouth the grooves are shallow and discharge onto the surface.

The stomach is long, thin, and tubular with delicate, threadlike, longitudinal rugae extending the length of the unfilled stomach (Fig. 7–4). These

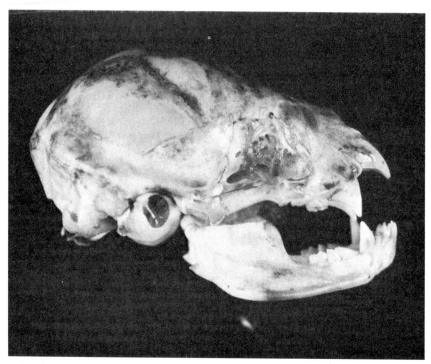

FIGURE 7-3 Skull of vampire bat *(Desmodus rotundus)*, showing the bladelike molars and the razor-sharp upper canines and incisors.

disappear when the stomach is distended, after ingestion of a meal of blood. The pyloric region is restricted to a small area near the junction of the duodenum and the esophagus. The rest of this tubular stomach is principally a fundic or cardiac cecum with acinar and tubuloacinar glands quite unlike gastric glands found in other mammals; the stomach ends in a blind pouch. Forman (1971) has suggested that the gastric mucosa system of vampire bats may be either degenerate or only partially functional. The stomach is highly vascularized, an anatomic feature which may provide for rapid water absorption to reduce the fluidity of the vampire bat's food. The blood enters the intestine first and then overflows into the stomach. Mitchell and Tigner (1970) suggested that the vampire bat's stomach functions not for protein digestion, as in most mammals, but rather for the storage of large amounts of blood and for water absorption. The small intestine, thin-walled and rubberlike, is twice as long as the 12-cm-long stomach. There is no microscopic indication of a large intestine. The efficiency of food utilization—73 percent—is amazingly high.

Vampires in captivity do not consume free water, nor are there records of observations of wild vampires drinking water. So, in reality, these mammals survive "in a behavioral desert of the severest kind" (Horst, 1969).

The vampire's diet contains 80 percent water, so the excretion of urea should present no severe problem. But a different turn of events does present a problem. A flying mammal of 30 g may nearly double its weight while feeding. Any early reduction in weight would seem to have survival advantages. Copious urinating begins almost immediately after the bat begins feeding. This is an isotonic urine with very low urea content. Later, back at the roost, digestion of the partially dehydrated blood necessitates the excretion of large amounts of nitrogenous wastes without losing excessive amounts of water. Since this bat does not drink water, it must achieve this excretion by effecting a high concentration of urinary urea. The remarkable efficiency of the vampire bat's kidney to concentrate urea is the highest known in any mammal, even exceeding that of many desert mammals (McFarland and Wimsatt, 1965; Horst, 1968). The urine-concentrating ability of mammalian kidneys is generally related to the relative length of Henle's loop, but in the vampire bat the loop is relatively short. Horst (1969) has postulated that the individual cells of the various tubules in the vampire are capable of either greater effort or more efficiency than is found in desert mammals.

That transmission of disease to livestock by vampires may result in economic losses has been emphasized, and is widely known, but the economic losses due to bloodletting are not so dramatic. As far as larger mammals are concerned it is probable that few of them die from this loss of blood, but the indirect economic drain of the vampire's food habits must nevertheless be great. Wimsatt and Gurriere (1962) made the assumption, considered conservative by them, that wild vampires drink on the average of 20 ml blood per day. A single bat would consume 7.3 l (15 p) per year. In its normal life span of 13 years (Trapido, 1946) it would consume 95 l (25 gal). Wimsatt and

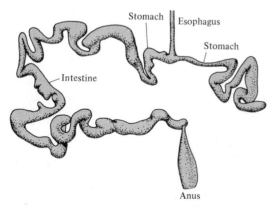

FIGURE 7–4 Stomach of vampire bat *(Desmodus rotundus). (After Grasse, 1967.)*

Gurriere (1962) wrote, "A moderate sized colony of vampires consisting of 100 adult animals, individuals would drain from the local livestock population approximately 730 liters (186 gallons) of blood each year. This amount is roughly equivalent to that contained in 20 horses, 27 cows, 465 goats or 14,600 chickens of average size." They wrote further, "In countries where vampires are common (in Mexico they presumably occur in the tens of thousands) the annual 'tribute' in gallons of blood drawn from livestock and other prey species certainly runs up into the six-figure category, or higher." More recently Thompson, Mitchell, and Burns (1972) have placed this estimate at $250 million annually for the Latin American countries. This would reduce the vitality of the animals considerably, causing a decrease in production of meat and in reproduction, as well as increasing susceptibility to disease.

An ingenious method has been devised for controlling vampires by treating cattle systemically with a single intraruminal dose of an anticoagulant, diphenadione. Roosting vampire bats, in their shuffling around, inflict minor wounds on each other, and the anticoagulant causes a drain on their blood supply. There is a differential sensitivity of cattle and vampires to diphenadione, which is great enough so that even calves consuming the milk of treated cows experience no hazards. So effective is this technique that in the cattle on three ranches near San Luis Potosi, Mexico, bites were reduced from 1.2 per animal per night to 0.07, a reduction of 93 percent. The cattle exhibited no ill effects. This type of control has several advantages: (1) there need be no physical contact between human beings and bats, potential carriers of rabies; (2) ranch personnel are more likely to have experience handling the cattle than in handling bats; and (3) the technique, even on a large scale, can easily be fitted into normal livestock management.

Rodents

Rodent digestive tracts exhibit great variability. There are major differences in the amount and extent of gastric glandular mucosa, in the kind of lining at the pyloric sphincter, and in muscular control of the pyloric sphincter (Dearden, 1969). In the five species he examined, two types of stomachs were demonstrated, on the basis of the type of epithelial lining. The first type was based on two kinds of stomach areas, a forestomach of cornified, stratified squamous epithelium and a pyloric or fundopyloric stomach of glandular mucosa through the pyloric sphincter. The second stomach type consisted of three stomach areas—a forestomach, a narrow glandular stomach, and a pylorus with a forestomach type of lining in the pylorus and at the pyloric sphincter. Muscle action at the sphincter also appeared to be of two kinds. In the meadow vole *(Microtus pennsylvanicus)* and collared lemming *(Dicrostonyx groenlandicus)*, the larger mass of sphincteric muscle was on the lesser curvature, indicating an almost symmetrically circular muscle action. In the brown lemming *(Lemmus trimucronatus)*, steppe vole *(Lagurus lagurus)*, and sagebrush vole *(Lagurus curtatus)*, the greater mass of the muscle was on the greater curvature, and the action appeared to be more like a milking action, the contraction progressing from proximal to distal. One can agree with Dearden (1969) when he wrote, "The diversity . . . would seem to indicate the need for additional studies in these areas in other mammals."

Carleton (1973), in a survey of gross stomach structure of the New World (and a few Old World) Cricetidae included 148 species in 42 genera. To these stomachs he accorded one or another of two anatomic designs. One of these he called unilocular-hemiglandular and the other he called bilocular-discoglandular (Fig. 7–5). The unilocular stomach is single-chambered with distribution of cornified and glandular linings corresponding to the basic stomach divisions, the cornified epithelium in the cornus (sometimes called the cardiac

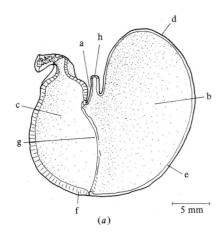

(a)

5 mm

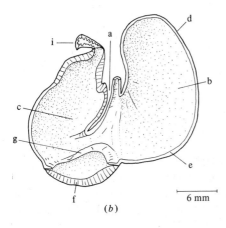

(b)

6 mm

FIGURE 7–5 The stomach morphology of rodent stomachs is usually of one or another of two anatomical designs, termed unilocular-hemiglandular and bilocular-discoglandular. Diagrams of *(a)* the unilocular-hemiglandular stomach of a South American cricetine *Oryzomys nigripes* and *(b)* the bilocular-discoglandular stomach of a neotomine-peromyscine *Neotomodon alstoni.* Anatomical features indicated are: a, incisura angularis; b, corpus; c, antrum; d, fornix ventricularis; e, cornified squamous epithelium; f, glandular epithelium; g, bordering fold; h, posterior end of esophagus; i, anterior end of duodenum. *(After Carlton, 1973.)*

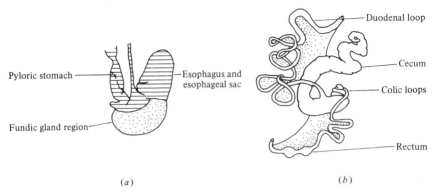

FIGURE 7–6 Digestive system of a meadow mouse, *Microtus pennsylvanicus. (Golley, 1960.)*

part), and the glandular epithelium limited to the antrum (sometimes called the pyloric region). The bilocular-discoglandular stomach is more strongly bipartite because of the deep incisura angularis (Fig. 7–5). Carleton (1973) has suggested that the unilocular-hemiglandular is probably the primitive condition and has discussed the taxonomic implications of stomach evolution in New World cricetines and other muroids. His reasons for suggesting that the discoglandular stomach is more recent is the absence of pyloric glands, which, together with fundic and cardiac glands, are found in most mammals, including those cricetines with hemiglandular stomachs. Toepfer (1891) originally inferred such an evolution. Two species of rodents with discoglandular stomachs, one a simpler type (*Microtus,* a microtine) and the other a complex type (*Onychomys,* a cricetine), will be discussed.

The digestive tract of the meadow vole, *(Microtus pennsylvanicus),* has been studied in detail by Golley (1960) (Fig. 7–6). He found that the esophagus, extending from the mouth to the stomach, averages 30 mm in length. The U-shaped stomach, very similar to that described for the wood rat by Howell (1928), is divided into an esophageal sac, fundic gland region, and pyloric gland region, or pyloric stomach. The largest portion of the stomach is the esophageal sac. The fundic gland region occupies the greater curvature of the U; it appears darkened and very thick and fleshy. The pyloric region makes up the distal portion of the stomach.

The small intestine averages 360 mm in length; it is divided into a duodenum and a Meckel's tract, and internally (by histologic examination) into duodenum, jejunum, and ileum. The dominant structure of the digestive tract of *Microtus* is the cecum, found to be 110 mm in length and with a series of complex folds. The colon-rectum averages about 156 mm in length.

In feeding experiments with alfalfa, or lettuce, carrots, and oatmeal, Golley (1958) found that only 10 to 14 percent of the gross food energy was recovered in the feces. Presumably the rest of the food energy was used by the mouse. This is a very efficient use of plant material. Morrison (1946) found that sheep *(Ovis)* and cattle *(Bos)* have only a 50 percent efficiency when fed a diet of alfalfa. Although Golley (1960) concerned himself only with the digestive structure, he felt that "the large cecum with elaborately folded walls, is probably a contributing factor, but the primary cause of the high efficiency will probably be discovered in a study of the intestinal flora and fauna."

It has been known for a long time that the food of the grasshopper mouse *(Onychomys torridus)* is almost entirely of animal origin, and further that most of the animal food consists of arthropods, including scorpions (Bailey and Sperry, 1929; Horner, Taylor, and Padykula, 1964). Even the whip scorpion, which sprays a highly concentrated secretion of acid on its opponent, is overcome by *Onychomys* (Eisner, 1962). *Onychomys* differs from most rodents, therefore, in being insectivorous and carnivorous rather than herbivorous. One of the changes that has occurred in the stomach of *Onychomys,* as well as in some other rodents, is an increase in the nonglandular area of the stomach, along with a decrease in the glandular area.

The appearance of the stomach of *Onychomys torridus* is quite typical of rodent stomachs except for a distinct bulge about halfway down the length of the great curvature (Fig. 7–7). According to Horner, Taylor, and Padykula (1964), a conspicuous blood vessel comes down from the lesser curvature near the esophagus. This blood vessel serves as an arbitrary boundary marker between the cardiac region and the pyloric region of the stomach. The bulge is the fundic or glandular region, which is very thick-walled, and all the gastric glands are confined to this region. There is no other glandular region. Pyloric glands are not present in the stomach of *Onychomys torridus longicaudus*

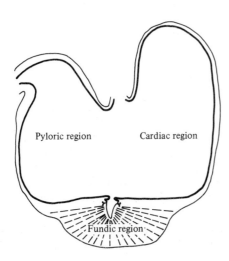

FIGURE 7–7 Stomach of a grasshopper mouse, *Onychomys torridus.* (Horner, Taylor, and Padykula, 1964.)

(Horner, Taylor, and Padykula, 1964). Of great interest also is the fact that the cardiac and fundic glands are housed in a distinct pouch, so that the soft tissue of the gastric glands is safeguarded from abrasion by the chitinous material of its insectivorous diet. There seems to be no enzyme which acts upon the chitin, and the only breakdown of chitin occurs during mastication. Horner, Taylor, and Padykula (1964) wrote, "The stomach of *Onychomys* is an example of evolutionary conservation in that small remodeling in basic architecture of a bilocular stomach highly efficient in handling plant material can be of adaptive value in the utilization of a completely different kind of food."

The stomach anatomy and diet of *Lophuromys sikapusi,* a murid, are almost identical with those of *Onychomys,* but Genest-Villard (1968) thought the anatomic arrangement was not connected with the insectivorous diet. Rather, he believed, the rapid growth of the young was connected with this peculiarity. In the neonate the distension of the stomach pouch allows the absorption of a large volume of milk. Horner, Taylor, and Padykula were not quoted by Genest-Villard; he may not have been aware of their interpretation. Carleton (1973) finds neither explanation acceptable. Both species lack pyloric glands, leaving a greater area for intensive mixing and reworking of the food bolus. In addition the enclosure of glands in a diverticulum permits the maintenance of a high pH in the main part of the stomach, which allows salivary amylase to continue its action of degrading glycogen and starch.

Whales

Of all the animals that are living or have lived, the whales are the largest. There is a record of a blue whale which weighed 123,000 kg (272,800 lb, or nearly 136½ tons). This was a female of 27.7 m (90 ft) in length, and since whales up to 30 m (100 ft) have been caught, they must weigh even more. A blue whale can weigh as much as four brontosauri, or 30 elephants, or 1600 people, which would be a fair-sized village.

Even though large mammals eat proportionally less (i.e., per pound of body weight) than small mammals, it still takes a tremendous amount of food to support a whale. The whale's digestive system is somewhat more complicated than that of a carnivore's or a rodent's but not as complicated as that of the subruminants, ruminants, and a few other mammals with ruminantlike digestion.

Plankton comprises the bulk of the diet for most whales, although they do feed on cuttlefish and squid; some, such as the killer whale, even feed on other whales. Plankton is a world of minute suspended organisms found floating in open water. The dominant organisms are the phytoplankton, including filamentous green algae, and single-celled desmids and diatoms. These tiny plants carry on photosynthesis as all other plants do. Suspended in the phytoplankton and feeding on it are the animal or zooplankton organisms.

Most numerous are the rotifers and the smaller shrimplike copepods and cladocerans, but there are also representatives of almost every phylum of the animal kingdom, including fingerlings of many kinds of fish. A major characteristic of these organisms is their fecundity, as exemplified by "krill," a 2-in.-long shrimplike crustacean of the Antarctic. It forms the major part of the diet of the blue whale during the 6 months that this whale spends eating in Antarctic waters. Young blue whales, which gain 39 kg (90 lb) a day, eat 2720 kg (3 tons) of krill every 24 h. This would add up to 453,600 kg (500 tons) of krill per whale per year, but the Antarctic is estimated to spawn 1.36 $\times 10^{12}$ kg (1½ billion tons) of krill per year! And this from organisms whose life span is reckoned in days or at most weeks.

The production of plankton is seasonal. The winter storms which stir the ocean bring the rich bottom nutrients to the surface. This, together with the increasing number of daylight hours in spring, begins the chain of events which will cover hundreds and thousands of square miles with plankton, upon which the baleen whales feed.

The cetaceans are divided into two suborders, those with teeth (Odontoceti) and those with baleen (Mysticeti). The baleen strains plankton out of the water. This is cleaned off the baleen and brought back for swallowing with a tongue which is composed mainly of spongy connective tissue, and is extremely soft and pliable. Toothed whales use their teeth for capturing prey, not for chewing. The salivary glands in whales are either very rudimentary or even lacking. The killer whales (with up to 56 teeth) eat squid, dolphins, seals, sea lions, narwhal, belugas, young walruses, and penguins. Slijper (1962) records the finding of 13 porpoises and 14 seals in the stomach of a 7-m (24-ft) killer whale. The food enters the stomach through the esophagus, a relatively small opening. In the mammals discussed so far the stomach is a single pouch lined with noncornified epithelium containing fundic and pyloric or other glands.

In all whales (except the beaked whale *Mesoplodon*) the stomach consists of three compartments. This is true of both the Odontoceti and Mysticeti. An exception is the beaked whale, in which the first compartment is absent and the pyloric stomach is broken up into as many as 12 compartments. The cetacean forestomach is anatomically identical with the forestomach of the ruminants, which consists of the rumen, the reticulum, and the abomasum.

Whales swallow their food whole, and the "chewing" is done in the stomach. The forestomach in whales is highly muscular, with a tough lining. It may also contain pebbles, stones, sand, or other hard objects, such as vertebrae or the chitin of crustaceans. Forceful contractions of the forestomach containing these materials must act much as the gizzard of a bird does in breaking down food. Even though the first and second compartments of big whales can hold over 760 l (200 gal) or 907 kg (a ton) of plankton, the relative proportion of a whale's stomach is not especially great compared with that of a cow, whose four stomach compartments can hold up to 91 l (35 gal).

The functions of the first compartment seem to be that of mechanically breaking down and storing food. The epithelium contains no glands and therefore secretes no materials. The lining of the second compartment contains fundic glands. Pepsin, hydrochloric acid, and small amounts of lipase (a fat-digesting enzyme) are found in the second stomach compartment. The third compartment, the pyloric stomach, contains a large number of pyloric glands.

Although whale stomachs resemble ruminant stomachs in their complicated anatomy and in their function for storage, there are several differences, because of their differing diets. Since whales are primarily carnivorous not herbivorous, there is no need for microfauna, whose function is to digest cellulose.

The intestine of cetaceans is not greatly differentiated. The Mysticeti and some dolphins (Odontoceti) have a very short cecum, which is absent in all other whales. The large and small intestines are similar, as in most other mammals with a carnivorous diet. The sperm whale *(Physeter catodon)* has an unusually long intestine, about 15,250 cm (500 ft) in an animal 1677 cm (55 ft) long. Expressed as a percentage of body length, carnivorous animals have a short intestine; omnivorous animals, a medium-length intestine; and herbivorous mammals, a long intestine. The cetacean pancreas differs little from that of the other mammals. The cetaceans have no gallbladder.

In any discussion of whale digestion the subject of ambergris cannot be left out, even though the origin of this substance is still vague. Ambergris was at one time worth its weight in gold as a perfume base. Although many synthetic substitutes have been found, it still can bring around $100 a pound. Ambergris is a dark brown, waxlike substance which is quite pliable. It is probably a pathologic substance produced in the intestine of sperm whales *(Physeter catodon)* and similar to intestinal stones sometimes found in terrestrial mammals such as cows.

Ruminants

The digestive system and metabolism of cattle *(Bos taurus)* have been studied more extensively than those of any other mammal, with the exception of man. The cow is a ruminant, a word meaning "cud-chewing." The cow, like other related mammals, can fill its stomach rapidly with vegetation, then spend the rest of the day working over the meal in some peaceful (or for wild animals, safe) refuge. This must have survival value for any kind of mammal in a wild state.

The machinery that makes the extraordinary metabolic feats of the ruminants possible is a compartmentalized, or multiple, stomach with the microorganisms which live in it (Fig. 7–8). The stomach is characterized by its great size and its division into four distinct compartments. It is believed that the first three compartments develop from the primitive foregut, as in other ani-

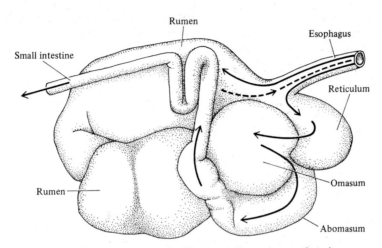

FIGURE 7-8   A ruminant stomach. *(Redrawn after various authors.)*

mals. They are (1) the *rumen* (paunch), (2) the *reticulum* (honeycomb, tripe), (3) the *omasum* (psalterium), and finally (4) the *abomasum* (the stomach of other mammals). They are connected by an opening so large that they are almost one very large compartment whose capacity reaches 91 l (35 gal).

The rumen, reticulum, and omasum are lined by a stratified squamous epithelium, whose outer layer is keratinized. The food is retained in the rumen and the reticulum until it is churned into a fine consistency. The contractions of the reticulum cause fluid to be washed backward into the rumen. This is followed by contractions of the rumen which return the material to the reticulum and then to the omasum. Dzuik, Fashingbauer, and Idstrom (1963) indicated that in the white-tailed deer *(Odocoileus virginianus)* the primary rumen contraction follows two successive contractions of the rumen. The duration of this cycle was 20 to 30 s.

Regurgitation of a bolus occurs with no contraction of the rumen, since the skeletal muscles do the work.

In the omasum, contractions of the muscle compress and pulverize the food. About 60 to 70 percent of the water is absorbed. The food is then passed to the abomasum, where gastric juice and digestive enzymes are secreted. The hydrochloric acid of the gastric juice destroys the microfauna. The abomasum is the glandular portion of the ruminant stomach and functions as do the simple stomachs of other mammals (Sisson and Grossman, 1969).

The abomasum also communicates by a groove, the *esophageal groove,* directly with the aperture between the rumen and the reticulum, where the esophagus enters. This groove is bounded by a pair of muscular ridges which can convert the groove into a canal. The food can be returned to the mouth in a rounded bolus, the "cud," where it is further masticated, while the individual is resting, and then swallowed again. This time it does not go into the rumen or reticulum but is strained (moisture removed) between the leaves of the omasum and enters the abomasum.

In calves feeding on milk, the milk goes directly via this groove to the abomasum for digestion. The ruminant type of digestion develops gradually. Calves are not born with rumen microfauna, but are believed to acquire them gradually. The saliva of the cow is spread around by grazing. The calf feeds on the same vegetation and also licks the cow itself. In this case slobbering all over is necessary for survival.

It was originally believed that the only function of the microorganisms in the rumen was to break down cellulose, because there is no cellulase in the ruminant stomach. But they do much more than this. For instance they synthesize vitamin B, so that the cow does not need to have these vitamins supplied in its diets, as man does. These microorganisms also make amino acids and proteins from simple materials, and further, they transform a considerable part of the low-grade plant material, with inferior protein and indigestible roughage, into milk and whey. Further, the propionic acid formed by fermentation seems to be the chief precursor of the glycogen. It is true that milk cows produce more if they are fed high-protein diets, but the ability of ruminants to make proteins from simple substances is one of the unique features of ruminant digestion. Some microbes in the rumen can split urea into ammonia; other microbes then fix the nitrogen of the ammonia to make it available for synthesizing amino acids. Some urea is transported back to the saliva, is swallowed and again recycled. Thus cows can even produce proteins from urea which is usually considered wasted material. K. Schmidt-Nielsen, B. Schmidt-Nielsen, Jornum, and Haupt (1957) have shown that camels also do this.

The microbial synthesis of foods is only one part of their function. Another is the breakdown of material to provide energy by decomposing the complex carbohydrates into large quantities of short-chain fatty acids, principally acetic acid. The salts of these acids go through the rumen wall into the bloodstream. The ruminant's metabolism of carbohydrates differs from that of the nonruminants. The nonruminants metabolize carbohydrates to form simple sugar, mainly glucose, and produce short-chain fatty acids later in the tissues. Thus, the nonruminant's blood contains much glucose and little of fatty acids, while the ruminant's blood is just the opposite. Most of a cow's energy requirements are supplied by the acetic acid produced in the rumen.

Ruminants cannot get along without glucose, because they must store their energy in the form of liver glycogen, which is made from glucose. They must have glucose for their nerve tissue, and they must have glucose to produce milk. Since sugars formed in the rumen are fermented, the animal must synthesize sugar from short-chain fatty acids in its tissues.

The rumen microorganisms are composed of bacteria and of ciliate protozoa (Oxford, 1958). It has been estimated that the rumen population of bacteria is about 10^{10} and that of protozoa is about 10^6. Under favorable conditions bacteria can divide every 20 min, but the protozoa divide only about two or three times a day. The protozoa are contained in two main groups, the holotrichs (two genera and three species) and the oligotrichs (many species in many genera), and are found only in the rumen. Oxford (1958) wrote, "The adaptation of rumen life is very highly developed, since the ciliates of the antelope of the tropical African bush veldt are in the main, identical with those of the reindeer of the Arctic tundras, as well as with those of domestic cattle, sheep, and goats." Some of the rumen oligotrichs possess a real elegance of form. For a detailed information on the microfauna the reader is referred to Hungate (1956).

The relationships between the nutritive processes of the animal and its range constitute a field which has not as yet received a great deal of attention. Food habit lists are a beginning, but "feeds are merely the carriers of the nutrients and the potential energy," according to Crampton and Harris (1969).

One might wonder about the use of antibiotics to combat disease in the ruminants. They are of course used, but in doses which would obviously not kill the entire rumen population of microorganisms.

Though ruminant digestion has many advantages, it also has some disadvantages. Fermentation in the rumen generates large quantities of carbon dioxide and methane. The animal rids itself of these gases by eruction (belching), but if something interferes with the elimination of these gases the rumen swells and the animal becomes bloated. Plants responsible for bloating contain saponins (foaming agents). This froth traps the gas, a process which can now be alleviated by dosing the animal with an antifoaming agent.

Another digestive ailment of cattle is ketosis, or excessive formation of ketones in the body. The disease corresponds to diabetes in human beings. In the cow a deficiency of sugar in the blood seems to be the cause; this deficiency may be due to a defect in the metabolism of certain fatty acids, a shortcoming in the diet, or a shortcoming in the microorganism population.

The cow synthesizes glucose from short-chain fatty acid units, acetate, propionate, and butyrate. The ruminant's energy and materials come from short-chain fatty acids produced in the rumen. Nonruminants depend on glucose and amino acids provided by the food.

The camels and llamas are usually described as "primitive cud chewers," and while their digestive system is ruminantlike, there are differences. Camels and llamas do chew their cud; the differences are in the anatomy of the stomach.

Ruminantlike digestion has been found in some marsupials and leaf-eating primates (Fig. 7–9). Moir, Somers, and Waring (1956) studied the anatomy of the stomach of the quokka *(Setonix brachyurus),* a marsupial, and found that the forestomach, or "rumen," is sacculated. An esophageal groove

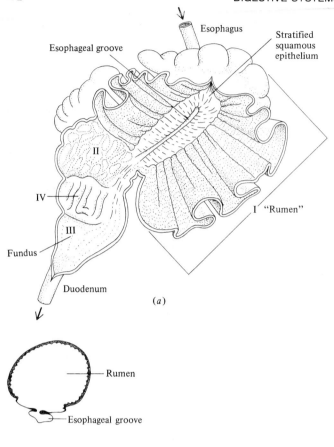

(a)

(b)

FIGURE 7–9 A ruminantlike digestion has arisen by parallel evolution in the grazing mar-supials (e.g., the quokka, *Setonix brachyurus*) although the stomach is not divided into four compartments like that of eutherian mammals (Fig. 7–8). *(a)* Stomach opened with a lon-gitudinal cut. *(b)* Stomach in cross section. In *Setonix* the esophagus leads into a putative rumen (I). An esophageal groove, by-passing the extensive sacculated main fermentative region (of tubular glands), passes through an area of stratified squamous epithelium like that of the rumen, reticulum, and omasum of sheep. Region III may be analogous to the aboma-sum; the functions of II and IV are still unknown. These and other complexities are associated with the possession of a dense bacterial population and a very small cecum. *(After Moir, Somers, and Waring, 1956.)*

bypasses the main fermentation region. The third region is analogous to the abomasum of the ruminants. The functions of two other well-defined but unnamed regions are still unknown. This capacious stomach, like that of the ruminant, allows the food to ferment. In *Setonix* there is a dense bacterial population.

Cellulose digestion, nitrogen metabolism, and vitamin synthesis are the significant features of ruminant and ruminant-type digestion which set this type of digestion apart from nonruminant digestion.

Chapter 8

The Internal Environment
and Its Maintenance

Previous chapters have emphasized morphologic and behavioral adaptations of mammals to their external environment. This chapter will deal with some of the internal adjustments which have to be made to external changes of the environment. Claude Bernard (1813–1878), the famous French physiologist (often called the father of physiology), pointed out that the cells of higher organisms live in an internal environment differing from the environment surrounding the organism. "All the vital mechanisms, however varied they may be, have only one object, that of preserving constant the conditions of life in the internal environment," is a translation of Bernard's statement.

Think for a moment of the range of external variables that mammals have to tolerate: air temperatures ranging geographically from −68°C (the Arctic in winter; the Antarctic is colder but there are no terrestrial mammals there) to 45°C (shade in the Sahara in summer) and 49°C (without shade in the Kara-Kum, east of the Caspian Sea), or annually from −16 to 40°C (central Great Plains of North America), and daily from −1 to 38°C (deserts of Africa); water temperatures ranging from −2°C (antarctic and arctic waters) to 30°C (at the equator); atmospheric pressures ranging geographically from 1 atm (at sea level) to less than half that on mountaintops and many times that at depths to which whales dive; salt concentrations of water from near zero to that of seawater. And add to that the problem of the giraffe whose brain must remain in a constant environment under extremes of blood pressure ranging from very low, when the animal is drinking, to very high, when the animal is grazing. Against all this the relative proportions of the different ionic constituents, the pH, and the total salinity, and the internal temperature are some of those constants which must remain within very narrow limits during the life of the mammals.

OSMOREGULATION

Water is a prime necessity for survival. It constitutes 70 percent of the body weight of mammals. Water serves these functions: (1) It acts as a solvent for the chemicals and chemical reactions in the body, (2) it is the fluid in which materials are transported, (3) it aids in the stabilization of body temperatures, and (4) it helps rid the body of wastes, especially nitrogenous wastes. Water loss must be balanced by water intake (or production).

Water Loss	Water Gain
1 Evaporation	1 In food, from 15 percent or less
a Respiration and sweating	in air-dried grain to 90 percent
b Insensitive water loss	in succulent vegetation (cactus)
2 In urine	2 Drinking water
3 In feces	3 Water of oxidation, formed
	when food is oxidized to form energy

Mammals are said to be osmoregulators, since they have well-developed mechanisms to control both body water and body solute content. The development of an energy expenditure system to control the extra-and intracellular fluids is presumably correlated with evolutionary history and the origin of life in a marine environment. The multicellular terrestrial animals of today are the products of nearly a half billion years of evolution along divergent paths, yet their bloods are remarkably alike in ionic composition. The evolution of a regulatory mechanism has been toward a maintenance of the earlier composition in the relevant parts of the organisms.

The kidney, whose special function is to excrete nitrogenous waste, is of great importance in this regulatory mechanism. For detailed discussions of kidney and kidney function the reader is referred to Smith (1961), Baldwin (1964), and Prosser (1973).

The most common end products (nitrogenous wastes) of protein metabolism in the animal kingdom are ammonia, NH_3, water-soluble and highly toxic; uric acid, $C_5H_4N_4O_3$, not water-soluble and not toxic; and finally urea, $CO(NH_2)_2$, water-soluble and moderately toxic. A plentiful supply of water is necessary for a rapid and complete elimination of ammonia, and so it is often the end product in, for instance, small aquatic organisms. Terrestrial vertebrates are divided. The reptiles and birds produce uric acid, but the mammals produce urea. Why does not the same method occur in all terrestrial vertebrates? An answer may be found in the mode of reproduction, which seems to have determined the type of end product produced. Enclosed (cleidoic) eggs, such as those of reptiles and birds, have a limited water supply. Ammonia or urea would be toxic, but uric acid is not and is the end product in reptiles and birds. The term for this type of nitrogen metabolism, when the end product is uric acid, is *uricotelic*. In mammals the embryos are connected with the maternal blood and the water supply is more plentiful than in the cleidoic egg. The urea diffuses across the placenta and is excreted by the maternal kidneys. The term for this type of nitrogen metabolism, in which the end product is urea, is *ureotelic*. Uricotelic metabolism is associated with the cleidoic egg and oviparity, and ureotelic metabolism is associated with viviparity.

Kidney Anatomy and Function

Each kidney is made up of a large number of units, basically similar in structure, called *nephrons*. Each nephron consists of a closed bulb, called a Bowman's (renal) capsule, and a long, unbranched tubule running a tortuous route through the cortex and medulla, finally ending in the pelvis of the kidney.

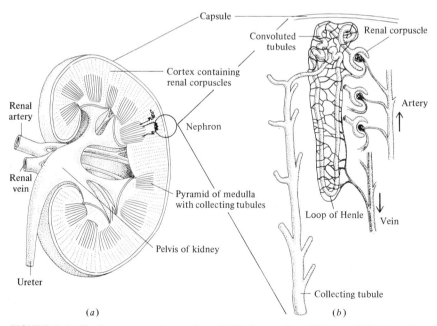

FIGURE 8-1   The human excretory system. *(a)* Median section of kidney. *(b)* Relationships of renal corpuscles, tubules, and blood vessels. *(From Storer, Usinger, and Nybakken, 1968.)*

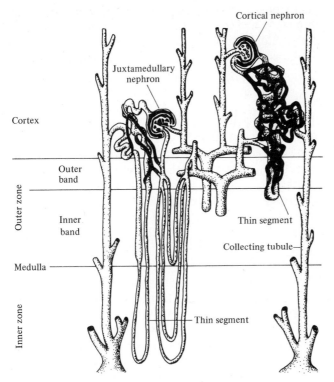

FIGURE 8–2 Major structural differences between a cortical nephron and one located in the juxtamedullary region. *(After H. W. Smith, 1951.)*

In most mammals it doubles and twists. The localization of the different segments of the nephrons in the kidney is the basis for dividing the organ into cortex, medulla, and papilla. The cortex contains the Bowman's capsule, proximal and distal convoluted tubules, the beginning of the collecting tubules, blood vessels, endocrine organ tissue, nerves, and supporting tissue (Fig. 8–1). The medulla consists essentially of the loops of Henle, which vary in length and location. The position of the nephron in the cortex varies. It is sometimes near the base of the cortex (juxtamedullary nephron) (Fig. 8–2). There is increasing evidence that nephrons may differ not only in position, but in structure and function as well. Cortical nephrons appear to function primarily in regulating salt excretion, while juxtamedullary nephrons regulate water excretion. The papilla is the innermost part of the medulla, or core, of the kidney and supports the openings of the collecting ducts that drain urine from the nephrons.

Arterial blood enters a convoluted network in Bowman's capsule, then a network surrounding the proximal and distal convoluted tube and the loop of Henle, finally emptying into a branch of the renal vein. Exchange of substances takes place through active transport and osmosis almost exclusively between blood capillaries and nephrons. Most of the mass of the medulla contains loops of Henle, parts of collecting ducts, blood vessels, and supporting tissues.

Water Conservation

Water and solute metabolism in mammals is not separated from other physiologic functions. Mammals, as warm-bodied organisms, are always producing heat by their own metabolism, and when the environment becomes too warm, one of their primary mechanisms for cooling themselves is evaporative water loss from the skin and respiratory surfaces. Mammals lack some of the water-conservation mechanisms such as the cloaca (except monotremes) of birds and reptiles or the salt glands of some sea birds and sea turtles, which are capable of excreting salt in very concentrated solutions. Characteristic water economies of mammals include the production of a concentrated urine and a lowered respiratory evaporation under extreme conditions. These may be combined in a number of ways with more specialized means of conserving water. These include behavioral adaptations for nocturnal activity, the use of burrows, and feeding on succulent vegetation, or, at times when moisture is unavailable, physiologic adaptations such as torpidity, and still more. In the following pages specific examples of these measures will be discussed in greater detail.

We now know that many mammals are able to get along without preformed water, and in the last several decades many studies have been conducted to learn how this is accomplished. The first North American mammal to be recorded as getting along without preformed water was the prairie dog *(Cynomys ludovicianus)*. The shipment of zoologic specimens by Lewis and Clark from Fort Mandan to the Executive Mansion on the Potomac in early 1805 included a live prairie dog. Antoine Tabeau, apparently one of the caretakers, wrote that he had found that the prairie dog never drank and that "Cap. Lewis had already kept it three months without being able to make it swallow a drop of water . . ." (Abel, 1939).

How the small desert-dwelling rodents acquired their water was especially interesting to earlier naturalists and biologists. One of the first to publish his ideas and speculations was Charles T. Vorhies (1945) of the University of Arizona. He found that some mammals, such as the wood rat *(Neotoma albigula)*, do not drink but acquire sufficient water in the succulent food they eat. Wood rats, also called pack rats *(Neotoma* spp.), are found most commonly where cactus is present. Cactus contains about 90 percent water, which helps solve the water problem, but cactus also contains oxalic acid, which is poisonous to most mammals. Oxalic acid is highly poisonous because it binds the calcium to form calcium oxalate, thus reducing blood calcium. This leads to the development of tetanic cramps in the muscles, as well as an impairment of nerve function. Thrombin is not formed, and the calcium oxalate may damage the kidney tubules. But pack rats can metabolize oxalic acid, thereby destroying its toxic effects. Ryan (1968) did find a population of *Neotoma lepida* in Deep Canyon, near Indo, California, living in an area where cactus was almost entirely absent. He wrote, "It is not yet known how this, the largest population of pack rats in the Deep Canyon area, satisfies its requirement for water in the absence of succulent foods." For many smaller mammals, succulent vegetation and dew meet their water requirements.

But eating succulent vegetation is only one of many ways of acquiring water in an arid region. Vorhies (1945) wrote that "the kangaroo rat *(Dipodomys spectabilis)* eats mostly air-dry food, mainly seeds with a minimum water content, yet the animal seldom if ever takes water." This was descriptive but not explanatory. Does the kangaroo rat store water? Does it live through dry periods at the expense of its body water, thus becoming desiccated? Or does it economize by not excreting urine for a long time, as, for instance, the African lungfish does? These were questions which led the Schmidt-Nielsens (1952–1953) to begin experimenting with kangaroo rats *(Dipodomys spectabilis, Dipodomys merriami)* in a field laboratory in Arizona. They found that the animals' water content (about 65 percent) did not vary between the rainy and dry seasons. The animals did not save water from the rainy season for the dry season. Kangaroo rats taken into the laboratory and given a diet of dry barley for 8 weeks had the same water content at the end of the 8 weeks as at the beginning. But they had increased in weight! Some rats were given watermelon along with barley, but the body water content remained the same. The urea and salt content of the blood also remained constant whether the diet was moist or dry.

Two sources of water are available to the kangaroo rat, preformed (drinking) water and metabolic water. Metabolic water, as the name suggests, is a product of metabolism and occurs whenever hydrogen is oxidized. The amount of water is dependent upon the hydrogen content of the food. Oxidation of a gram of protein yields about 0.3 g of water; and a gram of fat produces the greatest amount, about 1.1 g of water.

Obviously not a great deal of moisture is available to an animal which feeds heavily on seeds and grassy vegetation. The Schmidt-Nielsens, feeding the kangaroo rats on barley, found that each ate about 100 g in 5 weeks. If there was moisture in the air, the barley would contain some absorbed water. The total amount of water available to a rat during the 5 weeks varied between 54 and 67 g, a rather small amount with which to meet its physiologic function. What, then, are the adaptations for conserving water? There are several physiologic and behavioral adaptations, foremost of which is a very efficient kidney. The concentration of urea in the urine of the kangaroo rat can be as high as 24 percent, compared with a high of 6 percent for human beings. Turning this around, the kangaroo rat needs only one-fourth as much water as a human does to excrete the same amount of urea. The rat can excrete urine about twice as salty as seawater (K. and B. Schmidt-Nielsen, 1950). Anatomic alterations of the kidney have increased its efficiency to conserve water. Vimtrup and B. Schmidt-Nielsen (1952) wrote, "It is significant that the distal tubules and the collecting ducts in the kangaroo rat kidney are longer and differ in structure from those of the white rat kidney, while the glomeruli and the proximal tubules closely resemble those of the white rat kidney." Hypertonic urine is formed by active water reabsorption, presumably confined to the distal system. An exceptionally long papilla renis projecting down in the ureter seems to be important to animals which must excrete a concentrated urine. Another route of water loss is the feces. In the kangaroo rat the feces are nearly dry and therefore little water loss occurs through this pathway. A third route of water loss is evaporation from the skin and respiratory surfaces.

All mammals lose water directly through the skin, in addition to that lost through the sweat glands. This direct water loss is called insensible water loss. There is of course some water loss through the sweat glands. Quay (1965) has shown that rodents, including kangaroo rats (*Dipodomys* spp.), have sweat glands other than on the toe pads. The kangaroo rat has behavioral adaptations which reduce the water loss through the skin, sweat glands, and respiratory surfaces. Kangaroo rats are nocturnal mammals, avoiding the daytime high temperature and low relative humidity of the arid habitat. Tests showed that at 24°C the relative humidity would have to be no less than 10 to 20 percent if the rat were living on a dry diet. So the animal spends the day in its burrow. Here the relative humidity ranged from 30 to 50 percent and the temperature between 24 and 32°C. At night, desert temperatures varied from 16 to 24°C, and relative humidity from 15 to 40 percent. Respiratory surface water loss in kangaroo rats, and perhaps in other small desert-dwelling mammals, is reduced in a unique way. The inhaled dry air warms to body temperatures, and the moisture from the passage walls evaporates into it, cooling the passage walls. The cooled, humidified air proceeds to the lungs. When it is exhaled, the humidified air comes into contact with the cooled nasal passage walls, causing a considerable fraction of the water vapor to condense out and be absorbed by the nasal passages (Schmidt-Nielsen, Hainsworth, and Murrish, 1970). MacMillen (1972) has suggested that small desert rodents with high surface-to-volume ratios and poor insulation produce more water metabolically than they lose through evaporation, when ambient temperatures drop below the thermoneutral zone of the individual.

Sperber (1944) examined the kidneys of 141 species of mammals living in widely differing environments and found a clear relation between the environment and the size of the renal papilla. Mammals living in a wet climate, such as the platypus *(Ornithorhynchus),* and the mountain beaver *(Aplodontia)* in the humid Pacific Northwest, have kidneys without papillae. *Dipodomys, Gerbillus, Meriones, Jaculus, Ctenodactylus,* and *Hapalomis* (all desert rodents) have long papillae. The African sand rat *(Psammomys obesus)* has the most efficient kidney known, but the necessity for efficiency is not a lack of water but an abundance of salt. Sand rats are desert rodents, but their habitat is restricted to edges of saline and brackish waters. The sand rats feed on plants which contain 80 to 90 percent of water with about twice the salinity of seawater. The lack of water is not the problem. The plentiful salt content makes it necessary for the kidney of the sand rat to be able to concentrate its urine.

The grasshopper mouse *(Onychomys)* can get along without preformed water, for it lives on insects which contain from 60 to 85 percent water. The echidna, or spiny anteater *(Tachyglossus aculeatus),* also seems well adapted to survive in arid areas. From data published by Griffiths (1965) and Schmidt-Nielsen, Dawson, and Crawford (1966), it appears that the termite-eating echidna, living in dry air at 25 to 33.5°C, can survive without preformed water.

No discussion of mammalian ability to survive without water would be complete if the camel *(Camelus)* were ignored. Camels are denizens of the hot desert. It has often been said that there is more water in a camel and the man riding him than in the surrounding desert for a radius of 80 km (50 mi). The habitat of the camel provides little shade. This mammal occupies a habitat of maximum heat and minimum water, and has extraordinary attributes as a riding and pack animal in hot, dry regions of the world. It is reputed to be able to go up to 21 days or nearly 970 km (600 mi) without water. Man has wondered through the centuries about the camel's ability to get along without water; nearly 2000 years ago Pliny believed the "water sacs" in the rumen (stomach compartment) to be responsible. Recently, research has shown that they are not.

Although the camel originated in the central plains of North America, the last of them had migrated to Asia and South America a million years ago. During the middle of the nineteenth century some were brought back to the arid regions of the Southwestern United States, not because of anyone's burning desire to see a return of the native, but because of the camel's phenomenal ability to go without water for a long time and to survive on the coarsest of vegetation. This operation must rank as a most unusual practical experiment in the domestication of mammals.

For men to reach California during the gold rush meant either crossing the Southwestern deserts and the Rockies, mostly by foot, or a very hazardous trip of several months around Cape Horn. In 1850 there was little immediate prospect of a railroad connecting the East and the West. Edward Beale (1822–1893), a veteran of frontier fighting, supposedly convinced Jefferson Davis, then Secretary of War, that camels could be used for this desert transportation. In March 1855, Congress appropriated $30,000 to import camels for the transportation of military materials. By February 10, 1857, a total of 77 camels had been brought from Egypt, Arabia, and Asia Minor. The animals were taken to Camp Verde, 60 miles northwest of San Antonio, Texas, for acclimatization. They were a source of endless curiosity, amusement, and newspaper stories. Lt. Edward F. Beale was assigned to head the first and last Camel Corps. He was to select a new route from New Mexico to California, and to test the fitness of the camel as a beast of transportation in the arid Southwest. The test journey started on July 5, 1857, and ended on February 21, 1858. Beale (Lesley, 1929) wrote of the journey:

> A year in the wilderness ended! During this time I have conducted my party from the Gulf of Mexico to the shores of the Pacific Ocean, and back again to the eastern terminus of the road, through a country for a great part entirely unknown, and inhabited by hostile Indians, without the loss of a man. I have tested the value of the camels, marked a new road to the Pacific, and travelled 4,000 miles without an accident.

It should also be added that Beale lost no camels and was continuously extravagant with his praise of their performance. In December 1858, Secretary of War B. Floyd reported, "The entire adaptation of camels to military operations on the plains may now be taken as demonstrated." The Civil War dealt the death blow to an intriguing experiment, but some camels escaped or were sold and later released, and it was nearly a quarter of a century before the last of them perished. If this dramatic experiment was unsuccessful in its primary aim, that of providing efficient transportation of men and materials from Gulf ports to the goldfields of California, it fully confirmed that the camel could indeed get along on coarse vegetation and without water for long periods. How it does this was not revealed until a century later.

One way in which the camel is able to conserve water is by excreting a highly concentrated urine, just as the kangaroo rat does. As a matter of fact, camels are able to tolerate plants with a high salt concentration, as well as salt water. Camel urine has been found to be twice as concentrated as human urine (K. and B. Schmidt-Nielsen, Jornum, and Haupt, 1956 and 1957). Urine production in the camel is low but variable. The camel's bladder is very small, and in winter when the water content in vegetation is high, the urine output is increased and the bladder is emptied frequently. The volume varied from 1.2 to 8 l/day. Read (1925) reported that he found practically no urea in camel's urine.

The Schmidt-Nielsens (1952) and their colleagues (K. and B. Schmidt-Nielsen, Haupt, and Jornum, 1957), trying to attain maximum urine concentration, fed a young camel hay and dates, with no water, for 17 days. To their surprise they found a decrease in urine concentration. At the end of the period the total amount of urea excreted in one day was less than a gram! The excretion of less than a gram of urea would correspond to a metabolism of less than 2.5 g of protein per day. This seemed unreasonable. It would be more reasonable to attribute this deviation to the same kind of peculiar nitrogen metabolism that is well known in domestic ruminants. Stockmen have long known that urea and ammonium salts can be synthesized into amino acids and proteins by the microbes in the rumen of domestic livestock. Haupt (1959) has shown that in sheep, the bacteria of the rumen can synthesize protein from urea. In most mammals, the amino acids of the protein are deaminated and the nitrogen is excreted as urea, but the camel, like sheep and other ruminants, can utilize urea for synthesis of protein, and so excrete less. Even fresh camel dung is nearly dry.

Cooling is also accomplished with a minimum of water. The camel's daily temperature can vary from 34.0°C in the morning to 41.0°C in the evening without any harmful effect. By allowing the temperature to rise during the day, water used to keep down the temperature is saved. The excess heat can then be dissipated to the cooler and more humid environment which exists in the evening and night. The higher body temperature during the day also reduces the heat flow from a warm environment to the body. The concentration of fat in the humps rather than in a subcutaneous layer is also of value in heat dissipation. The camel's fur is thick enough to give some insulative effect from the hot desert air, but not so thick as to interfere with the evaporative cooling effect of sweating, although sweating is very limited in the camel.

Interesting as is the camel's ability to tolerate dehydration, a more surprising physiologic phenomenon is its ability to withstand almost "instant rehydration." To begin with, a camel can withstand dehydration exceeding 25 percent of its body weight. Many persons have had the experience of becoming sick when, after great physical activity on a hot day, they have tried to recoup their water loss too quickly. The use of salt tablets somewhat alleviates the problem, but the human being requires several hours for comfortable rehydration. Horses that have exercised vigorously and suffered a water loss are usually not permitted to drink freely at first, but are only slowly allowed to recoup their water losses. Camels can make up a 20 percent loss of body weight in 10 min. Schmidt-Nielsen (1964) records several instances of camels drinking more than one-quarter of their body weight of water in a very short period of time. This would be comparable to a 90-kg (200-lb) person drinking 28 kg (50 lb) of water in a very short time, probably 15 min! Why the camel can do so is not completely understood, but as mentioned earlier, the camel does not sweat much and therefore does not lose as much salt as a human being or a horse. The camel's legendary ability to get along without water, earlier shrouded in mystery, has gradually become untangled by researchers.

The large herbivores seem to be the most successful invaders of semiarid and arid regions. The giraffe, antelopes, and gazelles are the large ungulates most frequently seen in these regions of Africa. Schoen (1969) has studied three bovids from East Africa, the bushbuck *(Tragelaphus scriptus)* from moist bushland, the Uganda kob *(Adenota kob)* from dry savannah, and the little dik-dik *(Rhyncotragus kirkii)* from semidesert habitat. When those animals were dehydrated up to 85 percent of their initial body weights, their urine output fell, and their urine concentration rose. Measurements of the kidneys of the three species showed that the medulla occupied 31, 38, or 47 percent, respectively, of the kidneys. If the volume of the medulla is an indication of the lengths of loops of Henle, then the results suggest that the capacity to concentrate the urine is related to the aridity of the tropical habitats of these bovids.

Water balance in terrestrial carnivorous mammals has not received a great deal of attention, perhaps because it presents no great problem. Although protein metabolism provides the least amount of metabolic water, the original food (animal material) contains about 68 percent of water.

What about the whales and seals that are nearly always surrounded with water? At first glance they would seem to have no problem, but a second thought would soon lead to the realization that their problem is similar to those of the desert mammals—the unavailability of drinking water. Seawater could add a salt load to their kidneys, and they could conceivably also become desiccated. There is no evidence that they can drink seawater or that they have superior renal concentrating power, although the concentrating power of both whales and seals is slightly superior to that of humans and dogs. Whales and seals feed largely on marine fish. Marine fish contain as much body water as terrestrial animals and no greater concentrations of salt, so that whales and seals face no real desiccation problems. The salts that marine fish obtain by drinking seawater are removed by means of secretory cells in the gill membranes (Baldwin, 1964). In harbor seals *(Phoca vitulina)*, urine formation is directly correlated with feeding. As digestion proceeds and water (metabolic and preformed) becomes available, the urine flow greatly increases. Eight to ten hours after a meal the urine flow decreases to about 10 percent or less of the maximal. At the minimal urine flow, the maximal osmotic concentration of the urine is about 5.6 times that of the plasma, which is greater than that of a dog (4.5) or man (4.2) but certainly far less than that of the kangaroo rat, which is 17 times that of the plasma, or even higher. Thus, this seal, which can excrete a somewhat concentrated urine, gains more water from the fish it eats than it loses through the urine or by evaporation from the lungs. Therefore, it does not have to drink water.

While many of the whales eat mostly fish, thereby satisfying their water requirement, the whalebone whales (suborder Mysticeti) live on plankton and small fish which they strain out of the seawater with the baleen (whalebone). The baleen hangs from the palate. At the inner edge it frays into long tough hairs which bend back when the mouth is shut. When the mouth is open, these hairs reach from top to bottom. All the whale has to do is to open its mouth, fill the cavity with water, and close the mouth, to filter out up to 50 gal of plankton. Since most invertebrates are in osmotic balance with the seawater, the whale's diet contains more salt than that of seals. The whale kidney is capable of producing urine a little more concentrated than seawater, and this margin is sufficient to maintain the balance (Coulombe, Ridgway, and Evans, 1965).

Other mammals in which unusual water balance problems have been studied are the salt marsh–dwelling populations of *Microtus, Peromyscus,* and *Perognathus* mice (Fisler, 1961, 1961a, 1963; Haines, 1964); the wild rabbit of Australia (Hayward, 1961); the antelope ground squirrel *(Spermophilus leucurus)* of the American Southwest (Bartholomew and Hudson, 1959); and two nondrinking East African antelopes (C. R. Taylor, 1969). This by no means exhausts the list, but gives the reader some idea of the varieties of osmoregulation mechanisms as well as the kinds of mammals in which unusual or extreme kinds exist.

THERMOREGULATION

Animals are divided into cold-blooded or warm-blooded groups. Though these were the original terms, efforts to achieve a more precise definition have resulted in a proliferation of terms. A cold-blooded animal may not necessarily be cold to the touch. If it is in a warm environment, it may feel warm. So the term *poikilothermic* was substituted, which signifies a variable temperature for cold-blooded animals. *Homeothermic* came into common use for warm-blooded animals. It is surprising that with adoption of the term homeothermic for warm-blooded animals, the term *heterothermic* was not adopted for the cold-blooded animals, but that is now reserved for mammals with variable core temperatures. The term homeothermic is also spelled "homothermic" and "homiothermic." When the temperature of an animal is determined by the heat obtained from the environment, i.e., comes from the outside, the animal is *ectothermic.* When the temperature is obtained largely from its own oxidative processes—from within—the animal is *endothermic.* The only continuously endothermic animals are the birds and mammals, but not all birds and mammals are continuously endothermic (for instance, during hibernation).

Warm-blooded animals are described as those maintaining a constant and high body temperature, roughly 10 times higher than that of reptiles. The idea

Table 8-1 Equivalent Terms in Thermoregulatory Terminology

Cold-blooded	Warm-blooded
Poikilothermic	Homothermic (homeothermic, homiothermic)
Ectothermic	Endothermic
Adjusters	Regulators

of a constant body temperature is reinforced by our early training. If our body temperature was not 98.6°F (37.1°C) we were sick. But not all humans have exactly the same body temperature, and not all parts of the human body are the same temperature. Deep body or core temperatures of a mammal are quite constant, while the temperatures of the surface and appendages more closely approximate the external (ambient) temperature. Heat energy is produced when food is metabolized, and heat is lost when the environment is colder than the mammal. Mammals regulate the balance between heat production and heat loss so as to maintain a relatively high and constant body-core temperature. In spite of the great variations of environmental temperatures, from the icy Arctic to the fiery deserts, the polar bear *(Ursus maritimus)* and the pronghorn *(Antilocapra americana)* maintain nearly the same body temperature day and night, summer and winter, as do those mammals which experience more moderate temperature changes.

In some adult mammals the body temperature is more constant than in others. Some are not warm-blooded until nearly adult. Hibernating mammals apparently regulate their body temperatures but at a very low set point.

The regulation of body temperatures in endotherms is done by the hypo-thalamus, located at the base of the brain, together with heat receptors in the skin throughout the body. This system can be compared to the main thermo-stat and satellite thermostats operating in a modern home. The sensitivity of the hypothalamus is said to be so great that 0.01°C above the set point is sufficient to increase the dissipation of heat, through sweating, by one calorie per second. For more detailed discussions of temperature regulation the reader is referred to Benzinger (1964) and Hardy (1971), among others.

The cottontail rabbit *(Sylvilagus floridanus)* seems to have a very constant body temperature. James (1957) took rectal temperatures of 169 cottontails and found a variation of only "103.3 to 103.7°F" (41.1 to 41.3°C). In other species of mammals the "constant" temperature can vary by several degrees. This condition is called heterothermy. Of the opossum *(Didelphis marsupialis),* McManus (1969) wrote, "Although daily variation in mean body temperature of cage opossums was not detected, seasonal variation was observed." The average body temperature of 18 individuals exposed to an ambient temperature ranging from 18 to 32°C during June, July, and August was 35.55°C. The average of 11 individuals exposed to temperatures ranging from 7 to −10°C during November and December was 33.76 (the lowest was 30.8°C). In the nine-banded armadillo *(Dasypus novemcinctus),* Johansen (1961) found a daily variation from 34.0 to 35.4°C in 13 individuals.

Two sloths of tropical South America are also heterothermic. Britton and Atkinson (1938) recorded temperatures varying from 33.8 to 35.7°C in the two-toed sloth *(Choloepus hoffman)* and from 32.2 to 34.8°C in the three-toed sloth *(Bradypus griseus).* Then there is the well-documented situation in the camel *(Camelus)*, whose temperature can vary from 34.0°C in the morning to 41.0°C in the evening.

The greatest variation in temperature control occurs in the bats. Lyman (1970), who reviewed metabolism and thermoregulation in bats, wrote, "There seems to be a progression in the temperature regulation of bats from a reasonably well regulated homeothermism of the large Megachiroptera of the tropics, through the less adequate temperature regulation of tropical Microchiroptera, to a rather special form of hibernation in the microchiropterans of the temperate zone." McNab (1969) suggested that the insectivorous forms of tropical Microchiroptera are apt to show heterothermy, while nectar- and fruit-eating forms are efficient thermoregulators. Bartholomew, Dawson, and Lasiewski (1970) have shown that in the Megachiroptera heterothermy appears to be related to body size. In the species they studied, 35°C or above was considered the homeostasis state, but the bats were actively effective at body temperatures of 25 to 30°C in the Madang area of New Guinea.

The temperature zone in which a mammal requires the minimum amount of energy for temperature regulation (heat production or dissipation) is called the *thermal neutral zone, thermoneutral zone,* or *zone of thermal* neutrality. The zone may be as narrow as a single degree, as it is for the blossom bat *(Syconycteris australis),* a tropical nectar feeder living in New Guinea and nearby Australia and the smallest of the order Megachiroptera (flying foxes). Its thermal neutral zone is 33°C (Bartholomew, Leitner, and Nelson, 1964). The arctic fox has one of the widest thermal neutral zones, ranging from −40°C to well above 4.4°C. The human being's thermal neutral zone is relatively narrow, spanning about 4°C.

Heat Transfer between a Mammal and Its Environment

Heat is exchanged between an organism and its environment by radiation, conduction, convection, and evaporation; these processes may occur in the internal thermal region, the boundary region, and the external thermal region (Fig. 8–3). All objects exchange thermal radiation from their environment. The direction is always from the higher to the lower temperature. One of the problems mammals face is that in many circumstances they must lose heat against this gradient, i.e., lose body heat to an environmentally higher temperature. The measurement of energy exchanges is extremely complicated, because there are many components, such as behavior, variation of activity, and habitat selection. Special morphologic and physiologic mechanisms have evolved which diminish or increase the effects of these components. These components of energy flow, when summarized in cal/cm$^2 \cdot$ min, are sometimes expressed as $T = S + M + R - LE + C + A$, in which T is gain or loss in energy, S is solar radiation, M is heat of metabolism, R is thermal radiation (to or from the environment), L is latent heat of evaporation for water, E is rate of

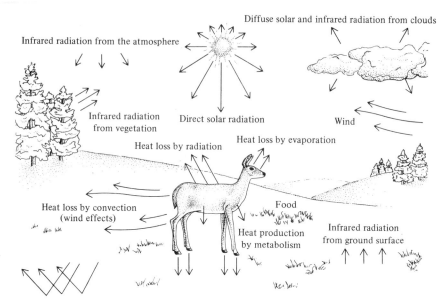

FIGURE 8–3 The thermal energy exchange between an animal and its environment. *(From Wildlife Ecology by Aaron Moen. W. H. Freeman and Company. Copyright © 1973).*

evaporation, G is conduction, and A is convection. Absorbed radiant energy is important in homeothermy and can be dissipated by any of the modes of heat transfer. Hair plays an important part in thermoregulation in mammals. Each hair functions as a little convection cylinder exposed to both free and forced convection by air currents. Whether the pelage is thick or thin, whether hairs are parallel to or at right angles to the skin are important in thermal exchanges. Air movement and the humidity are also important. More recently wind speed and temperatures have been combined (in the colder parts of the world) in a formula to give an "equivalent temperature" or "wind chill" index.

How mammals maintain a fairly uniform internal temperature against a constantly changing external temperature will be discussed in the next two sections.

Adaptations of Mammals to Low Temperatures

One might think that in a homeotherm the metabolic rate (oxygen consumption) would increase directly with increasing body weight, but this is not the case; the energy demand of a unit weight of tissue depends on the size of the mammal from which it came. The energy metabolism of a gram of mouse *(Mus musculus)* tissue is 20 times that of bison *(Bison bison)* tissue. This phenomenon has often been demonstrated on a "mouse-to-elephant curve" for mammals (Fig. 8–4), which shows that standard basal metabolism per unit weight decreases markedly as body weight increases. The *basal* or *standard metabolic rate* represents an approximation of the rate of metabolism of the mammal at rest and under no thermal stress.

In homeotherms heat production must be equal to heat loss, and it was early thought that heat loss was proportional to surface areas. Described as the "surface law," this led Bergmann to formulate *Bergmann's rule,* which stated that in any closely related group of birds or mammals, those that live in areas of lower temperatures will be larger than those living in areas of higher temperatures. This generalization has become very controversial and is by no means all-pervasive. Not all mammals living in cold climates are large. The generalization is based on the mathematic relationship of the surface area (which is squared) to the volume (which is cubed), a relationship expressed by the formula $M = aW^b$, where M is the metabolic rate; a, the constant of proportionality; W, the body weight; and b, the surface-to-volume ratio, frequently given as 0.75. Pearson (1948) suggested that a small shrew represents the lowest size limit possible for an endotherm. In a smaller mammal the heat would be lost faster than it could be produced, because of the great surface-to-volume ratio.

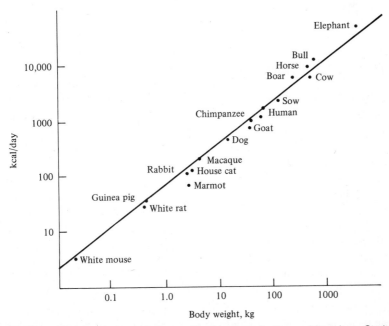

FIGURE 8–4 The mouse-to-elephant ratio of heat production and heat loss. Straight-line relationship is derived by plotting the logarithm at body weight to the logarithm at energy metabolism. *(From Kleiber, 1932.)*

Although $b = 0.75$ is a convenient rule-of-thumb figure, there is little consensus that it is precise. Variations in insulative features, fat, effective surface areas (postural adjustments and body conformation) are among those factors that influence this ratio. The theme that runs through cold adaptation is the circumvention of the relationship between size and thermal conductance.

Mechanisms for combating cold may be divided into two broad categories —increasing heat production and decreasing heat loss. Heat can be produced by exercise, shivering, and nonshivering thermogenesis. The most conspicuous of these is exercise, which is useful to the larger mammals, but in smaller mammals vigorous exercise does not necessarily increase the body temperature. There may be several reasons for this. One may be the higher surface-to-volume ratio in smaller mammals. Another may be that exercising disturbs the insulative value of the fur. Still another may be an increased peripheral circulation, which also increases the heat loss. Shivering is also an augmentation of muscular activity and is evoked by a decrease in surface or core temperatures. In several kinds of mammals heat production can be increased without increased muscular activity. Called *nonshivering thermogenesis,* this plays an important part in acclimation of mammals to low temperatures and in arousal from hypothermia. Rats acclimated to 30°C cannot replace heat loss at temperatures below 10°C by exercise, and their body temperature decreases. But when they are acclimated to 6°C, nonshivering thermogenesis can produce enough heat so that exercise can replace shivering down to − 20°C.

One of the sources of nonshivering thermogenesis is brown fat. This material is found in the young of mammals and also in the adults of hibernating species. Brown fat is concentrated around the neck and between the shoulder blades of newborn rabbits *(Oryctolagus).* It is found between the shoulder blades of marmots *(Marmota monax).* A human infant has a thin sheet of brown fat between the shoulder blades, and around the neck, as well as small deposits behind the sternum and along the spine. Bats have deposits of brown fat, especially between the shoulder blades. Dawkins and Hull (1965) found that at an ambient temperature of 35°C, newborn rabbits *(Oryctolagus)* produced a minimal amount of heat; at 25°C, they trebled their heat production; and at 20°C, they produced 400 cal/kg of body weight per minute. This would be enough to maintain their body temperature at a normal level even at 30°C if they were fully furred.

Brown fat cells have an abundance of mitochondria and generate heat by oxidation of fatty acids. Nerve impulses are sent to the brain from temperature receptors and from the brain along the sympathetic nerves to the brown fat. The nerve endings release norepinephrine, a hormone which activates the enzyme lipase, which splits triglyceride molecules into glycerol and free fatty acids, which in turn trigger the heat-producing cycle which turns chemical-bond energy into heat energy. Location of deposits of brown fat around the neck, thorax, and major blood vessels is strategic, since the heat produced can be quickly transported to the heart. This is where temperature control is most critical.

Decreasing heat loss is the other broad category of mechanisms for combating cold stress. There are many ways of accomplishing this. One of the best known in mammals is increasing effectiveness of the insulating shell (fur or blubber). It is well known that the thickness and quality of fur of mammals is greater in winter and in northern latitudes. Seasonal changes of the fur of arctic mammals varies from 12 to 52 percent in its insulative effectiveness, the greatest magnitude being found in the larger species, the smallest in the smaller mammals.

The rate at which a mammal loses heat to water is 10 times as great as the rate at which it loses heat to air. Human beings cannot survive in icy water for much longer than 10 min. Nevertheless there are some seals and porpoises which spend most or all of their lives in icy water. Many seals have an insulative layer of fur and of blubber, but the whales have only a thick insulative shell of blubber, inside their skin, as do seals of the family Phocidae, and the walrus. In the smaller whales, such as *Phocoena phocoena,* the metabolic rate of 3500 kcal/24 h is about three times that of terrestrial mammals. The lower surface-to-volume ratio of the larger whales reduces the problem of heat loss.

In those species of marine mammals without fur, the skin functions in a manner similar to the functioning of the skin on sparsely covered or naked appendages of terrestrial mammals. The skin temperature of harbor seals *(Phoca vitulina)* varies with the ambient temperature down to 0°C, resulting in only a small difference between the temperatures of the environment and the body surface. The extent of peripheral cooling is under physiologic control. Seals without fur are not as dependent on the appendages for loss of excess heat, but can employ local or general vasodilation. The superficial parts of these marine mammals function very well and are adequately nourished at temperatures ranging from 35.0°C down to near 0°C.

The muskrat *(Ondatra zibethica),* the water shrew *(Sorex palustris),* the beaver *(Castor canadensis),* Alaskan fur seal *(Callorhinus ursinus),* and other aquatic mammals can operate in waters of 0°C because of structural modification of their fur, which traps air. Their skin is never wetted, and they are surrounded by a layer of air. This not only helps their thermoregulation but also decreases their specific gravity, as discussed for the muskrat under the functions of fur in Chap. 6. The legs, tail, and ears cannot have as effective insulation as other parts of the body, and, like the appendages of seals and whales and the skin of furless seals, must remain viable at extremely low temperatures. Fat from the distal parts of the legs of arctic mammals such as the Alaskan red fox *(Vulpes vulpes)* and the caribou *(Rangifer tarandus)* has a melting point of less than 0°C, which is 30°C lower than the melting point of fat from core areas.

In the arctic mammals, the extremities and appendages are usually shorter than those of similar species living in warmer areas. This helps decrease thermal conductance. The dark coloration of the paws, nose, and tips of ears of mammals from colder climatic regions is still another adaptation to cold.

The aim of adaptation to low ambient temperatures is to reduce the lower limit of the thermal neutral zone and the rate of increase of energy metabolism when the ambient temperature decreases below the thermal neutral zone. The lower critical temperature of the red squirrel *(Tamiasciurus hudsonicus)* is 20°C, and it shows a rapid metabolic increase below thermal neutrality. It changes behaviorally when cold weather arrives by becoming less arboreal and spending more time on the ground or even in tunnels under the snow, where temperatures rarely go much below freezing (Irving, Krog, and Monson, 1955). Below-surface activity is a way of life for most small mammals in winter. They spend their time under the snow, under the vegetation, or below the surface. The deeper the tunnels, the less the deviation from fluctuating and often colder above-surface temperatures.

Nests and burrows reduce heat loss. The southern flying squirrel *(Glaucomys volans)* seems to maintain the same body temperature summer and winter, but survives the winter by hoarding food and huddling in well-insulated nests. The opossum *(Didelphis marsupialis)* uses denning, vasoconstriction, and shivering (McManus, 1969). Larger mammals such as dogs *(Canis familiaris),* coyotes *(Canis latrans),* and foxes *(Vulpes vulpes)* curl up to conserve heat. Bartholomew, Leitner, and Nelson (1964) studied several species of Australian "flying foxes," i.e., bats (*Pteropus poliocephalus* and *Pteropus scapulatus*). At low temperatures they enclose their hanging bodies in their wings and thus surround themselves in a layer of air at 10°C above the ambient temperature.

Specialized adaptations for reducing heat loss are the countercurrent heat exchange mechanisms (Fig. 8–5). Heat loss in seals and whales can be facilitated or reduced by "windows" of arteriovenous countercurrent heat exchange in the thinly insulated flippers and flukes. These heat exchange systems exist in a few other mammals also, and are not restricted to those living in the Arctic. An example of this principle used by engineers is the use of exhaust gases to preheat air coming into combustion chambers. In animals the heat in the arterial blood can be shunted into the venous system and carried back to the core. This direction can of course be reversed when the animal is warmer than the environment, for heat always moves from regions of high to low temperatures. In the simplest arrangement the artery and the vein lie side by side. In a more complicated arrangement the arteries and veins anastomose in a tangled bundle in the proximal parts of the appendages. They are then

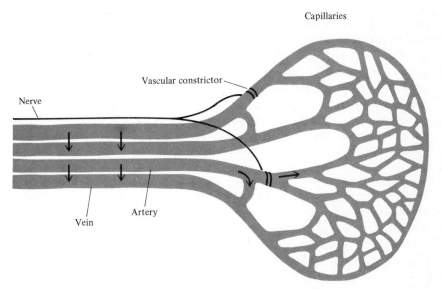

FIGURE 8–5   Schematic drawing of countercurrent heat exchange. *(Laurence Irving. Adaptation to Cold. Copyright © 1966 by Scientific American, Inc. All rights reserved.)*

called *retia mirabilia* (singular, *rete mirabile*). In the flippers and flukes of marine mammals there is a maximum surface for heat exchange, but the flippers contain alternate routes of separate superficial veins that are remote from the arteries. In this way the countercurrent system can be bypassed. Countercurrent systems of heat exchange are found in monotremes, sloths, armadillos, anteaters, beaver *(Castor canadensis),* and loris. Man has no elaborate nets, but the armadillo *(Dasypus novemcinctus)* has a simplified version which functions well for moderately cold conditions. These nets are absent in well-insulated arctic mammals such as foxes *(Alopex lagopus)* and huskies *(Canis familiaris),* whose effective insulation poses more of a problem in heat dissipation than in heat conservation.

Adaptation to low temperature has been accomplished by (1) increased heat production, (2) decreased thermal conductance, (3) behavioral avoidance of low temperature, and (4) daily or seasonal relaxation of thermal homeostasis, which should be thought of as a special resetting of the thermostat, not a return to poikilothermy. This last adaptation will be dealt with at length under its own heading—hibernation.

Adaptations of Mammals to High Temperatures

The problems of temperature regulation for mammals living in heat and drought are more difficult to solve than for those living in extremely cold environments. Under conditions of heat and aridity, heat is moving from the environment into the organism, so that endogenous heat produced by the mammal, must be moved against a temperature gradient. Heat loss is usually achieved by the loss of water, a difficult feat when water is scarce. Thermoregulation and osmoregulation are not separate physiologic functions. The physical factors that favor high environmental temperatures in deserts also favor heat loss during the night and during the winter.

The responses of mammals to high temperatures include behavioral adjustments, special modifications, and relaxation of precise limits of control. In examining various groups of mammals, we will see that combinations of these responses are the rule rather than the exception.

Marsupials Some of the larger kangaroos are excellent heat regulators and use a variety of mechanisms, such as panting, sweating, or increased salivating and licking of appendages in avoiding heat stress. In the North American marsupial the opossum *(Didelphis marsupialis),* heat stress occurs when ambient temperatures go higher than 37°C. McManus (1969) found that the main responses to high temperatures were the vasodilation of peripheral blood vessels and increasing the evaporative water loss by saliva spreading (licking its fur). Sweat glands are said to be lacking in the opossum, but highly developed bronchial glands (Sorokin, 1965) supplement the amount of fluid available to the opossum for licking its fur.

Bats Bats can use their wings for radiation in extreme heat. Bartholomew, Leitner, and Nelson (1964) found that Australian flying foxes have unique adaptations beyond the usual mammalian ones for thermoregulation. They can dissipate heat by panting, wing flapping, salivating, and licking their wings and chest. These authors wrote, "When the animals are hyperthermic the wings and interfemoral membranes become conspicuously engorged with blood, and the normally abdominal testes become scrotal."

Lagomorphs Jackrabbits *(Lepus)* live in plains and deserts of the Western and Southwestern United States. They do not have burrows. So far it is believed that jackrabbits do not have sweat glands. They meet heat stress in several ways. First of all, they are nocturnal and during the day seek whatever shade is available in shallow hollows under a bush or near a rock where the ground temperature is less than in sunny areas. In addition the large ears serve as a radiator to the sky where the microclimate is cooler than is that of the immediately surrounding area (Schmidt-Nielsen, 1964).

Desert rodents Many rodents are nocturnal, and the water economy regimen of the kangaroo rat *(Dipodomys)* and other rodents, discussed in detail earlier, is to some degree also involved in thermoregulation, since the two are not entirely separate functions.

Unlike many other rodents, ground squirrels are diurnal, but they still spend much of their daylight time in burrows. Vorhies (1945) recorded the temperatures in the burrows of the round-tailed ground squirrel *(Spermophilus tereticaudus)* for an entire year in southern Arizona and found that it never exceeded 29.0°C, even when surface temperatures reached 75.0°C. Another diurnal ground squirrel of the desert, the antelope ground squirrel *(Ammospermophilus leucurus),* was studied extensively by Hudson (1962). This ground squirrel neither hibernates nor estivates. The adaptation for desert living that evolved in this mammal was a tolerance of hyperthermia (above normal temperature) to the extent that metabolic heat is lost passively by conduction, radiation, and convection, and without the use of water. Antelope ground squirrels can tolerate body temperatures as high as 43°C. They exhibit intense periods of activity, interrupted by retreating into the burrow, flattening themselves out, losing the heat by conduction to the floor of the burrow and by radiation to the walls of the burrow. Thus the heat load accumulated by their activity is unloaded in the burrow. Living in the desert this animal does not have ready access to drinking water and would therefore be unable to lose its heat by evaporation, which would require a quantity of water loss equivalent to 13 percent of its body weight per hour.

Whales and Seals   Thermoregulation in these mammals is primarily one of keeping warm rather than keeping cool, for the cooling power of icy water is 10 times more effective than that of air, and some of these mammals live in −2°C water. But there must be an effective means of increasing thermal conductance during strong exertion, while in warmer water, or when the seals come onto land, as they must, to breed. Whales and seals have either atrophied sweat glands or none.

The thinly insulated extremities when flooded with blood during vasodilation serve as thermal windows for heat loss. During the harvesting of the northern fur seal, *Callorhinus ursinus,* care must be taken to avoid death from heat prostration, even though air temperatures remain below 10°C. Even the normal activities of breeding and territory maintenance of the males of the Pribilof Islands, where temperatures rarely rise above 12°C, may cause overheating, and activity decreases sharply when the temperature exceeds 12°C.

Carnivores   Panting and salivation are rather common forms of heat reduction. Panting is energetically more expensive than sweating. Panting is well known in the dog and also occurs in the other members of the family Canidae, as well as in cats, sheep, and antelopes. It avoids salt loss, but with the greater exchange of air there may be an excessive loss of carbon dioxide from the blood, causing severe alkalosis. This may be partly avoided by shallow respiration, so that gas exchange is not so great. In addition the dog also seems to be able to tolerate alkalosis better than many mammals, and this may be true of the other canids. Large serous-type glands, found in the nasal

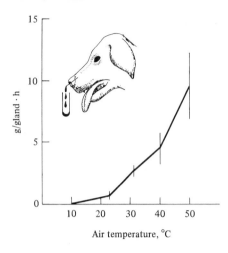

FIGURE 8-6 The lateral nasal gland in dogs as a source for evaporative cooling. Each point is an average weight of nine measurements from three dogs. Each vertical bar is twice the standard error at the mean. *(From Blatt, C. M., C. R. Taylor, M. H. Habal, Thermal panting in dogs: the lateral nasal, a source of water for evaporative cooling, Science* **177**: *804-805, 1 Sept. 1972. Copyright 1972 by the American Association for the Advancement of Science.)*

cavities of the dog, as well as in cats, sheep, and small antelopes, may be a major source of evaporative water during open-mouth thermal panting (Fig. 8–6). In this type of panting, most of the air enters the nose and leaves the mouth. Blatt, Taylor, and Habal (1972) found that (1) the secretion from this gland increased with increasing thermal panting, (2) about 30 percent of the increase in respiratory evaporation between 30 and 50°C could be accounted for by secretions from both glands, and (3) desiccation of nasal mucosa during thermal panting was alleviated because the opening of the gland was anterior to the turbinates (air came in through the nose).

There is a lack of hair in the axilla and the groin of many mammals. These thermal windows can be "opened" or "closed" behaviorally by stretching out or curling up. Dog owners regularly see their dogs stretch out on a hot day and curl up on a cold day. Foxes, coyotes, and wolves all regulate temperature in this way. Latitudinal and seasonal changes in thickness and quality of fur increase or decrease its insulative effect; these pelage changes are widespread among carnivores. Carnivores eat food with a high water content (about 70 percent), and therefore water is more readily available for losing body heat.

Ungulates Among the ungulates are the mammals which must survive the greatest temperature variation—not just at either extreme, but at both extremes. Musk oxen *(Ovibos moschatus)* are exposed to extreme cold, rarely experiencing temperatures over 30°C. But the majority of ungulates live on plains where shade is scarce, and escape from the searing sun is impossible. During the seasons of highest temperature many large mammals have sleek, glossy, and light-colored pelages which reduce solar heating. The pelage of mammals, usually thought of as a protection against cold, can thus protect against heat also.

The ability of the camel *(Camelus)* to get along without water in the hottest climates of the world is legendary. Cooling in the camel is accomplished with a minimum of water. The camel's daily temperature may vary from 34°C in the morning to 41°C in the evening without any harmful effect. By allowing the temperature to rise during the day, water used to keep down the temperature is saved. The excess heat can then be dissipated to the cooler and more humid environment which exists in the evening and night. The higher body temperature during the day also reduces the heat flow from the warm external environment to the body. The concentration of fat in the humps rather than in a subcutaneous layer is also of value in heat dissipation. The camel's fur is thick enough to give some insulative effect from the hot desert air, but not so thick as to interfere with the evaporative cooling effect of sweating, although sweating is very limited in the camel.

The water buffalo *(Bubalus bubalis)* is believed to have very few sweat glands, and little detailed information on heat and water balance in this mammal is available. This makes Cockrill's (1967) discussion on water buffalo all the more interesting. He wrote:

> In hot climates swamp buffalos must have almost unlimited access to water. Buffalos are not noticeably tolerant of heat, and they can suffer extreme discomfort if they are exposed for any length of time to the direct rays of the sun. They need to wallow [in water] uninterruptedly during the heat of the day. Few animals convey an impression of such blissful contentment as a swamp buffalo immersed to the nostrils in a mud wallow or standing ecstatically in a downpour of tropical rain.

Decreased (or increased) thermal conductance in mammals can be accomplished by decreasing the effectiveness of the insulating shell (fur or blubber), as well as its thermal conductance. This is accomplished on a seasonal basis; many ungulates have a much heavier coat in winter than in summer. Musk oxen *(Ovibos moschatus),* living in the far north, have a very heavy coat covering the entire body. In the bison *(Bison bison),* living in the temperate zones, the heaviest coat is restricted to the forepart, where the heart, lungs, and other core organs are located. Many of the large mammals can control the angle of their pelage so that it will either lose or conserve heat. Lack of hair, or at least a sparseness of hair, occurs in the axilla, the groin, on the scrotum, and on the mammary glands of many mammals, as already mentioned for the carnivores. In bovids, additional temperature sensors occur on the scrotum and udder. These initiate general evaporative cooling before an increase in ambient temperature has had the opportunity of increasing the core temperature of the individual. This would be similar to external thermostats on a house, which prevent a time lag in heating a house as the outside temperature is decreasing.

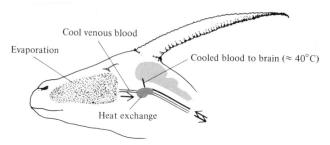

Cool venous blood

Evaporation

Cooled blood to brain ($\approx 40°C$)

Heat exchange

FIGURE 8–7 Countercurrent cooling for the brain of a gazelle (*Gazella* sp.). *(After Taylor, 1972.)*

That bovid horns serve as thermoregulatory organs may come as a surprise. Horns continue growing throughout the life of the individual, and a rich vascular plexus is necessary for this growth. Bone is carried away from the inner surface and is added onto the outer surface. This requires a rich blood supply. Taylor (1966) showed that in the horns of domestic goats *(Capra)* vasodilation takes place during exercise, and vasoconstriction when the animal is resting in cold weather. Barbary sheep *(Ammotragus lervia)* rams have been observed to cover their horns with moist sand or mud on a hot day (Ogren, 1965). Geist (1971) believed that the horns of mountain sheep *(Ovis canadensis)* also function to a degree in thermoregulation.

As is the camel, some antelopes which live in the desert and savanna areas of Africa are able to withstand an exceptional heat load. One of these, the oryx *(Oryx beisa)*, can withstand a body temperature of over 45°C for 8 h without injury (Taylor, 1969). Circulation specializations in the oryx and eland *(Taurotragus oryx)* protect the brain from heat damage. The external carotid artery (there is no internal carotid in ungulates) carrying blood to the brain divides into many branches in the cavernous sinus beneath the brain. These branches are in close proximity to veins carrying relatively cool blood returning from the nasal passages. Evaporative cooling of the mucosa of the nasal passages cools this blood supply. Temperature measurements have shown that in the gazelle (*Gazella* spp.), another of these African antelopes, the brain may be as much as 2.9°C cooler than the arterial blood leaving the heart (Taylor, 1969) (Fig. 8–7). Additionally, a decreased metabolism at high temperatures has been shown to contribute to heat tolerance in the oryx. The pale-colored pelage of these antelope also helps to reflect heat.

OTHER REGULATING MECHANISMS FOR MAINTAINING INTERNAL HOMEOSTASIS

This section will deal with two special questions: (1) How is it that some mammals, such as seals and whales, are able to stay under water so long? (2)

What happens to prevent the giraffe *(Giraffa)*, the world's tallest mammal, from figuratively "blowing its brains," even though while the animal is grazing, the brain may be 420 cm (10 to 12 ft) above the heart, and while it is drinking, the brain may be the same distance below the heart?

Adjustments during Diving

Muskrats *(Ondatra zibethica)* and beaver *(Castor canadensis)* can stay under water for 12 to 15 min; a seal, for 21 min or more; and whales, for an hour or two. Diving mammals do not generally have greater lung capacities than their nondiving counterparts. As a matter of fact, just the opposite is true in whales, which have about 50 percent of lung volume per unit weight compared to that of other mammals. It has even been suggested that because their lungs are not firmly attached to their thoracic walls, the lungs of whales may collapse during diving. Many modifications of whales for deep diving also probably greatly alleviate the problem of the "bends." A large proportion of the ribs lack any attachment, the lungs are dorsally situated above an oblique diaphragm, and the trachea are short and of large diameter. Whales sometimes dive to depths of over 915 m (3000 ft). Most diving mammals exhale before or shortly after diving, but this may be a buoyancy control rather than an oxygen storage maneuver. Although seals are not as completely adapted to marine life as are cetaceans, both groups have similar responses to prolonged submersion. In beaver *(Castor canadensis)*, muskrat *(Ondatra zibethica)*, and other semiaquatic mammals, these responses include (1) slowing of the heart rate (bradycardia), sometimes to one-tenth normal, (2) a drop in metabolic rate and body temperature, (3) a switch to anaerobic (without oxygen) glycolysis, (4) tolerance to high levels of lactic acid and carbon dioxide, (5) peripheral vasoconstriction, but a normal blood flow to vital organs.

Diving bradycardia has been reported for the beaver *(Castor canadensis)*, dugong *(Dugong dugong)*, hippopotamus *(Hippopotamus amphibius)*, porpoises, and seals among the mammals, and in such diverse animals as the crocodile, ducks, penguins, and turtles. Bradycardia even occurs in human beings and other nondivers to a limited extent. California sea lions *(Zalophus californianus)* and harbor seals *(Phoca vitulina)* have been trained to immerse their heads on command. When they are submerged, their heartbeat slows from 80 to only five or six times a minute (Elsner, Franklin, Van Citters, and Kenney, 1966). In the gray seal *(Halichoerus grypus)*, the heart rate is reduced from 120 to 125 per minute to 20 to 40 per minute (Harrison, Ridgway, and Joyce, 1972). Observations have confirmed earlier conclusions that peripheral vasoconstriction, i.e., a reduced renal, splanchnic, and cutaneous blood flow, together with a selective redistribution of the circulation so that an adequate cerebral and coronary blood flow continues, is a major result of this bradycardia.

But peripheral vasoconstriction is not the entire story. There is also a tendency for natural divers to have a higher blood volume and sometimes a higher oxygen-carrying capacity than nondivers. Considerable oxygen can be stored in the myoglobin (the hemoglobin of muscle), which contains a single heme group. Myoglobin has a much greater oxygen affinity than does hemoglobin at low partial pressures of oxygen. Nearly 50 percent of the oxygen reserves of diving seals is in the form of oxymyoglobin. Whales have about two to nine times the amount of myoglobin terrestrial mammals do, and about twice as many erythrocytes per blood volume. Vascular structure allows blood to bypass certain muscle tissue. There is a shift from aerobic (with oxygen) metabolism to anaerobic (without oxygen) metabolism in the muscles and a resultant buildup of lactic acid in the muscle tissue when an individual is submerged. Decreased metabolism is characteristic during diving. Seals lose body temperature quickly while submerged, because of a decrease in heat production. After long submergence seals often shiver on emergence.

An unusual method for allowing seals to remain under ice was discovered by Harrison, Ridgway, and Joyce (1972) while they were studying the gray seal *(Halichoerus grypus)*. They were using a tank in which the water was covered with a piece of glass containing a "breathing hole" to simulate natural conditions. They found the seals would release a bubble of air 250 to 750 ml in size and later retrieve it. Since CO_2 diffuses rapidly in seawater, this is a useful way of "purifying" the air, thus allowing them to stay under the ice longer.

A question that normally comes to mind is, "Do diving mammals suffer from the 'bends' or 'caisson sickness'?" This is a problem which plagues men diving in pressurized contrivances. The problem is caused by nitrogen, a rather insoluble gas, being forced under increasing pressure, into the circulatory system, during descent. During the ascent, as pressure decreases, the nitrogen leaves the blood system as bubbles, resulting in decompression sickness ("bends"). The danger of bubble formation in natural divers is minimal, for natural divers take below only the nitrogen dissolved at atmospheric pressure, and that in the lungs. In a diving suit or caisson there is a continual supply of nitrogen from air being pumped to the diver under pressure. That whales and other natural divers suffer from the "bends" under extreme conditions is at present debatable, although the consensus seems to be that they do not. Scholander (1940) and Harrison and Tomlinson (1963) have also suggested that excess nitrogen may be absorbed into the emulsified mucus and oil foam contained in the air sinuses, which is expelled during exhalation (the blow).

These responses, characteristic in conditions of reduced oxygen, are represented in many vertebrate groups and in their extreme forms are apparently based on mechanisms with a long evolutionary history. Some or all of these mechanisms also operate to prevent asphyxia during birth and hibernation.

Adaptive Hypertension in the Giraffe

A most unusual case of adaptation to pressure is that of adaptive hypertension in the giraffe *(Giraffa)* (Fig. 8–8). The giraffe competes successfully with some of the largest herds of grazing and browsing mammals still left, by browsing the tops of acacias 5.5m (18 ft) or more above the ground. The cardiovascular adjustment necessary to supply the brain with the proper amount of blood at proper pressures is doubly complicated. While the animal is browsing, the brain may be 420 cm (10 to 12 ft) above the heart; during drinking, the brain may be the same distance below the heart. While the giraffe is changing from drinking to grazing, undue blood pressures on the brain are prevented by another unique adaptation.

During fund drives for heart research, the Heart Fund has sometimes used a picture of the head and neck of the giraffe, with this caption, "My neck might save your heart!" The ad went on to say, "High blood pressure causes

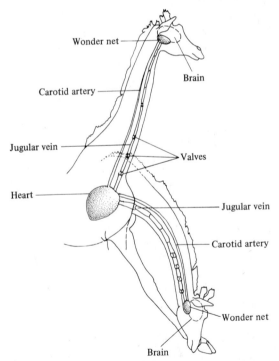

FIGURE 8–8 The giraffe (*Giraffa* sp.) may have the highest blood pressure in the world, to permit the blood to get from the heart to the brain, 4 m above the heart. Just beneath the brain is a spongy complex of arteries, which "dilutes" the pressure before it reaches the brain. The blood returns down the inch-wide jugular, past a series of valves, which narrow when the head is down, again reducing pressure to the brain. *(After Carr, 1967, and others.)*

stroke and contributes to heart attack in man. But giraffes aren't hurt by the sky-high pressure pushing blood up their ten feet of neck. Why?" The search for the answers makes fascinating reading, in addition to revealing the extent to which curiosity drives scientists (Goetz and Budtz-Olsen, 1955; Goetz and Keen, 1957; Goetz, Warren, Gauer, Patterson, Doyle, Keen, and McGregor, 1960; Van Citters, Kemper, and Franklin, 1966).

A giraffe may weigh over 907 kg (2000 lb), but it is a very sensitive animal and may faint and even die from shock when chased or captured. When struggling to escape it can easily break its long neck. Big game hunters when trying to capture giraffes alive, for zoos or other uses, followed the practice of immediately giving them Adrenalin to combat shock.

Goetz and Budtz-Olsen (1955) decided to try curare, or a derivative, to tranquilize their subjects. Dosage for an adult giraffe was calculated, correctly, to be 200 mg. But it was impossible to give so large a dose in liquid, so it would have to be administered in solid form. A suitable binding medium had to be found with which it could be attached to a bullet or an arrow. This binder had to be hygroscopic enough to dissolve rapidly in the tissues, yet solid enough not to disintegrate in flight. After many failures, the ideal material was found—icing sugar! Although Goetz and Budtz-Olsen (1955) first used .303 rifles, crossbows have been used more recently.

The giraffe's circulatory system is indeed huge. The heart of an adult animal examined by Goetz and Budtz-Olsen (1955) weighed 113.40 kg (25 lb), 40 times that of a human being, with walls up to 75 cm (3 in.) thick. The diameter of the jugular vein was over 25 cm (1 in.). It was equipped with a set of tricuspid valves which could withstand high pressures. Goetz and Budtz-Olsen felt that the size of the jugular vein was significant. They suggested that it acted as a reservoir when the head was down. The vein is collapsed when the giraffe is standing. The giraffe, like the llama *(Lama),* makes the oxygen exchange more efficient. The serum proteins and viscosity of the blood are the same as in humans. Since the giraffe has a high blood pressure, specimens were examined for arteriosclerosis. None was found.

The heartbeat of a resting animal (with a hood drawn over the head) was 100 per minute, and the respiratory rate was 15 per minute. The blood pressure in the carotid artery at the base of the brain with the head in an upright position was 200 mmHg. As the head was lowered it dropped to 175 mmHg. The hydrostatic pressure exerted at heart level by the column of blood in the carotid artery may exceed 250 mmHg. Van Citters, Kemper, and Franklin (1966) found that a temporary bradycardia occurred when the animal raised its head. Goetz and Keen (1957) found the carotid and vertebral arteries connected with a network, the *rete mirabile caroticum,* at the base of the brain, which "diluted" the blood pressure.

The massive heart sends the blood out at a pressure two or three times that found in human beings. By the time it has gone through the wonder net at the base of the brain, the pressure has been reduced so that no damage occurs to the brain. When the head is down the many arteries in the rete expand, and this temporarily reduces the pressure. The valves in the jugular vein close, holding the blood in the neck, so that it does not flow back into the brain. When the giraffe raises its head, the excess blood flows into the heart.

Chapter 9

Hibernation, Torpor, Estivation

The cost of homeothermy (warm-bloodedness) is great. The maintenance of a uniformly high body temperature in a cold environment requires sustained levels of food consumption. Many small to medium-sized mammals (and some birds) abandon homeothermy by daily or seasonal relaxation of thermal homeostasis; some animals are capable of such relaxation on both a daily and a seasonal basis. In these animals temperature change can range from only a few degrees of hypothermia, called *torpidity, seasonal lethargy,* or *partial hibernation,* to a pronounced drop in temperature, called *hibernation,* which is a seasonal period of dormancy in winter. When seasonal dormancy occurs in warm weather (with a less severe depression of core temperature) it is called *estivation.*

The most pronounced of these phenomena is hibernation (a translation of the German *Winterschlafen*). It is easier to characterize than to define. Some of the characteristics of hibernation are (1) body temperature reduced to within 1° or less of environmental (ambient) temperature, (2) oxygen consumption markedly reduced, (3) prolonged periods of apnea (cessation of breathing), (4) a torpor more pronounced than sleep, (5) arousal accompanied by activation of the major heat-producing mechanisms. There is an enormous literature on the physiology of hibernation, but we are as yet unable to fathom what triggers natural entry into or arousal from this state. Dawe and Spurrier (1969) reported that they were able to induce hibernation in thirteen-lined ground squirrels *(Spermophilus tridecemlineatus)* during spring and summer (March to August) by the transfusion of hibernation blood. Their results "indicate that a 'trigger' for natural mammalian hibernation in the ground squirrel is carried in the blood of the hibernating summer animals from the warm room causing them to hibernate after their introduction into the cold."

Their data have been questioned, (1) because they might not have correctly judged the onset of hibernation in the individuals used, if they had not been transfused, and (2) other researchers have been unable to repeat their results with *Spermophilus lateralis* or *Spermophilus tridecemlineatus* (Pengelley and Asmundson, 1972). In spite of this, further research may be very interesting.

Hibernation patterns are complex, and because they occur in a great variety of animals under varying circumstances, they must be polyphyletic in origin. The cardiovascular, metabolic, and thermoregulatory responses which allow hibernation are strikingly similar to those allowing deep or long submersion in diving mammals. These are (1) increased oxygen storage in the myoglobin, (2) bradycardia, (3) increase of lactic acid storage in the muscle (a partial switch to anaerobic metabolism), (4) reduced heat production, and (5) peripheral vasoconstriction.

Point four requires further elucidation in connection with hibernation. Hibernation in mammals is not a return to poikilothermy but rather a "resetting of the thermostat." The reasoning behind this will become more apparent as the discussion proceeds. If we accept as a part of the definition of hibernation a pronounced drop of the animal's temperature to near the ambient temperature, then such mammals as bears (*Ursus* spp.), badgers *(Taxidea taxus),* raccoons *(Procyon lotor),* and skunks (*Mephitis* spp.) cannot be called hibernators. Although they become lethargic for extended periods, their temperatures do not drop dramatically. The bear *(Ursus americanus),* because of its size, is unable to hide completely away from any disturbance, and though its lethargy allows some saving in energy, it does not leave the bear unable to take defensive action. That the bear's physiologic processes are quite well regulated, even in lethargy, is shown by the fact that the tiny young (weighing less than a pound) are born during this period and birth represents an exquisitely precise coordination of hormonal activities. The type of dormancy that occurs in bears has been variously called "seasonal lethargy," "ecologic hibernation," or "carnivorean lethargy" (Hock, 1960), as contrasted with "physiologic hibernation."

A characteristic of hibernators is that the major activities of their lives are telescoped into a relatively short spring-summer period of activity. Many biologic and physiologic processes are intensified. The level of oxygen consumption is higher in hibernators than in nonhibernators of the same size.

Preparation for reproduction occurs during hibernation in the male gonads, at least in the woodchuck, *Marmota monax* (Rasmussen, 1917) and in some ground squirrels (Shaw, 1926; McKeever, 1966). Hibernating mammals are usually said to have one litter per year, although this varies. In the northern part of its range the thirteen-lined ground squirrel *(Spermophilus tridecemlineatus)* has only one litter a year, while in Texas it may have two or three litters per year (McCarley, 1966).

TAXONOMIC DISTRIBUTION OF HIBERNATING MAMMALS

The orders Cetacea, Edentata, Carnivora, Tubulidentata, Lagomorpha, Perissodactyla, and Artiodactyla contain no hibernating representatives. In the order Monotremata, the echidna *(Tachyglossus aculeatus)* is a hibernator (Griffiths, 1965; Augee and Ealey, 1968). In the Marsupialia, the koala bear *(Phascolarctos cinereus)* is a hibernator, as are some of the smaller marsupials. Although Bourlière (1964) listed the North American opossum *(Didelphis marsupialis)* as a hibernator, this is doubtful, for McManus (1969) wrote, "No indication of large daily fluctuations in body temperature was noted, nor was torpor ever observed, although such phenomena have been reported in other didelphids." Opossums reaching the northern limit of their geographic range (in Minnesota, for instance) always seem to have their tail tips and ear tips frozen and are active throughout the winter, although they may become torpid for a short period of time.

Among the Insectivora the European hedgehog *(Erinaceus europaeus)* is a well-known hibernator; it has been used a great deal in hibernation research. Gould and Eisenberg (1966) have reviewed information on the tenrecs of Madagascar, although comparative data are still incomplete. *Echinops* has been reported as being lethargic during the day. *Centetes* and *Setifer* both hibernate. *Hemicentetes* may have two hibernating periods, one in June and one in December. In the laboratory Gould and Eisenberg (1966) found that individual *Hemicentetes* were irregularly torpid for several days or weeks; they made the interesting suggestion that "frequent periods of torpor may be an important adaptation to spermatogenesis." Their speculation was in part based on the fact that "the testes of all Tenrecinae are adjacent to the kidneys where body temperature is probably higher than in the pelvis."

Members of the order Chiroptera avoid the stress of intemperate climate in either of two ways, hibernation or migration, sometimes combining both. In addition, daily hypothermia occurs in some bats, when temperatures are low. Bats are tropical in origin, and many believe that hibernation is an adaptation of a tropical species which has expanded its range into temperate areas. Annual hibernation is found only in a few species of bats restricted to two families, the Vespertilionidae and the Rhinolophidae, which live in cooler climates. It is also in members of these same families that the reproductive peculiarities of copulation in the fall, storage of the sperm through the winter, and fertilization of the egg in late winter or early spring are found. This is an interesting evolutionary coincidence. I shall return to this discussion of daily torpor a little later. Among the primates the evidence is that some prosimians exhibit a mild hypothermia.

Bears *(Ursus americanus)* are absent from their usual range for 6 or 7 months during the winter, and were included as hibernators by the early naturalists. Observations and physiologic measurements have been made on denned bears in the last 20 or 25 years. Though there is little body temperature reduction, there is a drop of 75 percent in sleeping heart rate from 40 to 90 beats per minute in summer to 10 beats per minute in winter. However, the bear can have a relatively high heart rate for at least 30 min on most days, in contrast to hibernating rodents, whose heartbeat remains depressed for 4 to 28 days (Folk, 1967). Aldous (1937), Morse (1937), and Matson (1946) discussed dormancy in black bears and described denning sites, which evidently are not very elaborate. Bears apparently change their sleeping positions almost daily.

In the far north the presence of the polar bear *(Ursus maritimus)* on the ice floes throughout the winter has led to the belief that only the females are lethargic in winter, and that the males are active throughout the year, but this is not a true picture. Harington (1968) wrote, "Female polar bears with cubs over a year old and immature bears sometimes den between October and January. Adult males are occasionally found in natural shelters from late August to the end of September, but some den from October to January (rarely in March.)" In a footnote, Harington (1968) wrote, "There has been some confusion about whether adult male polar bears den. It has been definitely established that they do, although many are active throughout the year."

The greatest number of hibernators is found among rodents living in the Northern Hemisphere. The lack of hibernators in the Southern Hemisphere can easily be understood when one compares the land area south of latitude 40° in the Southern Hemisphere with that of the land area north of latitude 40° in the Northern Hemisphere. The woodchuck *(Marmota monax)* and the thirteen-lined ground squirrel *(Spermophilus tridecemlineatus)* (Fig. 9–1) have received the greatest attention, but the list of species studied would be a long one. In fact so much work has been done on rodents that Cade (1964) explored the possibility of interpreting the physiologic findings concerning torpidity within the framework of phylogenetic and evolutionary concepts based on paleontology and systematics, especially that of Wood's (1959) relationships. Cade (1964) speculated, "If so many morphological similarities among rodents are the products of parallel evolution, then it is reasonable to suppose that many physiological similarities are also the result of the independent development of adaptations for the same way of life."

Figure 9–1 Hibernation of the thirteen-lined ground squirrel *(Spermophilus tridecem-lineatus)* has been intensively studied.

THE PHYSIOLOGY OF HIBERNATION

Entrance into Hibernation

The first major change in the physiology of a mammal entering hibernation is a decrease in heart rate; this is followed by a drop in oxygen consumption and finally by a drop in the body temperature. Strumwasser, Gilliam, and Smith (1964) wrote, "body temperature is the clearest physiological index of the state of hibernation."

In the woodchuck *(Marmota monax),* the process of entering hibernation lasts about 9 h; though the slope shows an overall gradual downward trend, there are temporary peaks and valleys (Lyman, 1958). The heart rate decreases from over 100 times per minute to as low as 15 times per minute. The body temperature drops from about 35 to about 1°C, and the oxygen consumption is reduced from 1500 ml oxygen per kg/h to less than 100 ml oxygen per kg/h. In an active hedgehog *(Erinaceus europaeus)* the heart rate has been measured as 210 times per minute and sleeping values have been measured as 147 times per minute (Folk and Hedge, 1964). Although the heart rate of a hedgehog diminished greatly when it entered hibernation, the heart rate can be very erratic in hibernation, as shown by Kristofferson and Saivio (1964). In the thirteen-lined ground squirrel *(Spermophilus tridecemlineatus)* the heartbeat can drop from 200 to 400 beats per minute in nonhibernating individuals to five beats per minute in the hibernating individual.

Contrary to what one would expect, the feet of hibernating mammals are usually very pink, denoting a plentiful supply of oxygen. This pink color results from a high blood pressure, which in turn results from constriction of the peripheral blood vessels. In the diastolic phase when the blood is forced through the arteries, the blood encounters resistance. It has been interpreted as vasoconstriction while hibernating, but during arousal vasodilation occurs.

Although respiration is reduced during hibernation, the patterns vary. The most unusual pattern recorded is that of the hedgehog. At 4.7°C the animal did not breathe for as long as 56 min and then had a breathing period of about 4 min—a typical Cheyne-Stokes respiratory pattern. In the dormant hamster there may be three or four respirations per minute and then a period of apnea lasting 2 min or longer. In the birch mouse *(Sicista betulina),* a European member of the family Zapodidae, or jumping mice, the respiration and heartbeat may be synchronized at 7°C (Johansen and Krog, 1959). The range for active birch mice is 550 to 600 beats per minute.

There are many indications that some temperature control exists during hibernation. Hibernation should be thought of as a "resetting of the thermostat," rather than as a "turning off of the thermostat." A sudden drop in low temperatures usually results in awakening (Mayer, 1953). Twelve individuals of the arctic ground squirrel *(Spermophilus)* awoke when their core temperature went down to 0°C. At −15°C, a temperature never recorded in their natural burrows, seven individuals either awoke or froze. If the individual were "cold-blooded" it could not arouse itself. Heat regulation at low temperatures has been demonstrated in many mammals, among them bats (Kayser, 1939) and the birch mouse, *Sicista betulina* (Johansen and Krog, 1959). Generally during hibernation, the mammal is regulating at a body temperature of 2 to 5°C. This homeostasis is very poorly understood.

In addition to a reduced oxygen consumption during hibernation, Burlington and Wiebers (1965) have postulated that "anaerobically produced energy is associated with the increased hypoxic tolerance exhibited by infant mammals and adult hibernating mammals."

Hibernation in small mammals is a more drastic change than in the medium-sized ones such as the European hedgehog *(Erinaceus europaeus),* Arctic ground squirrel *(Spermophilus undulatus),* and woodchuck *(Marmota monax).* Of the small mammals, the jumping mouse *(Zapus hudsonius)* has been studied most intensively. It belongs to the family Zapodidae. Two other genera are included in this family; *Napaeozapus* has been studied by Sheldon

(1934 and 1938) and by Hamilton (1935); *Sicista* by Johansen and Krog (1959). *Zapus* has been studied by Morrison (1962), Sheldon (1934 and 1938), and Quimby (1951).

Morrison (1962) found the average body temperature for active jumping mice *(Zapus hudsonius)* was 37.3°C, slightly higher than that for larger species of mammals. Prior to entry into hibernation, jumping mice, like most rodents, gain weight. Quimby (1951) found that the hibernating group averaged 10.38 g more than the nonhibernating group (26.44 g as against 16.06 g) at entry into hibernation. He wrote: "Regardless of other conditions, these data indicate that the mice exhibiting marked weight gains early in the season entered into hibernation before those that did not." Specimens taken in the late fall (after October 16 in Minnesota) were either subadults or adults that were not fat. Others have reported similar conditions for other species: Wade (1930) and Johnson (1931) for ground squirrels; Hamilton (1934) for the woodchuck; and Sheldon (1938) for the woodland jumping mouse *(Napaeozapus insignis)*.

With all the experimental work done on hibernation, it is as yet impossible to induce hibernation experimentally, except perhaps by injecting blood of an already hibernating individual. Hibernation is probably induced by a combination of many factors—cold weather, food supply, increased fat, increasing serum magnesium levels, and a change in the respiratory control center, reducing sensitivity to carbon dioxide. In trying to supply seminatural surroundings in which individuals would naturally enter hibernation, at least two conditions must be met—a cold environment and extreme quiet.

The State of Hibernation

Hibernation is not a continuously torpid state lasting from sometime in the fall to sometime in the spring. Wade (1950), who worked on thirteen-lined ground squirrels *(Spermophilus tridecemlineatus)* and Franklin's ground squirrels *(Spermophilus franklini)* in Nebraska, reported: "Extensive observations in this region on hibernating ground squirrels under controlled conditions, and in nature, show that they will, as a rule, awaken from deep sleep at irregular and sometimes frequent intervals and move about." Mayer (1953) found that arctic ground squirrels awakened at intervals varying from 3 to 18 days. Since then this pattern has been recorded for many species of ground squirrels.

The eastern chipmunk *(Tamias striatus)* is primarily terrestrial, only partially arboreal. Panuska (1959) found a full spectrum of cold-weather activity in individuals of this species, who ranged from fully active to semitorpid to torpid. The lack of deep (marmotive) hibernation is compensated for by short periods of torpidity supplemented with the hoarding of food. Black (1963) believed that chipmunks are ecologically intermediate between tree squirrels and ground squirrels.

Yellowish or white fats are present in all normal, healthy mammals. In addition there is a "brown" fat which has been found most abundantly in

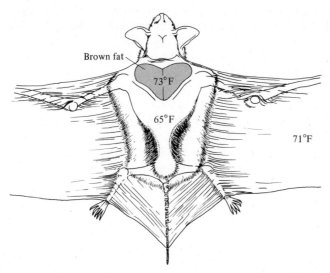

Figure 9–2 Temperature of intrascapular brown fat compared with temperature of other areas of body during arousal from hibernation in brown bat *(Eptesicus fuscus)* as shown by thermogram. *(After Hayward and Lyman, 1967.)*

mammals which hibernate, but small amounts are found in man and in laboratory rats and mice (Fig. 9–2). Oxygen consumption is higher in brown fat than in yellowish fat. Brown fat attains maximum size and weight just prior to hibernation, remains a large and dark gland during hibernation, and becomes small and pale just prior to arousal. Smith (1964) believed that brown fat tissue was strongly thermogenic and served to protect the animal by contributing heat to the cervical and thoracic region of the spinal cord. Smalley and Dryer (1966) have proposed a similar thermogenic function of brown fat in bats. Hayward and Lyman (1967) believed that the hibernating big brown bat *(Eptesicus fuscus)* had the highest capability for nonshivering thermogenesis of any species examined up to that time. They cautiously suggested that brown fat may contribute importantly to the production of heat for the process of arousal from hibernation. Brown fat apparently plays an important part in arousal from hibernation.

The role that endocrine glands play in hibernation is poorly understood. Hoffman (1964) has reviewed the information and interpretations. Usually morphologic and physiologic involution of the endocrine glands accompanies hibernation, but there cannot be a complete suppression of endocrine activity. There is a decline in such activity during the summer. The reproductive system becomes inactive with the onset of hibernation, but later in the winter hibernation continues even with increased activity of the endocrine organs, especially the pituitary. Thirteen-lined ground squirrels *(Spermophilus tridecemlineatus)* are producing sperm when they come out of hibernation.

It has been suggested that the different phases of entry into hibernation represent the stages of the general adaptation syndrome, or GAS (exhaustion of the body's adaptive abilities) and that the status of the adrenal glands during hibernation represents the exhaustion phase of the syndrome. Some investigators interpret the stages as resulting simply from pituitary activity. Basically there are two opposing views: (1) that hibernation is similar to GAS, and (2) that hibernation is entirely different from GAS.

It is obviously impossible to include even a brief summary of the voluminous literature appearing on hypothermia, or even to mention the many physiologic changes and processes that occur during hibernation. Rather it was the author's intent to alert the reader to the flexibility and variability of hibernation and hibernation studies and to suggest the enormousness of the literature.

Arousal from Hibernation

The time required for arousal differs among species, the larger species requiring more time than smaller ones. It varies from slightly over 30 min in some bats to several hours in the marmot *(Marmota caligata)* and around 20 h in the echidna *(Tachyglossus aculeatus).* The internal changes that trigger arousal (which is not the same as emergence above ground) have not yet been estimated, although several are suspected. One stimulus that has been suspected is the accumulation of metabolic end products, especially nitrogenous wastes. Although many hibernating mammals do empty their bladders when aroused, the golden-mantled ground squirrel *(Spermophilus lateralis)* apparently does not (Pengelley and Fisher, 1961). Fisher (1964), working with three species of ground squirrels and the common dormouse *(Glis glis),* found that "during these arousals individuals might defecate, or eat or drink, sparingly. There was, however, no regularity or consistency about these activities. On the other hand each animal was invariably found to urinate during an arousal." Fisher (1964) presented evidence which suggested that urine is not formed during torpidity and that the nitrogenous end products of metabolism accumulated in the blood between arousals, but cautioned, "It is not yet clear however that it is the gradually increasing level of NPN (non-protein nitrogen) in the blood which brings about an arousal."

Strumwasser, Gilliam, and Smith (1964), discussing the temporary arousals which occur during hibernation, wrote, "We have postulated that the animals arouse so as to reconstitute the storage of some chemical templates occurring particularly within neurons, which represent the storage of some orderly sequence of instructions (innate and learned)." Animals in deep hibernation are essentially somewhat denervated. Muscles of animals in deep hibernation are more sensitive to acetylcholine than muscles of nonhibernating mammals.

Another mechanism for arousal has been suggested by Baumber, South, Ferren, and Zatzman (1971). They found that in hibernating yellow-bellied marmots *(Marmota flaviventris)* metabolic imbalance occurred when there was a shutdown in carbohydrate mobilization and utilization. This was reflected in a rise in ketone bodies. Physiologic balance would be restored by an increased oxidative metabolism and glycogenolysis, a pattern which is typical during arousal.

The energy cost of arousal is great. Kayser (1965) has estimated that arousal from hibernation takes as much energy as is required for a 10-day period of hibernation. During arousal from hibernation, both nonshivering and shivering types of thermogenesis are used. Mechanisms for heat production are maximally activated, and the heat produced is concentrated in the anterior part of the body, serving especially the heart and the central nervous system. It is believed that this anterior concentration of heat results from the localization of the deposits of brown fat and the cardiovascular shunting.

Mayer (1960) described the arousal of the arctic ground squirrel under laboratory conditions. In the first 1½ h the temperature rose from 4.0 to 17.5°C and then from 17.5 to 32.0°C in the next 1½-h period. At hibernating temperatures there were three respirations per minute. These increased but at 11.0°C became obscured by the other chest movements. At 12.0°C shivering motions set in, first in the anterior part of the body and later in the posterior part. At 16.0°C the animal tried to right itself. At 17.5°C shivering stopped and the animal moved its tail. At 24.0°C the eyes were open and the animal sat up. At 25.4° the animal was essentially warm-blooded.

Wade (1950), working on the ground squirrels *(Spermophilus tridecemlineatus)* in Nebraska, noted that when they went into quarters prepared for hibernation they effectively closed any tunnels leading to these quarters from above ground with a considerable amount of earth firmly packed. He found that of the first emergents, most were males.

In the meadow jumping mice studied by Quimby (1951) the first emergents were also males, but the jumping mice emerged much later than other hibernators. Emergence dates for these mammals in south central Minnesota were during late April and early May. In this connection Quimby's (1951) comments are pertinent: "Other hibernating rodents were observed to be active long before the jumping mouse."

Beer and Richards (1956), working with bats, pointed out that fat reserves might determine the northern geographic distribution of the big brown bat *(Eptesicus fuscus)*. They did their work near St. Paul, Minnesota, just within the northernmost locality for the big brown bat in the central portion of North America. They suggested winter depletion of the fat reserves as one reason for this northern distributional limit of the big brown bat. In individuals of this species, hibernating in caves near St. Paul, Minnesota, fats amounted to 30 percent of the total weight at the beginning of hibernation and to about 10 percent near the end of hibernation. They calculated that the bats could survive an average of 194 days on their fat reserves. The frost period in this vicinity averages 195 days. The authors wrote, "The winters in this region are of approximately maximum duration for successful hibernation of big brown bats, and individuals of lighter than average weight would be expected to succumb unless the winter were shorter than average."

It has become obvious in this discussion that hibernation is an annual rhythm, probably programmed within the organism at birth and triggered by external conditions. Generally the pattern is one of progressively lengthening hibernating periods in the fall, a longer period during the winter ("deep hibernation" of some authors), and decreasing duration during the spring.

DAILY TORPOR

The phenomenon of daily torpor exists in insectivores, bats, and some of the smaller rodents. Herter (1962) reported that one *Echinops,* an insectivore from Madagascar, studied in his laboratory was active at night and torpid or lethargic during the day. Streaked tenrecs *(Hemicentetes),* also from Madagascar, were found to be irregularly torpid for several days or weeks in the laboratory (Gould and Eisenberg, 1966). As mentioned earlier, these authors suggested that frequent periods of dormancy may be an important adaptation to spermatogenesis, since in all the Tenrecinae (to which *Echinops* and *Hemicentetes* belong) the testes are adjacent to the kidneys, where body temperature is probably higher than in the pelvis.

In some small mammals daily torpor intergrades with seasonal dormancy. The birch mouse *(Sicista betulina)* undergoes periods of daily torpidity in spring, summer, and fall, then hibernates in winter.

Daily torpor seems more common in the bats (order Chiroptera) than in other groups of animals. Bats are a special group in which a wide range of hypothermia is exhibited. The order Chiroptera is divided into two suborders, the Megachiroptera and Microchiroptera. There seems to be little doubt that at least the larger Megachiropterans are true homeotherms (Burbank and Young, 1934; Morrison, 1959; and Bartholomew, Leitner, and Nelson, 1964). The Microchiroptera are generally regarded as having the poorest type of mammalian heat regulation. Reeder and Cowles (1951) and Kayser (1939), indeed, believed that *Myotis velifer* behaves like a poikilothermic animal.

But thermal instability is not uniform in all the Microchiropterans, and ranging all the way from the extreme instability of *Myotis velifer* to the stability exhibited by such species as *Desmodus* (the vampire bats). Reeder and Cowles (1951) obtained evidence that the leaf-nosed bat, *Macrotus californicus,* a species which remains active in the general vicinity of the Riverside Mountains of California the year round, is better able to regulate its internal temperature in the cold than is *Myotis velifer,* which migrates southward in the winter.

Many bats show daily torpor in summer and a prolonged torpor in winter. Hock (1951) stated that no physiologic distinction exists between the daily lethargy in summer bats and the lethargy of bats in hibernation. He further believed that bats differ from other heterotherms in that their body temperature when at rest at any season always approximates the ambient temperature. The western mastiff bat *(Eumops perotis)*, whose range extends along the southern coast of California and on both sides of the Arizona, New Mexico, and western two-thirds of the Texas border, shows daily torpor only during the winter.

Several adaptations to cold weather have been suggested in one species of bat, the Brazilian (Mexican) free-tailed bat *(Tadarida mexicana)*, which may migrate or hibernate, according to climatic conditions. Benson (1947), working in California, found that in *Tadarida* the body temperature approached that of the surroundings at any time of year. Twente (1956), studying the same species in south central Kansas and Woods County, Oklahoma, wrote: "Comparative fall and spring weights give an indication that *Tadarida* migrate southward to winter feeding grounds rather than hibernating throughout the winter in temperate regions without feeding."

The little brown bat, *Myotis lucifugus,* is a hibernating species but can maintain a high body temperature level, 29°C or above, when well fed and quiescent (Stones, 1964). There was some evidence that temperatures of pregnant and lactating females in summer were not so labile as in the others.

The great diversity of thermoregulation that is now known to occur in bats gave rise earlier to much controversy concerning the ability of bats to thermoregulate. The literature on thermoregulation in bats has been ably reviewed by Stones and Wiebers (1965) and by Henshaw (1970), among others. Henshaw (1970), in his introduction to his review, wrote, "In fact, as much diversity exists in thermoregulatory patterns in the Order Chiroptera as exists in all the rest of the mammals. Some species maintain their body temperature with a precision of ±1°C, whether active or asleep; while other species can remain in deep torpidity with the body temperature near 0°C for several months without arousal."

Among the other small mammals which undergo torpor at any time of year are the California pocket mouse *(Perognathus californicus)* and the cactus mouse *(Peromyscus eremicus)*.

The California pocket mouse, a small 20-g rodent, enters into torpor at any time that the temperature drops below the critical level of 32.5°C, by reducing its metabolism to the basal level. Equilibrium between heat loss and heat production is reached at or slightly above ambient temperature, the animal cooling much as would an ectotherm. During torpor above 15°C the body temperature remains slightly above the ambient temperature. If the body temperature drops below 15°C, the individual cannot arouse from torpor. This species undergoes torpor at any time of the year that the daily food ration is reduced. The daily cycle of torpor is inversely related to food intake. During arousal the rates of temperature increase are greater than that allowed by minimum heat loss and maximum aerobic heat production. It has been suggested that the animal depends partly on anaerobic metabolism and acquires an oxygen debt (Tucker, 1965). The energy saving for even a short period of torpor in this mouse is evidently enough to be of value. If an individual entered torpor at 15°C and immediately awakened, it would take 2.9 h for the body temperature to return to normal. At normal temperature the oxygen consumption would have been 11.9 ml of oxygen per gram. This is a reduction of oxygen consumption of 45 percent. A similar cycle exists in a desert-dwelling white-footed mouse, the cactus mouse *(Peromyscus eremicus),* which requires either a high environmental humidity or water. Individuals of this species can be induced into torpor by a negative water balance, even though food is plentiful.

ESTIVATION

Seasonal dormancy in response to prolonged periods of heat or drought (including lack of food) is usually referred to as estivation. It has also been characterized (for mammals) as dormancy at 20°C or higher ambient temperatures.

Estivation is a summer dormancy; hibernation is a winter dormancy. Some species become lethargic when deprived of food and water, others become lethargic when deprived of food. Still others become torpid spontaneously, i.e., when food and water are present. Estivation is probably not a direct response to lack of water but an evolutionary response to aridity. As in hibernation, heart rate and metabolism are reduced during estivation. Mammals which estivate generally have unusually low metabolic rates. Estivation is a way of minimizing environmental stress, not only from a lack of food and water but also from high temperatures, although indirectly this also is concerned with lack of water. Kirmiz (1962) wrote, "The jerboa can easily tolerate high environmental temperatures up to 45°C by entering into a state of lethargy (deep sleep)." Folk (1966) has said that the little pocket mouse *(Perognathus longimembris)* provides a classical example of estivation. Many but not all hibernators estivate.

Estivation is present among the marsupials (i.e., the pygmy opossum, *Cercaetus nanus*) and insectivores, but is most common among the rodents, and has been reported for the Piute ground squirrel, *Spermophilus townsendii* (Alcorn, 1940); the Columbian ground squirrel, *Spermophilus columbianus* (Shaw, 1925); and the thirteen-lined ground squirrel, *Spermophilus tridecemlineatus* (Wade, 1930, and others). Among the small rodents, estivation in the presence of food and water has been reported in the pocket mouse, *Perognathus longimembris* (Bartholomew and Cade, 1957), and the kangaroo mouse, *Microdipodops pallidus* (Bartholomew and MacMillen, 1961).

ADVANTAGES OF ADAPTIVE HYPOTHERMIA

Animals without external reserves seem to in some way conserve internal reserves to survive in climatic extremes. This has allowed many species to become widespread. The ground squirrels, including both *Spermophilus* and *Ammospermophilus,* are circumpolar from the arctic to the tropics and show all gradations of hypothermia: (1) where winters are cold, they hibernate; (2) in arid regions they estivate; (3) in northern arid regions with little summer rainfall they both hibernate and estivate; and (4) in warmer areas they neither hibernate nor estivate. Hypothermia—from the slight daily lowering of the temperature in summertime in some bats, rodents, marsupials, and the spiny anteater, *Tachyglossus* (a monotreme), to the extreme seasonal lowering of the temperature exhibited by many kinds of rodents and bats—is a saving of energy. Daily torpor and seasonal hibernation are physiologically very similar. Estivation differs from hibernation in that temperatures, heart rate, and respiration are not lowered as drastically in the former. It is also possible that dormancy may prolong the life of hibernators. The 24-year longevity record for *Myotis lucifugus* has been cited as evidence (Griffin and Hitchcock, 1965).

The immense effort spent on the study of hibernation has been stimulated in part by the hope that the unusual tolerance of hibernating mammals to low body temperatures may some day be transferable to nonhibernating mammals, including man. This would allow prolonged clinical hypothermia for surgery and even for space exploration. What makes this seem to come within the realm of possibility is the remarkable similarity of newborn animals, including humans, to hibernating mammals. Adolph (1951) showed that some newborn mammals survive low body temperatures that cannot be tolerated by the adults. The tolerance is equal to that of hibernation. Records of accidental cooling of premature human babies suggest a remarkable tolerance to low temperatures.

Though all the patterns of hypothermia differ in details, the conservation of energy during environmental extremes is the factor of adaptive significance shared by all.

Reproduction

Reproduction makes possible the continuity of the race. The characteristics of a species are preserved in its reproductive organization. In sexual reproduction the characteristics of two organisms can recombine in an infinite variety, providing a diversity of forms upon which natural selection acts, thus producing long-term evolutionary adjustments. These are the homeostatic functions of reproduction. The production of gametes and other reproductive activities cannot be viewed as a discrete process separate from the other life processes of the living mammals. The reproductive process lends continuity to life, allows evolutionary experimentation, extends the geographic range of species, and permits the repopulation of former habitats, as well as the invasion of new habitats.

REPRODUCTIVE PERIODS

The females of all mammals, with the possible exception of some primates, show rhythmic changes in the intensity of the sex urge. Sexual reception at its height is called *estrus* (adjectival form, estrous). In this estrous cycle all stages are determined by interaction of female sex hormones and those of the anterior pituitary. In mammals the reproductive periods are to a great extent seasonal. Evolutionary advantage favors those individuals that reproduce when the environmental conditions are most favorable for the pregnant mother and for the newborn young. What is important is not so much the time of mating as the time of birth, which must take place in the season that provides optimum conditions for the survival and growth of the young. Many kinds of mammals that have no specific mating season (humans, some of the nonhuman primates, elephant, rhinoceros, giraffe) are defined as continuous breeders. Most mammals mate at specific times of the year and are defined as seasonal breeders. Mammals belonging to this classification may show one (monoestrous) or several (polyestrous) breeding seasons per year, separated by periods of complete sexual inactivity. During this inactive period the gonads of the female and sometimes of the male resemble the gonads of sexually immature individuals, and gametogenesis stops completely.

Although seasonal cycles are common in the tropics, they are more clearly seen in animals living in temperate and arctic regions. In these regions, the larger mammals generally have short breeding seasons, timed so that the young will be born during the spring months when temperatures are getting milder and when there is a plentiful supply of food for parent and young. A notable exception is the yak *(Poephagus grunniens)* of the high mountain country of Tibet, whose young are born in the fall.

Mammals with fixed mating seasons are restricted in range by the season in which the young are born. The ocelot *(Felis pardalis)* and bobcat *(Lynx rufus)* are about the same size and have a carnivorous diet. Bobcats are born in the spring. Ocelots are born in September and October (which is spring in the Southern Hemisphere) and are evidently able to retain the mating habits from their place of origin in the Southern Hemisphere. The range of bobcat extends into Canada, while the northern limit of the range of the ocelot is in Texas. Young ocelots could not survive the winters farther north.

Mammals with more flexible breeding seasons, such as the mountain lion *(Felis concolor)* and the white-tailed deer *(Odocoileus virginianus),* have a wider geographic range. The mountain lion at one time ranged from Alaska to the southern tip of South America and from the Atlantic to the Pacific. In the north, mountain lion cubs are born in the spring months; and in South America the time of birth is September and October. The modern peccary *(Dicotyles tajacu),* whose range barely extends into the Southern United States from the south, has a 6-month mating period. In the tropics the mating season can be independent of temperature, as, for instance, in the tapir *(Tapirus),* and the mating season is often correlated with the coming of the rainy season.

The events of reproduction are subordinated to the brain. A whole series of events must occur in the proper sequence if reproductive efficiency is to be attained. Courtship, maturation and shedding of gametes, copulation, fertilization, gestation, and parturition require not only proper timing but also synchronization between individuals. Reproduction responds, as do all physiologic processes, to external stimuli, none of which alone can explain the phenomena of seasonal breeding and reproductive rhythm. Among the more important of the external stimuli are light, temperature, nutrition, neural stimuli, and social factors.

Most mammals, especially the smaller ones, start to breed as the days get longer, but the larger ones, with longer gestation periods, begin to breed as the days get shorter. Generally speaking, temperature affects reproduction in mammals rather indirectly. Nutrition affects reproduction in many ways, and its effect is difficult to measure. Amount and quality of food directly influence attainment of puberty, length of breeding season, proportion of females undergoing estrous cycles, and litter size.

Pheromones (ectohormones) have been described as external agents (odors) or "carriers of excitation," in contrast to the internal agents (hormones), and usually induce responses within members of the same species. This odor can act as a stimulus on the central nervous system to affect profoundly the reproductive behavior of the individual and play an important role in the sexual integration of animals. This phenomenon is receiving the attention of more and more researchers, and is certainly a very productive avenue of research, because of its relation not only to sexual integration but also to communication in general. One of the more interesting social factors which has come to the fore recently, and in which pheromones play an important part, is the effect of high density in delaying puberty and depressing litter sizes in the house mouse *(Mus musculus)*. When small groups of four or five females are kept together, there is an increase in pseudopregnancy and estrus is delayed. In larger groups estrous cycles may become irregular or stop altogether. If males are introduced, the average length of estrous cycles returns to normal, and may even become synchronized (Whitten, 1966). Whitten also provided some evidence that the odor of the male was one of the immediate causes of the modification of the cycle. If the male to which a female has been mated is replaced by a strange male, she reverts to estrus—a phenomenon called the Bruce effect, after its discoverer. Whitten, Bronson, and Greenstein (1968) have shown quite conclusively that the pheromone from male mice is volatile and "that it acts almost certainly through the olfactory receptors." Richmond and Conaway (1969) question that high density exerts an "oestrus-suppressing" effect, and believe that in *Microtus* disturbance is as much of a factor as changes in relative density. Although there has been little investigation on humans, indirect observations have indicated that menstrual synchrony occurs in all-female living groups. McClintock (1971), investigating this synchrony at a dormitory in a women's college, found a statistically significant increase in synchrony among roommates and closest friends from September until April. The cause of this has not yet been investigated, and McClintock (1971) cautioned, "Whether the mechanism underlying this phenomenon [is] mediated by awareness or some other process is a question which still remains open for speculation and investigation." Olfactory stimuli play a large part in the male's detection of the estrous female in marsupials, in rodents, in the perissodactyls, and the artiodactyls.

ANATOMY OF THE REPRODUCTIVE SYSTEMS

The basic blueprint of the reproductive systems is similar in both sexes. In mammals the reproductive and excretory systems are often, because of their intimate connection, referred to as the urogenital system. This association is very striking in their early development and the use of common ducts.

The Male Reproductive System

The male reproductive system includes paired testes, paired accessory glands, a duct system, and a copulatory organ. The undifferentiated gonads of the early embryo develop into testes in males and ovaries in females.

Testes The testes develop in the vicinity of the kidneys near the primitive ridge. They have a twofold function, to produce the male sex hormones, such as androgen, testosterone, and others, and to produce sperm. Testes are composed for the most part (90 percent) of seminiferous tubules, in which the sperm are produced. When a sperm which has the haploid or n number of chromosomes, fertilizes the egg which also has the haploid number, the fertilized egg will have the diploid or $2n$ (complete) number of chromosomes for that species. Chromosomal numbers vary between species of mammals; for a discussion of this, Tobias (1956) and Hsu and Benirschke (1968) should be consulted. The random manner of division and segregation of chromosomes during maturation and fertilization provides for the variation in the characters that appear in the new generation.

In nearly all mammals the testes descend into a sac, the *scrotum*. Exceptions are the monotremes, some insectivores, the edentates, sirenia, all the seals (except seals of the families Otariidae and Odobenidae), whales, hyraxes, rhinoceroses, and elephants. In the rodents the testes descend through the inguinal canal into the scrotum only during the breeding season, and are abdominal at other times. In some bats they are abdominal, and in others scrotal.

One function of the scrotum seems to be to provide an environment for the testes which is 1 to 6°C cooler than the body temperature. The separation from the body assures a cooler environment than that of the body cavity, but a double muscle system in the scrotum draws the testes close to the body wall for warmth or lets them fall away for cooling. Another mechanism, equal in importance to the scrotum, for providing temperature regulation is the *pampiniform plexus,* a looped system of scrotal veins and arteries which lie on the surface of the epididymis and which follow the spermatic cord (Fig. 10–1). Arterial blood entering the plexus at the temperature of 39.0°C has cooled to

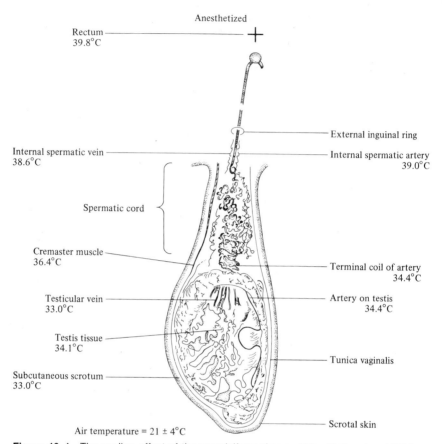

Figure 10–1 The cooling effect of the pampiniform plexus. *(After Nalbandov, 1969.)*

34.8°C by the time it enters the testes. This plexus probably plays a major role in cooling the testes in those mammals whose scrotum is tight and not variable and pendulous. It has long been known that domestic rams *(Ovis aries)* are subject to a summer sterility, which disappears with the coming of cooler fall weather. Temporary sterility has been experimentally produced by high temperature in laboratory animals. Darling (1937) has suggested that one of the reasons for the wallowing of red deer *(Cervus elaphus)* stags is to cool the testes. In hibernating mammals the body temperature approximates the ambient temperatures. The testes reach a peak of activity just before terminal arousal, and the males seem to be in breeding condition when they emerge from hibernation (McKeever, 1966, and others). Obviously, in mammals which retain the testes in the coelomic cavity, sperm are viable at higher temperatures, or other mechanisms, unknown at present, are in operation.

Sperm   The male gamete, or sperm, consists of a head, with a neck, a middle piece, and tail. A protoplasmic cap, the *galea caput,* a normal component of the head, is usually dissolved when sperm are strained with fat solvents. The neck, middle portion, and tail do not consist of a single piece, as generally pictured, but are composed of several strands of fibrils which are covered by a sheath. At the tip of the tail, the fibrils flare out into a brush. The shape of the head varies with the species. In some marsupials the spermatozoa are paired for a time (until fertilization), a phenomenon not yet understood. Though sperm survive a long time in the epididymis, their life span, with a few exceptions, is very short after ejaculation into the female reproductive tract. Usually this is only about 24 h, but there are variations, the most extreme occurring in some bats, where copulation occurs in the fall and the sperm are stored in a seminal receptacle until spring, when fertilization takes place. This phenomenon is called delayed fertilization.

Male Duct System and Accessory Glands   The duct system in the male includes the epididymis and the vas efferens, which empties into the larger vas deferens, which in turn empties into the seminal vesicle, the ejaculatory duct, and then the ureter. The accessory glands are the prostate (not always, but most often, a single structure), paired seminal vesicles, which are sometimes considered lobes of the prostate gland, and paired bulbourethral or Cowper's glands. There is considerable variation of these glands among species. In the cat and dog the prostate is well developed and the seminal vesicles are absent; in a domestic bull the seminal vesicles are enlarged and the prostate is small. All glands contribute seminal plasma to the sperm during ejaculation. The seminal plasma serves two important functions: it is an activating medium for previously nonmotile sperm cells, and it furnishes the sperm cells with essential nutrients, fructose, and electrolytes. The pH of fresh semen approximates 7. In rats and mice the semen coagulates as a result of the secretion of the coagulation gland, a lobe of the prostate. This results in the formation of a plug in the vagina of the female. The plug, which is often used as an indication that the female has mated, drops out about 24 h after mating. Vaginal plugs are also formed in baboons and in some squirrels.

The Penis   The penis consists of two to three cylindric bodies, the *corpora cavernosa penis* (Fig. 10–2). These spaces become filled with blood during sexual excitation. In some species the penis contains a bony structure, the *os penis* or baculum, whose structure and size have been used as taxonomic characters (Burt 1960; Hamilton, 1946; and many others) and as an age indicator (Elder, 1951; Sanderson, 1964; and others). The os penis appears to be the ossification of a corpus cavernosum and is found in most carnivores and aquatic mammals and also in a variety of other mammals. In some mammals the penis forms an S-shaped loop called the sigmoid flexure. The end of the penis is capped by the *glans penis,* which is variously shaped. For instance in the opossum *(Didelphis marsupialis)* the 2-cm terminus is bifid, as it is in the platypus *(Ornithorhynchus),* and in the pig *(Sus),* it is spiral or corkscrew-shaped.

The Female Reproductive System

The reproductive organs of the female consist of the paired ovaries and the duct system. The duct portion of the reproductive system receives the ovulated eggs and conveys them to the site of implantation, the uterus. The same duct portion also receives the sperm and conveys it to the site of fertilization, the oviduct.

Ovary The ovaries, paired in mammals, have a twofold function: (1) to produce female sex hormones, the estrogens; and (2) to produce the female gametes, or ova. The ovaries remain near the kidneys, where they first differentiated. The growth of the ovary is controlled by hormones from the pituitary gland. The size and shape of the ovaries vary with age, reproductive stage, and species. If the female has many young in each litter, she is called *polytocous* and the ovary is usually berry shaped. If one young is born at a time, the female is called *monotocous* and the ovary is ovoid. At ovulation the follicular wall ruptures, releasing the ovum into the coelom, where it is picked up by the oviduct. In monotremes the left ovary is larger than the right. The egg of the monotreme contains a large quantity of yolk, as do the eggs of birds and reptiles. In marsupials large numbers of ova are shed at one time. The female

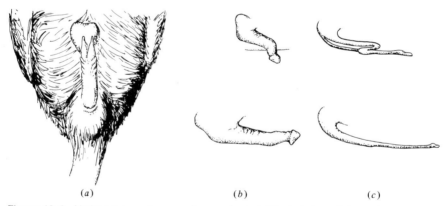

    (*a*)   (*b*)   (*c*)

Figure 10-2 Various types of mammalian penises. *(a)* Ventral view of the male external genitalia of the opossum *(Didelphis marsupialis).* The penis is caudal to the scrotum and its glans is bifurcate. The ureter opens between, not at the tips, of the bifid prong. A groove extends along the medial side of each part of the glans penis for more than half its length. *(b)* Diagrams of the vascular-muscular type penis of horses *(Equus caballus)* and *(c)* the fibroelastic type penis of cattle *(Bos taurus)* in the nonerect (above) and erect (below) positions. *[(a) After McGrady, 1938. (b) and (c) After Hafez, 1969.]*

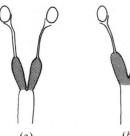

(a)   (b)   (c)   (d)

Figure 10–3   Several types of uteri (stippled) found in placental mammals, showing degrees of fusion of the two "horns" of the uterus. A duplex uterus *(a)* occurs in the orders Lagomorpha, Rodentia, Tubulidentata, and Hyracoidea; a bipartite uterus *(b)* is known in the order Cetacea; a bicornate uterus *(c)* is found in the order Insectivora, in some members of the orders Chiroptera and Primates, and in the orders Pholidota, Carnivora, Proboscidea, Sirenia, Perissodactyla, and Artiodactyla; a simplex uterus *(d)* is typical of some members of the orders Chiroptera and Primates and of the order Edentata. *(From Evolution of Chordate Structure: An Introduction to Comparative Anatomy by Hobart M. Smith. Copyright © 1960 by Holt, Rinehart and Winston, Inc. Reproduced by permission of Holt, Rinehart and Winston, Inc.)*

European mole *(Talpa europea)* is intersexual. During the nonbreeding season the medulla (the embryonic remnant of the testes) becomes very large and pushes the cortex to the anterior part of the gonad. During the breeding season the cortex becomes active and produces ova. For a review of this process, see Deanesly (1966). In many bats, only the left ovary is functional. Seals and whales ovulate from one ovary one year and from the other the following year. Chromosome reduction to the haploid or *n* number occurs in the female gamete as it does in the male gamete, although the process differs.

Female Duct System and External Genitalia   The female duct system consists of paired oviducts (or fallopian tubes), the uterus, where the embryo develops, including the uterine body and horns, the single cervix, which is a sphincter muscle between the uterus and the vagina, and the single vagina, which receives the male's penis. The external genitalia include the clitoris, the labia majora, and the labia minora. The clitoris is the embryologic homologue of the penis. The labia minora are present in only a few mammals.

Uterus   The uterus has thick muscular walls, many blood vessels, and a specialized lining, the *endometrium,* in which the developing embryo is embedded (Fig. 10–3).

ESTROUS CYCLE AND OVULATION

All female mammals, except the higher primates, permit copulation only during a definite period of the sexual cycle. The period of the proper physiologic state and psychologic receptivity which permits copulation is called the *estrus* (the adjectival form is spelled "estrous"). The time span from the beginning of one estrous period to the beginning of the next, during the reproductive period, is called the estrous cycle.

The physiologic processes that initiate and control reproduction are mediated by hormones. The hypothalamus produces about 10 releasing factors, only one of which is important in this discussion—the releasing factor for the production of follicle-stimulating and luteinizing hormones by the anterior pituitary. In biologic shorthand these are FSH-RH and LH-RH, secreted by the hypothalamic neurons which reach the anterior pituitary gland, or hypophysis. The pituitary is a composite organ attached by a stalk to the base of the brain. Its location, in a bony depression at the base of the brain, affords the greatest amount of protection that any location in the body can give. In the adult form, there are two main lobes, named simply the anterior and posterior lobes, which in most mammals are fused together. Although each part secretes a number of hormones, it is the anterior pituitary that is most important in our discussion.

The typical estrous cycle (Fig. 10–4) begins with the secretion of a gonadotropic hormone of the anterior pituitary, which stimulates an egg-containing follicle in the ovary to grow (Fig. 10–5). The hormone is appropriately called the follicle-stimulating hormone or FSH. In males the same hormone stimulates the germinal epithelium of the testes to produce spermatozoa.

Then the follicular cells secrete estradiol, which causes a thickening of the uterine lining (endometrium) but does not cause ovulation. This seems to be caused by a balance of gonadotropic hormones from the pituitary. Estradiol, while stimulating the growth of the follicle, inhibits the production of FSH, thus preventing the development of other follicles.

When the egg is mature, the enlarged follicle also matures and, bulging from the surface of the ovary, bursts, releasing the egg. Although it is often stated that the pressure of the follicular fluid causes the rupture, the pressure may actually decline slightly before ovulation, so the mechanism is not yet fully understood. The egg either passes into the oviduct, where it may be fertilized if sperm are present (this will be discussed later), or it goes into the uterus, where it will degenerate.

The development of the follicular cells is stimulated by several hormones from the anterior pituitary—the luteinizing hormone (LH) and the luteotropic hormone, originally labeled LTH, a symbol which permeates the literature. Later LTH was discovered to be the same hormone as prolactin, which acts on the mammary gland; it is now called PRL, sometimes PL. Under the influence of these two hormones the follicular cells fill the ruptured follicle with a yellow body, the *corpus luteum,* which in turn becomes an endocrine gland. The corpus luteum continues to produce estradiol in reduced quantities, and it produces the second female hormone, progesterone. This hormone

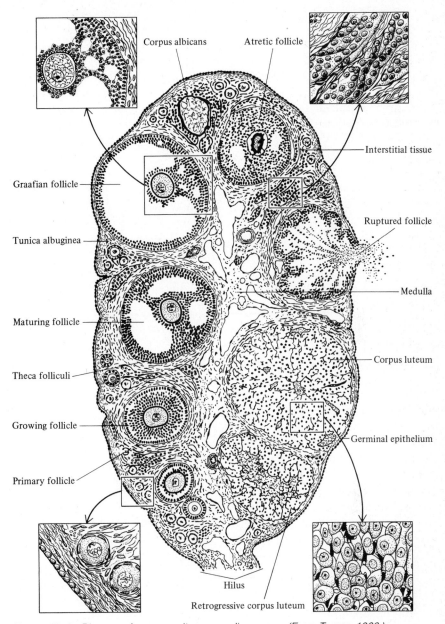

Corpus albicans

Atretic follicle

Interstitial tissue

Graafian follicle

Ruptured follicle

Tunica albuginea

Medulla

Maturing follicle

Corpus luteum

Theca folliculi

Growing follicle

Germinal epithelium

Primary follicle

Hilus

Retrogressive corpus luteum

Figure 10–4 Diagram of a composite mammalian ovary. *(From Turner, 1966.)*

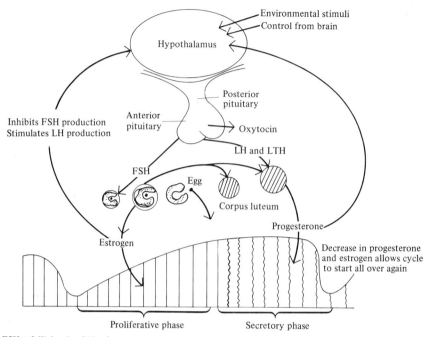

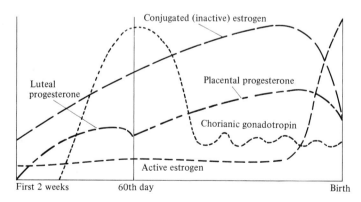

FSH = follicle stimulating hormone
LH = luteinizing hormone
LTH = luteotropic hormone influences on uterine lining = prolactin influences on mammary gland

Figure 10–5 Hormones influencing the estrous cycle and gestation.

promotes the growth of the uterine lining (the endometrium), makes possible the implantation of the fertilized egg, and stimulates the growth of the mammary gland. Estradiol and progesterone are also produced in males, as is the luteinizing hormone, which stimulates the growth of interstitial cells in the testes as well as in the ovaries. It is sometimes labeled ICSH when it occurs in males. The presence of estrogens and progesterone inhibits the secretion of FSH by the anterior pituitary. If the egg is not implanted in the endometrium, the corpus luteum disintegrates and no more progesterone is produced, causing the disintegration of the endometrium. If the fertilized egg is implanted, the activity of the corpus luteum is prolonged. The sequence of events under those conditions will be discussed later.

During the estrous cycle changes occur both in the uterine endometrium and in the vaginal epithelium of all mammals except the monotremes; in some marsupials it occurs but to a lesser degree. In mammals with short estrous cycles (the smaller mammals such as the mice and rats), the changes in the vaginal epithelium are more closely correlated with ovarian events than in mammals with longer cycles. The cycle is usually divided (by definition) into diestrus, proestrus, estrus, and metestrus. If there is a period of dormancy, it is called the anestrus. The changes in the vaginal epithelium during early and late estrus are caused primarily by estrogen.

Although there is a cyclic sloughing off and regeneration of the uterine endometrium in all mammals, it is most pronounced in the primates. Among the primates this process is accompanied by bleeding, possibly because a completely new endometrial lining replaces old endometrial lining each time. In other mammals, such as the rat, there is an almost continuous sloughing off of the endometrium.

In the nonprimates the estrous cycle is characterized by a short period of time when the female is receptive; this period is called the estrus or "heat." Ovulation occurs shortly before or shortly after this period. In primates, copulation can occur at nearly any time.

In primates (which includes humans), the estrous cycle is called the menstrual cycle, although the difference between the two cycles is arbitrary, rather than fundamental. Estrous behavior is not obvious in humans. In most mammals estrus is also used to denote a time when the female is receptive to the male. The word derives from a Greek word meaning "gadfly" and was used to describe the nervous, erratic behavior of cattle *(Bos taurus)* when being attacked by this fly. The sequence of events in the ovaries of primates does not differ greatly from the sequence in other mammals. Ovulation in the majority of mammals is *spontaneous* and occurs without coitus. Many reproductive phenomena seem to be controlled by nerve impulses which may signal the anterior pituitary that a hormone should be released. Nerve impulses reaching the hypothalamus may there be translated into neurohumoral agents, which reach the anterior pituitary by way of the hypothalamopituitary portal system. This may explain induced ovulation.

In *induced ovulation* the egg is shed within a few hours after copulation, i.e., copulation induces ovulation. Rabbits, many carnivores, thirteen-lined ground squirrels *(Spermophilus tridecemlineatus),* meadow mice *(Microtus pennsylvanicus),* and mink *(Mustela vison)* are induced ovulators. It would appear that in the larger species of mammals, the females that lead a solitary life, or at least remain mostly in female groupings, are spontaneous ovulators. Rutting season and the endowment of females in heat with an odor to attract distant males increase the chances of procreation. This is probably the case with cervids, bovids, and the larger carnivores. The smaller mammals that live singly seem to be in a continuous state of sexual receptivity. For a list of these smaller mammals the reader should consult Van Tienhoven (1968, p. 276).

OVULATION, FERTILIZATION, AND THE PLACENTA

After ovulation, the egg begins a seemingly hazardous journey to the site of fertilization in the oviduct. In some mammals [rodents, cats *(Felis),* and dogs *(Canis),* for instance] the ovary is surrounded by a tissue capsule which is continuous with the oviduct, so that the egg can hardly fail to find its way from the ovary to the oviduct. But in many mammals (e.g., rabbits and humans) there is no such tissue and it appears as if the egg might miss the oviduct and get lost in the body cavity. To prevent this the *infundibulum* of the oviduct diligently seeks the ovulated egg. The infundibulum is the funnel-shaped free end of the oviduct whose edge is developed as a fringe, the fimbria, or fingerlike projections covered with cilia.

The union of a spermatozoon with an ovum to form a zygote is called fertilization. This restores the original diploid chromosome number and combines the hereditary characteristics of both the male and female. Only one of the millions of sperm cells released into the vagina penetrates and fertilizes the egg. As soon as the egg is fertilized the outer membrane of the egg changes in consistency and becomes impenetrable to the other sperm. The zygote is carried down the oviduct by cilial currents and muscular contractions of the wall of the oviduct, under the influence of sex hormones. During this travel, cell division begins. About the time the zygote enters the uterus it has become a *blastocyst,* a hollow ball of cells filled with fluid. After further enlargement the blastocyst implants in the endometrium. During the interval between fertilization and implantation, the embryo is nourished by a limited supply of yolk in mammals and also by material secreted by glands in the female genital tract.

After implantation, embryonic membranes form the *umbilical cord,* through which blood vessels contributed by the *allantois* run to a structure formed by the embryonic membranes and from the adjacent uterine tissue.

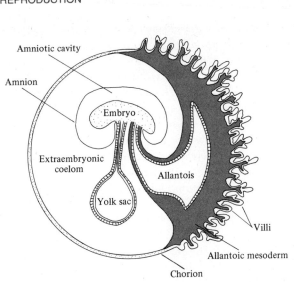

Figure 10–6 Embryo and membranes of a placental mammal. *(After Balinsky, B. I.: An Introduction to Embryology, 3d ed., W. B. Saunders Company, 1970.)*

This structure is the *placenta*. Within the placenta the blood vessels of the mother and young lie close together, but the maternal blood is separated from the fetal blood. Nutritive substances and oxygen diffuse from mother to embryo, while urinary wastes and carbon dioxide diffuse in the opposite direction. The placenta is a distinctive and important structure in therian (placental) mammals. Higher taxonomic categories such as subclasses and infraclasses have been based on differences in the placental types. The placenta also serves as an endocrine organ synthesizing hormones necessary for the maintenance of pregnancy (Fig. 10–6).

Placentas have been classified according to (1) the degree of intimacy between the embryonic and maternal parts of the placenta, and (2) the form of the placenta.

1 *Choriovitelline placenta.* This is the most primitive type of mammalian placenta; it occurs in all the marsupials except members of the family Peramelidae (bandicoots). The yolk sac acts as a placenta. The blastocyst does not embed in the uterine lining but simply sinks into a depression. The embryo receives nourishment from limited diffusion from the maternal blood in the depression, but mostly from "uterine" milk secreted by the uterine mucosa. The weak fetal-maternal connection and the poor nourishment seem to limit the length of the gestation period, which is very short in marsupials—12 days in the opossum.

2 *Chorioallantoic placenta.* This type is found in the marsupial bandicoots and all the eutherian mammals, in which the young complete their develop-

ment in the uterus (eutherians comprise all but the monotremes and marsupials). The basic structure is the same in all chorioallantoic placentas, but in the bandicoot, transfer of materials between fetal and maternal circulations is least effective. The allantois is fairly large and highly vascularized. The blastocyst rests against the endometrium on the side where the allantois contacts the chorion (the outer membrane of double membrane that surrounds the embryo). In the rest of the mammals the structure of the chorioallantoic placenta allows for a greater degree of diffusion of materials between the fetal and uterine bloodstreams. In these the blastocyst sinks into the endometrium. As implantation proceeds, chorionic villi grow further into the endometrium. The uterus is becoming highly vascularized in the area of the implantation. Such an arrangement increases the surface areas, allowing a more rapid interchange between the maternal and fetal circulations.

a   In the *epitheliochorial placenta* the embryonic tissue is in contact with the endometrium and the villi rest in pockets in the endometrium. This is the kind of placenta found in lemurs, pigs *(Sus),* horses *(Equus),* and whales.

b   In *syndesmochorial placenta* the uterine epithelium is locally eroded, reducing the separation of the maternal and fetal bloodstreams. This kind of placentation occurs in cows *(Bos),* sheep *(Ovis),* and goats *(Capra).*

c   *Endotheliochorial placenta.* A greater amount of erosion of the endometrium reduces the separation of maternal and fetal bloodstreams still more. This kind of placentation is found in the carnivores.

d   *Hemochorial placenta.* Destruction of the endometrium proceeds so far as to involve the endothelium of the uterine blood vessels, allowing direct contact between the chorionic villi and the maternal blood. This occurs in some insectivores, bats, apes, and some rodents.

e   *Hemoendothelial placenta.* The greatest destruction of placental tissues and least separation of fetal and maternal bloodstreams occur in this kind of placenta. It is the kind found in rabbits and some rodents.

The shape or form of the placenta varies with different mammals and is determined by the distribution of villi over the chorion. When the villi occupy one or two disklike areas on the chorion, the placenta is called *discoidal.* This kind occurs in insectivores, bats, some primates (including man), rabbits, and rodents. When the villi cover the chorion, the placenta is called *diffuse.* This kind occurs in lemurs, perissodactyls, and artiodactyls. In carnivores the villi are arranged in a bond encircling the equator of the chorion; this is called a *zonary placenta.*

The embryonic part of the placenta is expelled either with the young or shortly thereafter, as the "afterbirth." In mammals with the epitheliochorial type of placenta (which provides the least intimacy between maternal and fetal membranes), the placenta is termed *nondeciduous.* In mammals with a more intimate mingling of maternal and fetal tissue, bleeding occurs when the placenta is shed. This type of placenta is termed *deciduous.*

GESTATION AND PARTURITION

If the egg is fertilized, the course of the estrous cycle changes and the corpus luteum persists for a time rather than becoming atretic, i.e., degenerate. With few exceptions, each fertilized egg produces only one individual. A notable exception is provided by the armadillos (*Dasypus hybridus* and *Dasypus novemcinctus*), in which four identical embryos result from one fertilized egg. Being identical, the embryos are all of the same sex. Only one corpus luteum is formed. In females with long cycles, there is plenty of time after the egg is fertilized for the development of the luteal tissue to secrete progesterone, which in turn prepares the uterus for implantation of the embryo. In mammals with short estrous cycles the corpus luteum from the previous ovulation may be maintained beyond this normal regression time (as in pseudopregnancy) thus getting a "head start" if fertilization occurs.

In mammals in which pseudopregnancy does not exist, the presence of the embryo in the uterus is probably the signal that releases a substance from the pituitary gland which maintains the corpus luteum.

The uterus must also be maintained in a noncontractile stage during pregnancy; otherwise the fetus would be aborted. Progesterone blocks the contraction of the uterine muscles and helps maintain the quiescent stage.

The successful termination of pregnancy is yet another multifaceted and complicated interplay of hormones and neural hormones, and maybe even mechanical causes. Successful termination of pregnancy involves dilation of the cervix and the pelvis, contraction of the uterus, breaking the physical bond between mother and young, and initiation of breathing in the newborn young. Termination of pregnancy is poorly understood. At present some combination of all the following events is thought to be necessary, the process varying among species:

1 Decrease in progesterone secretion by the ovary, the placenta, or both and an increase in volume of uterine contents initiate uterine contractions.

2 Decrease in inactive (conjugated) estrogen.

3 Increase in active estrogen.

4 Increase in oxytocic substance, secreted by the posterior pituitary gland and possibly by the placenta in some species.

5 Decrease in relaxin shortly before or at parturition. Relaxin plays an important role in dilation of the cervix of the cow, pig, and laboratory rat. In the rat maximal effect is attained when progesterone and estrogen are present. In cattle and rabbits, as well as humans, relaxin concentration increases throughout gestation, and probably is necessary for the maintenance of pregnancy, but drops just before or at the time of parturition. Relaxin is produced in the uterus, placenta, and ovary, and so it is difficult to define its precise role. There is disagreement among endocrinologists as to the existence of relaxin. Some endocrinologists believe the results attributed to relaxin can be attributed to the action of estrogen and/or progesterone under varying conditions.

Recent evidence (Liggins, 1969) has led to speculation that the fetus indirectly controls parturition in all mammals, by stimulating (in an unknown way) the production of a prostaglandin, PGE_2, which could either function mainly as a luteolytic agent (in those mammals where the progesterone is secreted primarily by the corpus luteum) or have primarily an oxytocic action (in those mammals in which the progesterone is secreted mainly by the placenta). In some fossorial mammals (mammals living almost entirely underground) the pelvis is normally too narrow for the birth of young. In the pocket gopher *(Geomys bursarius)* the pelvis is rigid when the animal is maturing, but in the female, relaxin causes the pubic symphysis to relax, accommodating the birth of young.

The fetus must pass through the pelvis and the dilated cervix, and the uterus must contract. The precise interactions and hormonal levels which initiate parturition need much elucidation. The placenta as well as the fetus is expelled, and most wild mammals consume their own placenta.

THE MAMMARY GLANDS AND LACTATION

The growth of the mammary glands and the production of milk are also controlled by hormones. In some species of mammals estrogens alone control the growth of the mammary gland; in others, growth of the larger duct system is controlled by estrogen and STH (somatotropic or growth hormone produced by the anterior pituitary). The growth of the fine ducts and alveoli is under the influence of progesterone.

But neither estrogens nor progesterone causes the production of milk. The initiation and maintenance of lactation is accomplished partly by the hormone PRL, also called prolactin, coming from the anterior lobe of the pituitary gland. It was the first hormone isolated from the anterior lobe and, surprisingly, was first isolated in birds, not in mammals. Riddle, a Canadian biologist, found it in pigeon "milk," a material produced by the female for the newly hatched young. He named it prolactin. More recently it has been suggested that several other hormones [STH and adrenocorticotropic hormone (ACTH), and neural hormones] have an effect on lactation, but their function is, as yet, poorly understood. PRL induces maternal behavior in females and sometimes in males in both mammals and birds.

Increased amounts of oxytocin have been found in the blood of rabbits, sheep *(Ovis)*, goats *(Capra)*, cattle *(Bos)* during (but not before) parturition. In rabbits, cattle, and humans there is a "milk letdown" in lactating females when a stimulus such as sucking or milking occurs. These stimuli send a neural reflex to the hypothalamus and from there to the posterior lobe of the pituitary, which then releases oxytocin. The oxytocin causes contraction of myoepithelial tissue in the mammary gland.

The hormone requirements needed to prepare the mammary gland for lactation and to initiate and maintain lactation are not yet clearly understood. Complicating the picture is the fact that species differences exist.

Though the mammary gland and the production of milk are major char-

Figure 10–7 The mammary gland and milk are characters of major importance in the class Mammalia.

acteristics of the class Mammalia, little attention has been paid to this phase of mammalogy. The young of most mammals are born in a helpless state and remain for a period of time in this condition in a den, burrow, or lair. A female which is nursing young forages for her usual food, but a part of this food is converted to milk. The milk is stored and secreted by the mammary gland (Fig. 10–7). The hypothetical evolutionary advantage is that the chance of discovery of both the female and the young by predators is reduced.

Anatomy of the Mammary Gland

Although the mammary glands are one of the most distinguishing characters of mammals, they are by no means uniform in structure. Although the early stages of the evolution of the mammary gland are unknown, it is generally accepted that it evolved from the sweat glands. An abdominal incubation area probably evolved in the endothermic therapsids (Long, 1972; Hopson, 1973). Sweat glands, which seasonally hypertrophied as a result of hormonal activity, moistened the brood area to prevent desiccation of the eggs and young. Originally young may have licked the glandular secretions for their moisture. Eventually a shift to maternal secretions for the food of the young would have allowed the female to confine her exposure to optimal hunting times.

Mammary Glands of Cattle   Although there is no "typical" mammary gland, the mammary gland of cattle *(Bos taurus)* has been intensively studied and will be used to show the basic anatomy of a mammary gland (Fig. 10–8). Histologic examinations show mammary secretory cells of monotremes, marsupials, and placentals to be remarkably similar.

The udder of a cow is made up of four mammary glands, each with a teat. The duct at the lower end of the teat is called the streak canal; in cows it can range from 8 to 12 mm in length, and is closed by an involuntary sphincter which holds the milk in the gland against pressure of the milk in the storage system. An additional function of the streak canal may be to reduce the entrance of any foreign matter.

Above the streak canal there is a cavity, or cistern, in the teat called the *sinus papillaris.* Within this sinus there are primary and secondary folds, as well as numerous longitudinal and circular ones. These folds allow for expansion of the teat as it fills with milk. In addition there are strong strands of connective tissue and elastic fibers—structural folds—which give the teat its solidarity.

The upper end of the cistern of the teat opens into the cistern of the mammary gland, the *sinus lactiferus* or *galactophorus.* This cistern, a large cavern of various sizes and shapes, stores the milk as it is being secreted. The size has been variously estimated as 125 cm^3 and 400 cm^3. From 8 to 12 large milk ducts lead into the cistern of the mammary gland. These larger milk ducts are generally short and wide, and branch into smaller ducts which are very

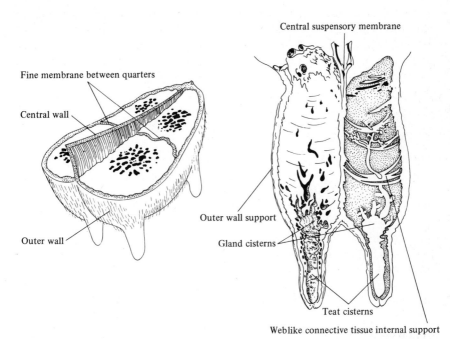

Figure 10–8 Gross anatomy of the mammary gland of a cow *(Bos taurus). (Courtesy of Babson Brothers.)*

irregular. These may again rebranch. All the ducts end in enlarged spaces called *alveoli* or *acini,* which serve as collecting spaces throughout the udder. Each alveolus is made up of two kinds of cells. A layer of secretory epithelium surrounds the lumen (Fig. 10–9). Peripheral to this is a coarse mesh of myoepithelial cells (Feldman, 1961). These layers are in turn surrounded by a thin basement membrane. Surrounding each alveolus is a network of capillaries which bring the blood in intimate contact with the secretory cells (Fig. 10–9). It is in the cells lining the alveoli (not the duct) that milk is secreted. The filling of the gland begins with the filling of the lumen of these alveoli; as milk secretion continues, the milk is forced into larger and larger dilations until it reaches the cistern of the gland.

Thus the alveoli, or acini, are drained by terminal (end, intercalary) ducts. A group of alveoli or acini form a lobule drained by intralobular ducts. Lobules are united into lobes drained by interlobular ducts and finally by lobular ducts. The mammary gland is thus composed of lobules and lobes connected by a series of ducts. These units of gland structure are separated by septa, or walls of connective tissue, called the *stroma.* The secretory portions of the gland are called the *parenchyma.* The stroma appears white, the parenchyma orange-colored. A central suspensory membrane separates the cow's udder into quar-

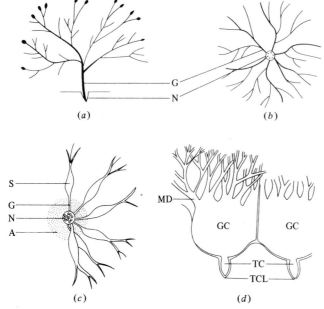

Figure 10-9 Examples of different arrangements of the mammary duct system. *(a)* Rat. Ducts all unite to form one main duct (G, galactophore, a milk-carrying duct), which opens at the tip of the nipple (N). *(b)* Rabbit. Ducts unite to form several main ducts. *(c)* Woman. Each main duct is dilated near the base of the teat to form a sinus (S); the nipple is surrounded by the dark-colored areola (A). *(d)* Ruminant. The main ducts (MD) open into a single large gland cistern (GC), which in turn opens into the teat cistern (TC). (TCL, teat canal.) *(From Cowie, 1972.)*

ters (Fig. 10–8). The mammary gland is served by branchings of the mammary artery and the mammary vein, a lymphatic system, and a nervous system.

Mammary Glands in Other Mammals The most rudimentary milk glands are found in the egg-laying mammals, the Monotremata—in the spiny anteater or echidna *(Tachyglossus aculeatus)* and the duck-billed platypus *(Ornithorhynchus anatinus)* of the Australian region. In the platypus, the mammary tissue is scattered over the ventral surface and consists of about a hundred or so lobules on each side of the median line. Each lobule opens separately on the skin surface. At each opening there is a stiff hair along which the milk oozes out and forms a drop. There is no nipple, and the young (with sucking noises) lick the milk off these hairs. No pouch is formed in the platypus.

In the female spiny anteater, or echidna, a pouch or incubatorium is formed during pregnancy. The mammary tissue is gathered into two glands located on either side of the abdomen. At their proximal ends the lobules of the glands are collected together and open onto the ventral surface at two small areas (6 X 3 mm) within the pouch. The terminal duct of each lobule opens into the invagination of the skin surrounding a hair, the follicle of which was situated in the invagination. The sebaceous glands associated with the follicle also open to the exterior through the same channels.

In the Marsupialia, the egg develops in the uterus, absorbing nourishment through its membranes, and the young are born in a very immature stage after a very short gestation period. An allantoic placenta is rarely present in this group. The young reach the brood pouch, or marsupium, under their own power and there attach semipermanently to the long tubular teats. Not all Marsupialia have pouches, one example of a marsupial without a pouch being the murine opossum of Central and South America. Before the animal reaches sexual maturity, the teats in a kangaroo resemble small circular apertures on the skin surface. Each has a channel, and at the bottom of the channel there is a small papilla in which the ducts of the milk glands open. During pregnancy these teat pouches evert like the finger of a glove. The young attach to these teats. In the pouchless opossums the young are also attached to the teats and are simply carried around in this manner.

The mammary glands of the marsupials differ from those of the monotremes in that the secretory ducts are brought together and lead to the surface through a teat or nipple. The number of teats varies considerably, totaling as many as 24 in the genus *Phascolaretos*. An unusual feature is that usually a single nipple is found between the two lateral rows; thus, the North American opossum *(Didelphis marsupialis)* has 13 teats.

The young of placental mammals are nourished for quite a long time in the uterus by means of a placenta. The mammary gland is very well developed in placentals, and the anatomy is basically similar to that of the domestic cow, in which the mammary tissue has been localized and secretory ducts are brought together and lead to the surface through a teat or nipple. The number of teats varies from two to 22, and their locations also vary greatly.

Milk Composition

It would be difficult to formulate a "waterproof" definition of milk, but for legal purposes the United States Department of Agriculture has defined it as "the whole, fresh lacteal secretion obtained by the complete milking of one or more healthy cows, excluding that obtained within 15 days before and five days after calving or such longer period as may be necessary to render the milk colostrum-free. The name milk, unqualified, means cow's milk."

Milk provides nutrition for the young of mammals, transmits passive immunity, and sometimes supports growth of symbiotic intestinal flora. It is not surprising that the mammal whose milk we know the most about is the domestic cow, *Bos taurus* (Jenness, 1974). Good general references on milk and mammary glands of domestic animals include Jenness and Patton (1959), Kon and Cowie (1961), and Schmidt (1971).

The main components of fats, carbohydrates, and proteins which make up the bulk of milk (except water) differ from most other fats, carbohydrates, and proteins.

Fats (Lipids)   Though fats are usually called lipids in the literature on milk, they most frequently are labeled fats in tables of milk composition, while in the dairy industry they are referred to as butterfat. Milk lipids are complex and variable. Fat is dispersed in milk in the form of droplets or globules (colloidal suspension), the size of which in cattle is an inherited characteristic. A flat globule membrane prevents the globule from being dissolved. The fatty acids may be saturated or unsaturated, volatile or nonvolatile.

The longer-chain fatty acids are taken from the blood plasma, while the shorter-chain fatty acids are synthesized in the mammary glands. In ruminants, much of milk fat is composed of a relatively high concentration of short-chain fatty acids, resulting from synthesis in the mammary gland of acetate and β-hydroxybutyrate produced by rumen organisms. The laboratory rat *(Rattus norvegicus)* and laboratory mouse *(Mus musculus)* are able to synthesize fatty acids from glucose. This is probably one reason for their great adaptability to a wide variety of habitats and diets. Ruminants, on the other hand, are unable to utilize glucose for the synthesis of fatty acids. Ruminant milk contains a large concentration of citrate, which increases the calcium-carrying capacity of milk. Rat and mouse milk is nearly devoid of citrate and depends almost entirely on casein to carry calcium (Jenness, 1974).

Carbohydrates   Lactose (milk sugar) is the principal carbohydrate of milk, except in monotremes, in which fucosyllactose and difucosyllactose predominate (Messer and Kerry, 1973), and in most pinnipeds (seals), in which lactose is lacking (Pilson, 1965; Schmidt, Walker, and Ebner, 1971). Chemically, lactose is a galactose-glucose sugar synthesized in the mammary gland. The function of lactose has not yet been clarified, but apparently it is not to supply a need for galactose, since mammals readily convert glucose to galactose. The unique values of lactose appear to be indirect. It appears to regulate the amount of water secreted in milk. Lactose persists further along in the alimentary canal than other sugars. This characteristic results in favoring an acid type of fermentation, with better utilization of calcium and strontium (Wasserman and Lengemen, 1960). Lactic fermentation also discourages the growth of putrefactive organisms. These properties are important in the developing young. Other indirect values of lactose are that it may discourage fat deposition, and may encourage gastrointestinal motility. Enzymes for the hydrolysis of lactose have been found in significant amounts in the mucosa of mammals, especially in domestic calves, and in some species of yeasts.

Proteins   Milk contains a number of proteins, the most important of which are caseins. Caseins are defined as proteins precipitated from milk by acidification to a pH of 4 to 5. Casein is believed to be composed of four variants, none of which has been crystallized but all of which can be separated by zonal electrophoresis. They are evidently synthesized in the mammary gland and have the property of forming micelles, in the presence of calcium and inorganic phosphate and milk k-casein (one casein variant) as a basic ingredient. These micelles range in size from 20 to over 200 nm (nanometer, or 10^{-9} meter) in diameter. The organization and structure of caseinate micelles are poorly understood, but they are a neat way of packaging amino acids and mineral salts essential for the growth of young (Jenness, 1974).

Other proteins found in milk include β-lactoglobulin (found in artiodactyls, function unknown) and λ-lactalbumins, found in artiodactyls, rodents, and humans. Two kinds of iron-binding proteins have been found in milk. One appears to be identical with the transferrin of blood; the other, a very different protein, named lactoferrin, is characteristic of milk and has been found in saliva, semen, and tears. Another protein found in milk is identical with the serum albumin of the species.

Milk also serves to carry antibodies and immunoglobulins to suckling young. Their concentration is especially high in colostrum, the first mammary secretion after parturition. Colostrum contains more mineral salts, total protein, casein, and serum proteins and less lactose than normal cow's milk. In the artiodactyls the placenta excludes the transfer of immunoglobulins, which are transferred by the colostrum. In primates and rodents, young acquire immunoglobulins by placental transfer. Immunoglobulins persist in milk throughout lactation. IgA-type immunoglobulins are produced in part in the mammary gland and are especially concentrated in the milk of lagomorphs and primates. In the ruminants, IgG, which appears to enter the colostrum from the bloodstream, is more concentrated than IgA.

At least 19 milk enzymes have been purified, isolated, or identified from milk (Shalhani, 1965). Vitamins and minerals are also important components of milk.

There are of course many miscellaneous substances which come directly from the blood in unaltered form or may be products of decomposition. Among these are glucose, galactose, uric acid, and leukocytes, as well as antibiotics, pesticides, and disinfectants.

The three phases of mammary glandular tissue activity may be described as (1) proliferation of glandular epithelium (early and midgestation); (2) colostrum formation, when great quantities of serum gamma globulin are characteristic (a short time before and after birth); and finally (3) synthesis of protein and fat (after the birth of young).

The composition of milk changes as lactation progresses. Silver (1961) showed that protein content in the milk of white-tailed deer *(Odocoileus virginianus)* increased from 8.83 percent on the second day of lactation to 11.50 percent at the end of the fifth month. Fat increased from 8 to 18 percent at the same time, but lactose decreased from 3.82 to 2.22 percent. The same trends occurred in caribou *(Rangifer tarandus)* milk, according to Aschaffenburg, Gregory, Kon, Rowland, and Thompson (1962), and the milk of mountain sheep *(Ovis canadensis)* showed the same pattern (Chen, Blood, and Baker, 1965).

The Ecologic Significance of Milk Composition

It can be argued that because milk differs greatly in composition and young are born at widely different stages of physiologic maturity, the nutritive needs of the young have not been an important selective force in the evolution of milk composition. On the other hand, there must be some adaptive significance. For example, we might choose two mammals which live in arid regions—the Merriam's kangaroo rat *(Dipodomys merriami)* and the two-humped bactrian camel *(Camelus bactrianus)* of Asia—and see how their milk is adapted (see Table 10–1). Both need water for cooling themselves. The camel's milk has a higher percentage of water (87 percent) than that of the kangaroo rat, whose milk is 50 percent water. The camel and its young are diurnal mammals, active during the heat of the day. The kangaroo rat and its young are nocturnal and therefore do not lose as much water to the surroundings as camels do. In addition, the fat content of both the camel's and the kangaroo rat's milk is somewhat higher than, for example, that of domestic cattle *(Bos)*. This would also be helpful in arid regions, for fat yields both water and more energy than either the carbohydrates or proteins do when metabolized. Fats yield more energy (8.4 to 9 kcal/g) than either protein (3.1 to 4.4 kcal/g) or carbohydrates (3.8 to 4.1 kcal/g) when metabolized. Fats also yield the greatest amount of water (1.1 g/g when metabolized, compared with about 0.6 g for carbohydrate and about 0.4 g for a protein). The correlation of milk composition with the climatic environment of the species is yet another example of the many adaptations which help mammals to survive in many climates. While looking at Table 10–1 one should be mindful again of the amounts of energy and water produced per gram of carbohydrates, fats, and proteins.

The fat content of the milk of arctic mammals and of aquatic and marine mammals, as well as those living in deserts, is also high.

The normal environment of the Pinnipedia (seals) is the sea, in which they spend a great deal of their time. The vital processes of birth, weaning, and mating occupy only a brief visit ashore—in the case of the Atlantic gray seal

Table 10–1 Selected Examples of Milk Composition of Various Species, in percentages

Species (common and Latin names, author, and no. of samples when more than one)	Fats	Proteins	Lactose	Ash	Total solids	Water
Monotremata Spiny anteater *Echidna aculeata multi-aculeata* (Marston, 1926)	19.62	11.3	3.81	0.78		
Marsupialia Virginia opossum *Didelphis virginiana* (Shaul, 1962)	4.7	4.0	4.5	0.77		
Insectivora Short-tailed shrew *Blarina brevicauda* (Shaul, 1962)	6.5	11.0	3.2	0.75		
Chiroptera Long-nosed bat *Leptonycteris sanborni* (Huibregtse, 1966)		4.37	5.39*	0.63	12.1	
Brazilian free-tailed bat *Tadarida brasiliensis* (Huibregtse, 1966)		11.07	3.70*	0.73	34.4	
Rodentia Merriam's kangaroo rat *Dipodomys merriami* (avg. of 11) (Kooyman, 1963)	23.48					50.42
Beaver *Castor canadensis* (Shaul, 1962)	19.8	9.0	2.2	2.0		
Cetacea Blue and fin whales *Balaeonoptera musculus* and B. *physalus,* (sample) (White, 1953)	17.1–51.6	3.6–13.1	0.7–4.5	0.8–1.8		
Fin whale *Balaeonoptera musculus* (avg. of 2) (Gregory et al., 1955)	42.34	12.16	1.29	1.42		
Humpback whale *Megaptera novaeangliae* (Pedersen, 1952)	38.5			1.48	39.53	46.7

Table 10–1   Selected Examples of Milk Composition of Various Species, in percentages (*Continued*)

Species (common and Latin names, author, and no. of samples when more than one)	Fats	Proteins	Lactose	Ash	Total solids	Water
Spotted porpoise *Sternella graffmani* (avg. of 6) (Pilson and Walker, 1970)	25.3	8.28	1.09			
Spinner's porpoise *Sternella microps* (Pilson and Walker, 1970)	26.2	7.09	1.03			
Carnivora Black bear *Ursus americanus* (Hock and Larson, 1966)	10.5	7.34	1.51	1.14	23.5	
Polar bear *Ursus maritimus* (Baker, Harington, and Symes, 1963)	30.6	10.05	0.48	1.15	43.45	
Lion *Panthera leo* (Shaul, 1962)	18.9	12.5	2.7	1.4		
Pinnipedia California sea lion *Zalophus californianus* 　Feb. (avg. of 3) 　June 　Dec. (Pilson and Kelly, 1962)	 31.1 36.5 37.0	 13.3 13.8	 0 0	 0.64	 52.7	
Atlantic gray seal *Halichoerus grypus* (= *Eschrichtius robustus*) (pooled sample) (Amoroso and Mathews, 1962)	52.2	11.2	2.6	0.7	67.7	
Alaska fur seal *Callorhinus ursinus* (avg. of 5) (Ashworth, Romaiah, and Keyes, 1966)	52.2	9.59	0.112	65.2		
Proboscidae African elephant *Loxodonta africana* (Sykes, 1971)	7.0	4.0	6.5	0.5		82.0

Table 10–1 Selected Examples of Milk Composition of Various Species, in percentages (*Continued*)

Species (common and Latin names, author, and no. of samples when more than one)	Fats	Proteins	Lactose	Ash	Total solids	Water
Artiodactyla						
Peccary						
Pecari tajacu (= Dicotyles tajacu)						
Milk	3.55	5.8	6.55		16.45	83.55
Colostrum	4.8	6.0	5.2	0.64	18.2	
(Sowls, Smith, Jenness, Sloan, Regehr, 1961)						
Pecari tajacu (= Dicotyles tajacu) (avg. of 5)						
Milk	6.61					
Colostrum	3.1					
(Brown, Stull, Sowls, 1963)						
Bactrian camel						
Camelus bactrianus	5.39	3.8	5.10	0.69		87.0
(Shaul, 1962)						
Mule deer						
Odocoileus hemionus	10.4	8.88	4.37	1.49	25.00	
(avg. of 3) (Kitts, Cowan, Bandy, and Wood, 1956)						
Odocoileus hemionus						
July	8.3			1.66	20.78	
Aug.	8.4			1.50	22.30	
Sept.	13.6			1.39	27.63	
(Browman and Sears, 1955)						
White-tailed deer						
Odocoileus virginianus	15.06	11.92	3.75	1.51	32.2	
(Ruff, 1938)						
Odocoileus virginianus	8.3			1.44	20.4	
(Hagen, 1951)						
Odocoileus virginianus	7.3	7.04	4.28	1.48	20.1	
(Murphy, 1960)						
Caribou						
Rangifer tarandus	13.05	9.54	3.31		27.36	
(avg. of 2) (Aschaffenburg, Gregory, Kon, Rowland, and Thompson, 1962)						

Table 10–1   Selected Examples of Milk Composition of Various Species, in percentages (*Continued*)

Species (common and Latin names, author, and no. of samples when more than one)	Fats	Proteins	Lactose	Ash	Total solids	Water
Musk ox *Ovibos moschatus* (avg. of 2) (Tener, 1956)	11.0	5.3	3.6	1.8	21.54	78.5
Mountain sheep *Ovis canadensis canadensis* (avg. of 4) (Chen, Blood, and Baker, 1965)	12.01	8.79	4.29	1.18	26.5	73.5
Pronghorn *Antilocapra americana* (Browman and Sears, 1955)	9.6			1.37	26.28	
Antilocapra americana (Shaul, 1962)	20.7	10.6	2.4	1.5		64.5
Bison *Bison bison* (avg. of 2) (Schutt, 1932)*	1.73	2.52	4.73	9.1		13.32

* Casein and albumin.
Source: Mostly from *Dairy Science Abstracts*, as listed by Jenness and Sloan, 1970.

(Halichoerus grypus), only 2 to 3 weeks. This short sojourn out of the water protects these essentially marine species from prolonged exposure away from their true element. It also means that the young must grow very quickly. For a long time it was assumed that the sugar present in the milk of all mammals was lactose, but Pilson and Kelly (1962) first demonstrated that lactose was absent in milk of the California sea lion, *Zalophus californianus.* Pilson (1965) later reported that lactose was absent in the milk of members of the other subfamily, the Phocoidea, of the Pinnipedia. Seals' milk contains about 52 percent fat and little or no sugar.

The gestation period of the Atlantic gray seal is nearly a year. One young at birth weighed 16.8 kg (37.5 lb), gained about 1.9 kg (3 lb) every 24 h, and at 10 days weighed 45.3 kg, or 100 lb. At 13 to 14 days it was weaned. Then it began fasting, "as well he might after a fortnight of intensive suckling on the richest known milk." The calf passed little or no waste matter during the fasting period, and "the scavenging birds went away disappointed" (Lockley, 1954). When the calf left his birthplace on land he was only 3 weeks old. Amoroso and Mathews (1951) also gave the average daily weight gain as 1.9 kg (3.3 lb), birth weight as 13.6 kg (30 lb), and weaning weight as 41.8 kg (92 lb). During this feeding period the mother seal's weight decreased from 168 to 129 kg (371 to 276 lb), a loss of 44 kg (97 lb). This quick growth of the young on rich milk is another advantageous adaptation, permitting them to spend the shortest time possible in a hostile environment.

The whales (order Cetacea) are the only mammals that never come on land, but spend all their time in the sea. The milk of blue *(Balaenoptera musculus)*, fin *(Physalus)*, humpback *(Megaptera novaeangliae)*, gray *(Eschrichtius robustus)*, and sperm *(Physeter catodon)* whales (Slijper, 1962, and others) contains about 40 to 50 percent fat, 11 to 12 percent protein, very little lactose, and 40 to 50 percent water. Pilson and Walker (1970) also noted that the average chloride concentration (62 meq/l of water) in the nine samples of *Sternella* milk was twice the average chloride concentration (32 meq/l of water) in bovine milk (Jenness and Patton, 1959). It is probable that the chloride replaces the lactose in the maintenance of isotonicity in whale's milk (Pilson and Walker, 1970). According to these authors, data are yet lacking to show that the composition of milk changes during the nursing period in whales as it does in other eutherian mammals. Cetacean milk usually has a creamy-white color, but sometimes has a pinkish tint. It has a slightly fishy smell, and its taste, according to Slijper (1962), is "reminiscent of a mixture of fish, liver, Milk of Magnesia, and oil." The milk of seals and that of whales are similar in composition, which seems reasonable when we consider that the combustion of fat releases a maximum of energy and salt-free water, which are major requirements of their life in the sea. For aquatic mammals, natural selection has placed a high premium on the secretion of milk containing a high proportion of energy-yielding material. It has been shown that the blubber of newborn whales contains little fat, and their relatively large body surfaces exposes them to great heat losses. Therefore it would be of great value for these animals to have a high metabolic rate and concentrated foodstuffs. The milk of blue and fin whales contains about the same proportional amount of vitamins A and B, potassium, and chlorides as the milk of terrestrial animals. But calcium and phosphorus occur in greater amounts. Whale calves gain a maximum of weight in a minimum of time. Blue whales *(Balaenoptera)* double their birth weight in 7 days, dogs *(Canis)* in 9 days, cows *(Bos)* in 47 days, and horses *(Equus)* in 60 days. Whale calves gain nearly 5 cm (2 in.) in length and nearly 90 kg (200 lb) in weight per day! They must be surfaced quickly after birth, and suckling calves rarely leave their mother's side until weaned.

To complicate the life of whale calves further they must get their milk underwater (as do the young of sea cows, sea otters, and at times hippopotami). Although the mammary glands of several species of mammals have been found to contain myoepithelial (muscle) cells (Feldman, 1961; Richardson, 1949–1950), these cells are especially important in the mammary glands of whales. A whale cow spurts her milk into the calf's mouth. The calf must obtain a maximum of milk in a minimum of time. The mammary glands are two long, fairly small (for a whale), fairly flat organs inclined toward each other, with their tips near the umbilicus. Each gland is divided into countless lobes and lobules. These all lead into narrow ducts, which in turn lead into a central lactiferous duct comparable to the cistern in a cow's udder. The average dimension of the mammary gland in a nonlactating fin whale is 2 m $\times$ 76 m $\times$ 5 cm thick (2½ ft $\times$ 7 ft $\times$ 2¼ in.), but during lactation the thickness increases to 30 cm.

It has been estimated that a whale calf, which gets 492 l (130 gal) of milk in about 40 feedings per 24 h, must then get about 13 l (3½ gal) of milk per feeding. The bottle-nosed dolphin *(Tursiops truncatus)* takes one to nine sucks of only a few seconds each at each feeding.

Whales have a comparatively short period of lactation. The bison *(Bison)* nurses its young for about a year and a half, the rhinoceros *(Diceros)* for 14 months, and the elephant (*Elaphas* and *Loxodonta*) up to 3 years. In contrast whales rarely nurse their young much over a year, although the bottle-nosed *(Tursiops)* dolphin has been recorded as lactating for 16 months. Right *(Eubalaena)* and gray *(Eschrichtius)* whales nurse their young about a year; blue and fin whales about 7 months.

The milk of arctic mammals such as the polar bear *(Ursus maritimus)* and the caribou *(Rangifer tarandus)* also has a high fat content, although not so high as that of seals and whales. Again this must be of adaptive significance, because the young of these mammals must need a food with high energy and high water production potentials.

Sharman (1970) has given a good review of lactation in marsupials, especially in kangaroos; the following information comes mostly from his review. The composition of marsupial milk, like that of eutherian mammals, changes during the growth of the young. Bailey and Lemon (1966) reported that the early milk of the red kangaroo *(Megaleia rufa)* contains little if any fat, but the amount gradually increases, so that the milk which the advanced young receive may contain up to 20 percent fat. The protein content increases but the sugar decreases as lactation progresses. The most notable fact about some marsupials, especially kangaroos, is that the separate mammary glands of the same female can produce two kinds of milk concurrently. In the kangaroos, one young (out of the pouch) may be over 220 days older than the pouch young (Bailey and Lemon, 1966). The two kinds of milk differ in several respects. The milk for the advanced young contains three times as much fat as the milk of the other gland. There are also specific protein differences (Bailey and

Lemon, 1966). Although the two milks differ in chemical composition, they are produced in the same endocrine environment.

A correlation between ecology, nursing behavior, and milk composition was devised by Shaul (1962) for the use of zoo personnel. She based her five groups on the nursing habits of the species. Mammals nursing on a demand basis tend to produce more dilute milk than those that nurse their young infrequently. Jenness and Sloan (1970) believed that this hypothesis has merit but that five groups may be excessively precise; they suggested three groups. Group A (Shaul's groups I and II) contains those species that nurse on demand and would include marsupials, primates, perissodactyls, and some artiodactyls. In this group sugar and ash make up more than 50 percent of the solids and nonfats, and sugar furnishes over 25 percent of total calories. Group B (Shaul's groups III and IV) includes those species nursing on a scheduled basis—lagomorphs, most rodents, most carnivores, and many artiodactyls. Group C (Shaul's group V) includes arctic, aquatic, and some desert mammals—*Castor, Myocaster, Lontra,* the Ursidae (bears); pinnipeds (seals); and cetaceans (whales). In this group, fat furnishes more than 75 percent of the total calories and also furnishes a great deal of water.

For good summaries of milk composition see Evans (1959), Shaul (1962), and Jenness and Sloan (1970).

Milk as a Taxonomic Character

Milk has been used to clarify or further substantiate already established opinions concerning the relationships or evolution of mammals, particularly among the higher taxons. Sloan, Jenness, Kenyon, and Regehr (1961) studied the whey and casein proteins of representatives of the Artiodactyla, Perissodactyla, Carnivora, Rodentia, Primates, and Marsupialia. They wrote, "Although no two species have precisely the same pattern, the patterns from several groups of species of tribal to superfamily rank exhibit definite similarities." In their discussion they wrote, "In general, primitive surviving stocks of any order of mammals have fewer quantitatively important milk proteins than phylogenetically more recent derivatives."

In another publication, Jenness, Regehr, and Sloan (1964) wrote, "The various generic and suprageneric taxa investigated (represented by fifty-two species) each show a more or less standard pattern of proportions of sugar components." Marston (1926) demonstrated the presence of a protein in echidna which is presumably a casein. This caused Sloan, Jenness, Kenyon, and Regehr (1961) to raise the interesting point, "Thus the ability to synthesize casein must have been acquired independently in the therian and monotreme ancestry or casein (and thus milk) was secreted by the therapsids themselves."

More recently Jenness (1974) has plotted the proportions of protein, fat, and lactose plus ash in the milk solids in ternary diagrams (Fig. 10–10). Such diagrams show that taxonomically related species tend to cluster at similar

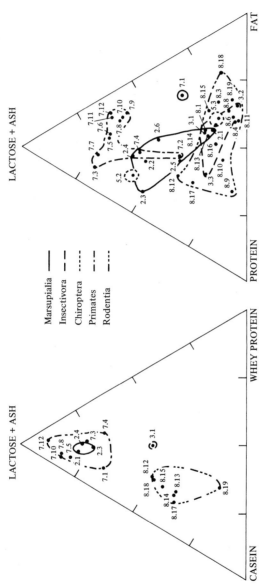

Figure 10–10 Ternary diagram of milk solids of marsupials, insectivores, bats, primates, and rodents. The numbers on left stand for nonfat solids, the numbers on right stand for total solids. *(From Jenness, 1974.)*

compositions of these solids. Primate and perissodactyl milks tend to cluster in the high-lactose-plus-ash area of the diagram. Milks of aquatic carnivores and cetaceans cluster in the high-fat zones. Rodents and rabbits have about equal proportions of fat and protein. Marsupials, carnivores, and artiodactyls cluster in the center of the diagram, showing equal proportions of the three classes of constituents. The composition of solids-not-fat in terms of casein, whey proteins, and lactose-plus-ash show pronounced clustering within mammalian orders. Thus milk can be of use, although limited, as a taxonomic character, at the higher taxons.

The study of milk, its adaptive significance, importance in population dynamics, use in evolutionary and taxonomic studies, and immunologic properties for the survival of young has just begun, and appears to be a productive area for future research.

PSEUDOPREGNANCY

At times female mammals behave as if they were pregnant, even though they are not. This is a commonly noted occurrence in certain pets, such as cats and dogs, and is commonly referred to as false pregnancy, or *pseudopregnancy*. It is characterized by the maintenance of the corpora lutea from the last ovulation beyond the time of normal regression when fertilization of the egg has not occurred. Most often pseudopregnancy lasts about half the length of time a normal pregnancy does.

REPRODUCTIVE DELAYS

Normally the fertilized egg becomes implanted in the uterus, and development continues without interruption until full term. The gestation period is designated as the time between the fertilization of the egg and parturition, but in some species of mammals modifications have arisen to lengthen the period from conception to parturition. These include delayed development, delayed implantation, embryonic diapause, and (though not strictly fitting the definition of a delay between conception and parturition) delayed fertilization.

Delayed Fertilization This unusual development pattern is known to occur only in some of the bats of the north temperate regions of both the New and Old World, and is seemingly a response to winter dormancy. In those species in which delayed fertilization occurs, the males become reproductively active in August. Spermatogenesis ends before winter, but the caudal epididymis retains viable sperm much longer. Most copulation occurs before hibernation, although it can occur during hibernation. The sperm are stored in the uterus. Ovulation and fertilization commonly occur after hibernation ceases, although it may occur while the animals are still at their winter roosts. In the

Table 10–2 Reproductive Data for Some Common or Well-known Mammals

	Approximate gestation period	Litter size	Number of litters per year
Order Monotremata			
Family Ornithorhynchidae			
Duck-billed platypus	From mating to egg laying 12–14 days		
Ornithorhynchus anatinus	Incubation period, 10–12 days		
Family Tachyglossida			
Echidna *(Tachyglossus aculeatus)*	27 days (d), then 8 weeks (wk) in pouch as an egg and young. Then young are cared for 3 months (mo) after they leave the pouch.		
Order Marsupialia			
Family Didelphidae			
Virginia opossum			
(Didelphis marsupialis)	12 d	9 (8–12)	1 or 2
Family Phalangeridae			
Koala *(Phascolarctos cinereus)*	35 d	1	1 (not known for sure)
Family Macropodidae			
Red kangaroo *(Megaleia rufa)*	33 d	1	Within a few days of giving birth, females mate. This second embryo remains dormant in the uterus as a reserve.
Order Insectivora			
Family Erinaceidae			
Hedgehog *(Erinaceus europaeus)*	34–35 d	3–6	1–2
Family Soricidae			
Short-tailed shrew			
(Blarina brevicauda)	21–22 d	3–8 (4.5 avg.)	Several (3–5)
Least shrew *(Cryptotis parva)*	21–23 d	4–6	

Family Talpidae			
Eastern mole (*Scalopus aquaticus*)	42 d	2–5	
Order Chiroptera			
Family Vespertilionidae			
Little brown bat (*Myotis lycifugus*)	50–60		1
Big brown bat (*Eptesicus fuscus*)	Delayed fertilization. Mating in August, young in May.	2	1
Order Primates			
Family Callithricidae			
Marmoset (*Callithrix jacchus*)	140 d	2 (1–3)	Every 3½ to 4½ yr
Family Pongidae			
Gorilla (*Gorilla gorilla*)	251–289 d		
Chimpanzee (*Pan troglodytes*)	196–260 d	1	3–4 yr
Orangutan (*Pongo pygmaeus*)	275 d	1	2-yr intervals
Gibbon (*Hylobates lar*)	210 d	1	
Order Edentata			
Family Dasypodidae			
Nine-banded armadillo (*Dasypus novemcinctus*)	Breed in July–September; egg fertilized; then dormant for 120–126 d; young born 120–150 d; young born after fertilization. 4 young, all from single egg.	4	1
Order Lagomorpha			
Family Ochotonidae			
Pika (*Ochotona princeps*)	30 d	2–6	1–2
Family Leporidae			
Snowshoe hare (*Lepus americana*)	30–38 d	3–4	1

Table 10-2 Reproductive Data for Some Common or Well-known Mammals (Continued)

	Approximate gestation period	Litter size	Number of litters per year
White-tailed jackrabbit (*Lepus townsendi*)	43 (41–47) d	5 (1–11)	1–3
Eastern cottontail (*Sylvilagus floridanus*)	30 d	3–6	3–4
Order Rodentia			
Family Sciuridae			
Woodchuck (*Marmota monax*)	30 d	2–6	1
Black-tailed prairie dog (*Cynomys ludovicianus*)	28–32 d	4.9 (3–6)	1
Thirteen-lined ground squirrel (*Spermophilus tridecemlineatus*)	27–28 d	4–10	1
Eastern chipmunk (*Tamias striatus*)	31 d	2–6	1–2
Gray squirrel (*Sciurus carolinensis*)	44 d	3–6	1–3
Fox squirrel (*Sciurus niger*)	45 d	2–5	1–2
Red squirrel (*Tamiasciurus hudsonicus*)	40 d	3–6	1–2
Southern flying squirrel (*Glaucomys volans*)	40 d	1–4	1–2
Northern flying squirrel (*Glaucomys sabrinus*)	30 d	3–6	1–2
Valley pocket gopher (*Thomomys bottae*)	30 d	4.6	1
Plains pocket gopher (*Geomys bursarius*)	40–50 d	1–9	1

	Gestation	Young per litter	Litters per year
Family Heteromyidae			
Merriam's kangaroo rat (*Dipodomys merriami*)	33 d	2–4	3
Family Castoridae			
Beaver (*Castor canadensis*)	100 d?	5	1
Family Cricetidae			
Deer mouse (*Peromyscus maniculatus*)	23–27 d	3–7	4–8
White-footed mouse (*Peromyscus leucopus*)	27–31 d	3–5	1
Wood rat (*Neotoma cinerea*)			
Hamster (*Mesocricetus auratus*)	16 d	1–12	3–4
Bog lemming (*Synaptomys cooperi*)	23 d	1–5	3–5
Red-backed vole (*Clethrionomys gapperi*)	18 d	5	3–7
Muskrat (*Ondatra zibethica*)	22–30 d	6–8	2–3
Meadow vole (*Microtus pennsylvanicus*)	21 d	1–9	Up to 17
Prairie vole (*Microtus ochrogaster*)	21 d	3–6	Many
Family Muridae			
Norway rat (*Rattus norvegicus*)	21 d	4–10	Many
House mouse (*Mus musculus*)	19 d	4–8	Many
Family Zapodidae			
Jumping mouse (*Zapus hudsonius*)	17–21 d	5 or 6	2–3
Family Erithizontidae			
Porcupine (*Erithizon dorsatum*)	112 d	1	1
Family Hydrochoeridae			
Capybara (*Hydrochoerus hydrochoeris*)	119–126 d	2–8 young	1

Table 10-2 Reproductive Data for Some Common or Well-known Mammals *(Continued)*

	Approximate gestation period	Litter size	Number of litters per year
Family Chinchillidea			
Chinchilla *(Chinchilla laniger)*	105–114 d	1–4	1–3
Family Physeteridea			
Sperm whale *(Physeter catodon)*	16 mo	1	
Family Monodontidae			
Beluga *(Delphinapterus leucus)*	14 mo	1	Every 2–3 yr
Family Balaenopteridae			
Hump-backed whale			
(Megaptera novaeangliae)	12 mo	1	
Blue whale *(Balaenoptera musculus)*	10–11 mo	1 (rarely 2)	Alternate years
Order Carnivora			
Family Canidae			
Coyote *(Canis latrans)*	60–63 d	5–10	1
Wolf *(Canis lupus)*	62–63 d ± 4 d	5–9	1
Dog *(Canis familiaris)*	53–71 d	1–15	
Arctic fox *(Alopex lagopus)*	51–57 d	4–11	1
Red fox *(Vulpes vulpes)*	51–63 d	4	1
Gray fox			
(Urocyon cinereoargenteus)	55–63 d	5	1
Family Ursidae			
Black bear *(Ursus americanus)*	Delayed implantation about 4 mo; in 7 mo young are born and weigh less than a pound	2 is usual litter	Every 2 yr

Species	Gestation	Number of young	Breeding frequency
Grizzly bear *(Ursus horribilis)*	180–250 d	1–3	2–3 yr
Polar bear *(Ursus maritimus)*	8 mo	2	Every other year
Family Procyonidae			
Raccoon *(Procyon lotor)*	60–73 d	3–4	1
Family Mustelidae			
Short-tailed weasel *(Mustela erminea)*	10 mo (Delayed implantation)	6–13	1
Least weasel *(Mustela nivalis)*	35 d (Little or no delayed implantation)	4–6	2
Mink *(Mustela vison)*	45–70 d (Delayed implantation with implantation about day 24)	3–10	1
Badger *(Taxidea taxus)*	Delayed implantation. Gestation is 6 wk. Total time 183–265 d.	1–5	1
Striped skunk *(Mephitis mephitis)*	63 d	4–10	1
Family Felidae			
House cat *(Felis cattus)*	52–69 d	1–6	1
Bobcat *(Lynx rufus)*	60 d	1–4	1
Mountain lion *(Felis concolor)*	90 d	2–4	Every 2 yr
Leopard *(Panthera pardus)*	3 mo	2–3	1
Lion *(Panthera leo)*	105 d	2–4	1
Tiger *(Panthera tigris)*	92–113 d	2–3	1
Order Pinnipedia			
Northern fur seal *(Callorhinus ursinus)*	Including period of delayed implantation 12 mo		1 (Alternate horns of uterus each year)

Table 10-2 Reproductive Data for Some Common or Well-known Mammals *(Continued)*

	Approximate gestation period	Litter size	Number of litters per year
California sea lion (*Zalophus californianus*)	342–365 d	1	1
Walrus *(Odobenus rosmarus)*	12 mo (No delayed implantation)	1	Every 3 yr
Order Proboscidea			
Family Elephantidae			
African elephant (*Loxodonta africana*)	22 mo (Puberty in 10–13 yr)	1	Every 2–2½ yr
Indian elephant (*Elaphas maximus*)	21 mo	1	Every other year
Order Sirenia			
Family Trichechidae			
Manatee *(Trichechus manatus)*	152–180 d	1–2	1
Order Perissodactyla			
Family Equidae			
Domestic horse *(Equus caballus)*	329–345 d	1	1
Zebra *(Equus hippotigris burchelli)*	12 mo	1	1
Family Rhinoceratidae			
Black rhinoceros *(Diceros bicornis)*	540 d	1 young	4–5 yr
Order Artiodactyla			
Family Suidae			
Domestic pig *(Sus scrofa)*	112–116 d	6–18	1
Collared peccary *(Dicotyles tajacu)*	142–148 d	1–4	1–2

	Gestation	Litter size	Litters per year
Family Camelidae			
Camel (Two-humped *Camelus dromedarius*)	370–440 d	1	Alternate years
Family Cervidae			
Elk (*Cervus canadensis*)	249–262 d	1	1
White-tailed deer (*Odocoileus virginianus*)	208 d	1–2	1
Mule deer (*Odocoileus hemionus*)	203 d	1–2	1
Moose (*Alces alces*)	240 d	1–2	1
Caribou (*Rangifer tarandus*)	327 d + 3 d	1	1
Family Giraffidae			
Giraffe (*Giraffa camelopardalis*)	450 d	1–2	1
Family Antilocapridae			
Pronghorn (*Antilocapra americana*)	7–7½ mo	2	1
Family Bovidae			
Bison (*Bison bison*)	9–9½ mo	1 (Rarely 2)	1
Musk ox (*Ovibos moschatus*)	157–165 d	1 or 2	1
Bighorn sheep (*Ovis canadensis*)	157–165 d	1 or 2	1

* Some of the more common domestic mammals are included as a familiar point for comparison. The short gestation periods in marsupials results in a "premature" birth, after which the young find their way to the pouch where development continues. In some species delayed implantation lengthens the time between fertilization and parturition. In delayed implantation the fertilized egg begins development, then becomes dormant for a period of time before implantation in the uterine wall, where development again continues until parturition.

It becomes obvious from this table that the number of young per litter and litters per year vary. Among those factors which cause this variation are age and physical condition of the female, quality of the food, and whether or not the female is lactating. In white-tailed deer, for instance, a young female produces one fawn, older females twins. Twins and even triplets are more likely when good food is abundant (Cheatum, 1947). Productivity has also been shown to bear a direct relationship to kinds of food available to the beaver. Huey (1956) working in New Mexico found the average litter size in aspen areas to be 4.20 young, in cottonwood areas 2.75 young and in willow areas 2.06 young. Still another factor shown to affect litter size is climate. Litter size has been shown to be correlated also with latitude in the genus *Sylvilagus* in North America (Lord, 1960). In this genus litter size has changed by approximately one for every 250 mi of latitude. Litter sizes of prairie voles (*Microtus ochrogaster*) studied in Kansas were larger at the height of the breeding season than those at the beginning and end (Jameson, 1947). Lactation lengthens the gestation time in some species, as for instance among deer mice (*Peromyscus maninculatus*). Svihla (1932) found the gestation period of four subspecies, other than *bairdi*, was prolonged for as many as 8 days, while in the meadow vole (*Microtus pennsylvanicus*) lactation does not seem to increase the gestation period (Hamilton, 1941).

big-eared bat *(Plecotus townsendii)*, the gestation period varies from 56 to 100 days, a variation probably due to regional temperature differences. Delayed fertilization supposedly confers an advantage because a part of the reproductive activities occur in the fall, allowing parturition to occur earlier in the spring, giving the longest possible time for the development of young before the following winter. For more detailed information the reader is referred to publications by Pearson, Koford, and Pearson (1952) and Wimsatt (1945 and 1960).

Delayed Development   This pattern, like the one just described, is also found in bats. The blastocyst does implant shortly after fertilization but develops very slowly. In the California leaf-nosed bat, *Macrotus californicus (= M. waterhousii)*, the embryo grows very slowly for a 4-month period (Bradshaw, 1962). In Schreiber's bat *(Miniopterus schreibersii)*, of Southern Europe, the egg is fertilized in the fall. The blastocyst implants soon after, but development is very slow and the young are not born until spring, after hibernation (Courrier, 1927). Gestation is from 3 to 4 months longer than in species of the same genus living in the tropics.

Delayed Implantation and Embryonic Diapause   In some species of mammals the egg develops to the blastocyst stage and then lies dormant for a time before it becomes implanted in the uterus to continue its development. Presumably this serves as an adjustment to the estrous cycle or to delay birth to a time of favorable conditions for the newborn young. If the delay is a consistent part of the reproductive cycle, it is termed *obligate;* if it provides a delay when an animal is nursing a litter, it is termed *facultative.*

The discovery of delayed implantation has been credited, by some, to William Harvey, who kept an excellent record of his hunting trips with King Charles I of England. His "field notes" were published in 1651, and translated into English by Willis (1847). Others (Short and Hay, 1966) believe that Ziegler (1843) and Bischoff (1854) should receive the credit.

Since roe deer *(Capreolus capreolus)* were known to mate in July and early August and the young are born in May or early June, quite a discrepancy exists between the gestation periods of the roe deer and those of the larger red and fallow deer, whose mating occurs during the middle of September but whose young are nevertheless also born in May or June. This fact has evidently intrigued both hunters and biologists since early times. From these observations the idea arose that the July-August rut was "designed purely for pleasure" and the true rut occurred in December.

Several interesting theories arose in the middle 1800s, but one early investigator (Bischoff, 1854) examined 150 roe deer uteri and recovered blastocysts throughout the period from August to December. This investigator believed the arrested development was due to the slow growth of the blastocyst once it had reached the uterus. It is said that Bischoff brought down upon himself the wrath of the citizenry surrounding Giessen, Germany, in accumulating the 150 uteri, for he virtually exterminated the local roe deer.

Stieve (1950), who pursued this same research nearly a century later, found that the majority of does mated during the rut of the last half of July and first part of August, while a few, especially the young ones, mated during the last part of November and the first part of December. In those which mated in November and December the eggs developed with little or no delay.

Delayed implantation occurs in many other species of mammals. In the black bear, *Ursus americanus* (Wimsatt, 1963), implantation of the blastocyst is delayed for 5 months. Young are produced in alternate years. For such a large mammal, the young are extremely small at birth—weighing from 9 to 12 oz. They are born in January or February when the female is in a semidormant (but not hibernating) state. This pattern of discontinuous development apparently is found in all species of bears (Craighead, Hornocker, and Craighead, 1969). Delayed implantation has also been found in many other carnivores, including seals (with the exception of the walrus); badger *(Taxidea taxus);* marten *(Martes americana);* fisher *(Martes pennanti);* wolverine *(Gulo luscus);* river otter *(Lontra canadensis);* short-tailed weasel *(Mustela erminea);* long-tailed weasel *(Mustela frenata);* and European badger *(Meles meles).* In these species the "gestation" periods vary, lasting in some cases nearly 2 years. Mammals other than carnivores in which delayed implantation occurs are armadillos *(Dasypus novemcinctus);* lactating rats *(Rattus norvegicus);* mice *(Mus musculus);* hamsters *(Cricetus);* and some kinds of marsupials, when lactating, although a different type of delay has been reported in some species of macropodids, which Sharman (1970) has called *embryonic diapause.* Delayed implantation has been reported for two species of bats—the tropical fruit bat, *Eidolon helvum* (Mutere, 1967), and the Jamaican fruit bat, *Artibeus jamaicensis* (Fleming, 1971).

The ecologic implications of delayed implantation are many, and certainly confer some advantages in those mammals in which it occurs. To northern forms such as the gray seal *(Halichoerus grypus)* it is essential to survival. Birth of young takes place out of the water, at the end of February to mid-March, along the northwest Atlantic shores. This is the only time of year that a suitable platform, in the form of dense pack ice, is available in this area. Mating is essential soon after birth, for later the females are scattered over a tremendous area. Prell (1930) believed that delayed implantation was most significant in mammals originating in the Arctic, where the young must be born immediately after the winter season if they are to survive. Paradoxically, delayed implantation has not been demonstrated in the least weasel whose range both in America and Eurasia penetrates deeply into the Arctic Circle (Wright, 1963).

Embryonic Diapause. As already mentioned, delayed implantation occurs in some lactating marsupials, but in 14 species of the family Macropodidae, a delay occurs whose pattern differs somewhat from delayed implantation. Sharman (1970) has called this pattern embryonic diapause and has pointed out the differences. (1) The quiescent marsupialian embryo (equivalent to the blastocyst of eutherian mammals) is covered by a shell membrane and an albumin layer in which numerous spermatozoa remain viable for several months. (2) This quiescent embryo is composed of a single layer, called protoderm, in contrast to the blastocyst, which is divided into an inner embryonic region which gives rise to the embryo and an extraembryonic region which gives rise to the extraembryonic membranes.

For many years there were reports of prolonged gestation in kangaroos, and in 1912 Carson reported the case of a female red kangaroo which had a young in the pouch 332 days after the death of the male. He believed that because of the double uterus in the kangaroo, double fertilization occurred and one fetus developed quickly, while the other developed slowly. Since 1912 there has been a diapause of about 50 years in our knowledge concerning the true situation of the apparently long gestation period in marsupials. Much of the research to unravel the reproductive cycles in marsupials has been reviewed by Sharman (1970), who also did much of the research. The remarks which follow are based mostly on this paper.

Marsupials ovulate spontaneously, and the gestation period is short, often shorter than the estrous cycle. If insemination occurs at estrus, the resulting pregnancy does not interrupt the estrous cycle. In those species whose gestation period is shorter than the estrous cycle, the sucking stimulus of the newborn pouch young prevents the female from returning to estrus at the proper time and the female will remain in anestrus while the young is in the pouch.

But in those species (14 known species in the family Macropodidae) in which embryonic diapause occurs, the gestation period is either equal to or

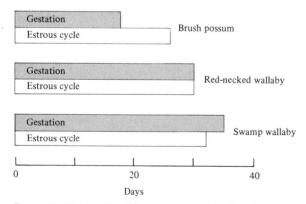

Figure 10–11 Relationship between gestation length and estrous cycle in three marsupials. *(After Short, 1972.)*

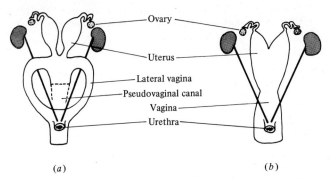

Figure 10-12 Female reproductive tracts of marsupials and placentals. *(After Sharman, G. B., Reproductive physiology of marsupials, **167**: 1221–1228, 27 February 1970. Copyright 1970 by the American Association for the Advancement of Science.)*

longer than the estrous period and the joey (pouch young) does not reach the teat soon enough to prevent estrus, so that a postpartum estrus occurs (Fig. 10–11). Because of the peculiar anatomy of the female reproductive tract, the sperm can descend in the central vaginal canal (Fig. 10–12). To simplify a potentially complicated event, ovulation alternates between the two ovaries. Copulation occurs during the postpartum estrus, and the fertilized egg begins development. When the first young is born in this situation the sucking stimulus of the young joey arrests the development of the embryo. It seems likely that the sucking stimulus of the pouch joey produces a release of oxytocin from the posterior pituitary which arrests the development of the corpus luteum, although it is not completely inactive for it is this corpus luteum which apparently inhibits the return to estrus. In the red kangaroo *(Megaleia rufa)*, this hollow sphere of about 100 undifferentiated cells remains dormant in the uterus for just over 200 days. Add to this a normal intrauterine gestation period of 33 days, and the "gestation period" in embryonic diapause can be 235 days! If the joey dies, normal embryonic development of the quiescent embryo resumes and a new individual can be born in 31 days.

In a red kangaroo a female can have up to three dependents at the same time (Fig 10–13). The first estrus of the breeding season produces a young after a gestation of 33 days. Estrus occurs 2 days later, resulting in a quiescent embryo, which will assume development when the sucking activity of the first young wanes. After the birth of this second young, estrus again occurs, resulting in a quiescent embryo. This also poses a paradox of having two young, widely separated in age, feeding on milk. As discussed earlier, milk of two different compositions is provided by the same mammary gland. In the eutherian mammals, delayed implantation ensures that the young are born at the most favorable time of the year. In kangaroos young are protected from the rigors of the external environment by the pouch. The reproductive delays

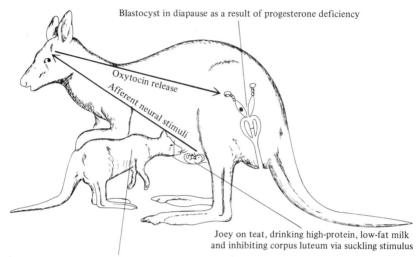

Blastocyst in diapause as a result of progesterone deficiency

Oxytocin release

Afferent neural stimuli

Joey on teat, drinking high-protein, low-fat milk
and inhibiting corpus luteum via suckling stimulus

Young kangaroo returning to drink low-protein, high-fat milk

Figure 10–13 Female red kangaroo *(Megaleia rufa)* with young of three different ages. *(After Short, 1972.)*

in kangaroos may be directed more toward ensuring the female's reproductive success. In Australia the irregular periods of drought could place an ever-increasing stress on the female by the ever-increasing demands of the milk supply. If the joey dies, it can be immediately replaced without the intervention of another estrous cycle. Evidently loss of pouch young during parental flight is another hazard in the life of kangaroos, at least of the larger species. Short (1972) has called the quiescent embryo a sort of reproductive "spare tyre," and Ealey (1963) has discussed the ecologic significance of reproductive delays in kangaroos.

Sharman (1970) points out that there have been many evolutions of viviparity; he believes that viviparity evolved separately in eutherian and marsupial stocks after derivation from a common oviparous stock.

Homes

Many mammals return to regular sites, seasonal or permanent, for hiding, resting, sleeping, hibernating, reproducing, and rearing their young. Most occupants of such sites are intolerant of others of their kind, and defend their sites from intruders. Larger living areas which are not defended have been defined as home ranges. The more sedentary mammals usually have a temporary or a permanent shelter within one or both of these areas. Young born in a permanent shelter, e.g., cottontails *(Sylvilagus)* or mice, are usually naked and helpless at birth, while mammals which produce young more readily able to take care of themselves at birth, such as jackrabbits *(Lepus)* and deer *(Odocoileus)*, do not have a shelter. The terms "altricial" and "precocial," widely used by ornithologists, though not so widely adopted by mammalogists are nevertheless useful in describing the young of mammals as well as the young of birds. Altricial mammals could be defined as those whose young are helpless at birth and require complete parental care for some time. Precocial mammals could be defined as those whose young are at birth able to move about and follow their mother, are usually well haired, and do not require a nest or other permanent shelter.

There are, of course, nomadic mammals that never have a shelter which serves as a home, although they may have a home area. Whales come quickly to mind, although there is now some evidence that at least some species of whales have a home area. The seals are another group which have no permanent shelters, although they come to land to reproduce. The young of both these groups of mammals are precocial.

The ungulates are still another group without permanent shelters; they wander throughout the year within their home ranges, which in some cases may shift with the season, thus producing a migratory pattern. Many species of cervids seek warmer areas in winter. Those that live in mountainous areas "come down" in the fall. In western North America, the deer *(Odocoileus),* elk *(Cervus canadensis),* and moose *(Alces americana)* winter in the lower areas. In summer the home ranges of matriarchal groups of the red deer *(Cervus elaphus)* of Scotland are near the mountaintops and may overlap. In winter their ranges are on the lower slope (Darling, 1937). The caribou *(Rangifer tarandus)* of the Northwest Territories in Canada seek the shelter of the trees in winter but move out to the open, windswept tundra for the summer, perhaps in part to escape the tormenting by insects. In the Great Lakes states of the United States the white-tailed deer *(Odocoileus virginianus)* move to "winter yards," usually swamps of white cedar.

A variety of burrows, lodges, dens, and tree cavities serve as homes for altricial mammals. They provide a relatively stable environment by reducing both the body-to-air temperature differential and widely fluctuating relative humidities. In some cases they serve only until the young are nearly grown. The timber wolf *(Canis lupus)* for instance, has a den in which the young are born and where they remain until summer or early fall, when nearly grown. In the winter months they simply "bed down" on the surface in some sheltered area. In the spring the pregnant female, with the help of a male, or other adults, starts preparing a den or cleaning out an old one. It would be difficult to categorize neatly the variety of homes used by mammals, and I have chosen rather to organize this discussion in a taxonomic sequence.

MONOTREMES

The egg-laying mammals use burrows for escape and nests for the young. The quills, or spines, covering the echidna *(Tachyglossus aculeatus)* protect it when the echidna digs vertically into the ground whenever it is disturbed. Echidnas also dig a nest burrow, but the information on this is fragmentary. They have a marsupium in which they can carry the egg and when it hatches, the young. Because quills develop on the young, females cannot long carry it in the pouch, and the evidence seems to indicate that young are then left in the burrow. Barret (1942) found a young echidna in a burrow dug in sandy soil. Hodge and Wakefield (1959) found a female and a baby echidna at the end of a 122-cm (4-ft)-long burrow. For 3 weeks the young remained in the nest, and although the mother was never found in the nest, the young remained healthy. Griffiths

(1968) reported an incubation period of 10 days. The newly hatched young weighed 378 mg, and "the egg tooth was evident."

The duck-billed platypus *(Ornithorhynchus)* does not have the stiff quills of the echidna, nor does it have a marsupium. Its burrows serve somewhat different functions. Burrell (1927), in his book on the platypus, devoted a complete chapter to "the nesting apartment." In general the burrows are of two kinds. One is a general living quarters, which is usually a semicircular excavation beneath roots of large trees. The other is the nesting burrow, built and occupied by the female, in which the eggs (usually two) are laid. The female plugs the burrow at intervals. The nesting burrow is a tunnel, anywhere from 38 cm (15 in.) to 16 m (60 ft) in length, about 38 cm (15 in.) below the surface. The nest is made of grass, gum leaves, reeds, and sometimes other materials. The incubation period is believed to be 14 days or less.

INSECTIVORA

The moles (like another completely fossorial mammal, the pocket gopher) dig two kinds of tunnels, the surface runways or ridges, which are used for collecting food, and deep passages which are used for thoroughfares and for everyday living (Hisaw, 1923). The surface runways are formed by pushing the earth up into ridges; the deep passages are formed by digging the earth loose and either carrying it to the surface or putting it in discarded tunnels. All manipulation of dirt is done with the front feet (W. B. Carpenter, 1857; Herrick, 1892; and Hisaw, 1923).

The shallow trenches of the European mole *(Talpa europaea)* were formerly believed to be made only during the breeding season, and were known as "rutting runs" or, with a French flair, *traces d'amour.* But Godfrey and Crowcroft (1960) pointed out that although they are often associated with mating, they are also formed at other times. These are also called "open runs," and together with the regular *surface runs* and *deep tunnels,* 10 to 20 cm below the surface, compose the two kinds of tunnels. The nest of the European mole is usually an enlarged part of the deep tunnel, about 24 cm (8 in.) in diameter, filled with grass or dry leaves. In the floor of this chamber, a vertical tunnel, called the bolt hole, serves for drainage or escape. A breeding nest is constructed by the female just before parturition.

Hamilton (1931) has described the runways and burrows of the swamp-dwelling star-nosed mole *(Condylura cristata).* The tunnels sometimes open under water, and the long tunnels may sometimes be completely under water. Mounds of excess dirt appear much like the work of a crayfish. Spherical nests for the reception of young are built well above the water, under logs and fallen trees. They are composed of nearby material such as dead leaves and dead grasses, which are not shredded.

Shull (1907) and Hamilton (1929) have described the nests of the short-tailed shrew, *Blarina brevicauda* (Fig. 11–1). Shull (1907) found that the nests, 12 to 15 cm in diameter, were usually made of leaves, of grass, sedge, nettle, goldenrod, and ash, which were not shredded or torn into smaller pieces. One nest was composed entirely of the hair of a meadow mouse. Hamilton (1929) found two breeding nests and several resting nests. Both were under rotten logs. In one nest, two runways led in opposite directions and a third went

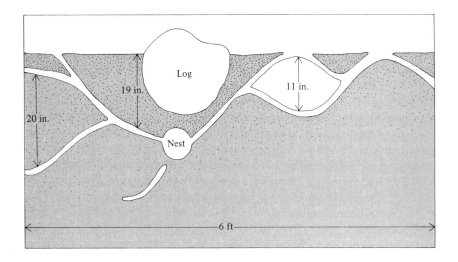

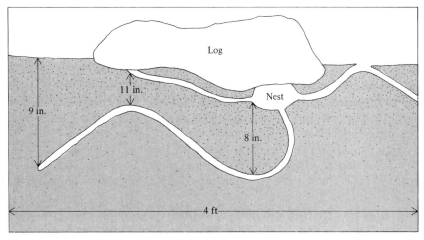

Figure 11–1   Two diagrams of nests and runways of short-tailed shrew, *Blarina brevicauda.* *(From Hamilton, 1929.)*

directly down. The nest was composed of dead elm leaves, 24 cm (8 in.) across the top and 12 cm (5¾ in.) deep. The female left the nest when it was uncovered, and seven young about 9 days old were found in it. The second nest, 30 cm below the surface, was composed mostly of unshredded maple leaves.

CHIROPTERA (Bats)

The habitations of bats are paradoxically either simple or unique. There are no specialized habitations for the individual young. Many are carried about by the mother, to whose fur and breasts they cling. The young are helped in this by the specialized and backwardly hooked milk teeth and of course by the claws. As they grow older and heavier, they are left behind to cling to the walls of a building or cave. Solitary bats, such as the red bat *(Lasiurus borealis)* and hoary bat *(Lasiurus cinereus),* carry their young until they are quite large and then leave them to cling to twigs or bark. Photographs of red bat females carrying young make it appear as if they are carried until at least two-thirds grown. The young of *Myotis lucifugus* are carried by the mother for the first 3 or 4 days after birth, but after that they seem to be left in the roost (Griffin, 1940).

In many species of bats which roost in colonies the young are never carried, and colonies consist exclusively of adult females and young. These roosts are generally called maternity wards, or nursery colonies. The little brown bat *(Myotis lucifugus)* and big brown bat *(Eptesicus fuscus)* are common bats which leave their young in roosts or small clusters (Griffin, 1940). In Maryland the big brown bat maternity colonies were found in sheltered places such as chimneys and hollow trees (Christian, 1956). The young of the pallid bat *(Antorozous pallidus)* cling to the females for the first 2 or 3 days and then form nursery colonies. The most dramatic and widely known maternity wards are those of the Brazilian free-tailed bat *(Tadarida brasiliensis),* whose range extends across the southern half of the United States and into Mexico and includes the famous Carlsbad Caverns of New Mexico. There is little parental care beyond nursing of the young in these bats, and females are rarely found with young attached. The young cling together in large masses, separate from the adults (Krutsch, 1955). An estimated 100 million females migrate north from Mexico each season. Bracken Cave, near San Antonio, Texas, is estimated to have a population of 20 million adults. Each female has only one offspring and returns to the nursery after her nightly foraging. The males move north later than the females.

Though the habitations selected for young bats may not be too unusual, resting places other than trees, caves, or buildings can be quite specialized. G. M. Allen (1939) in his book on bats has a chapter entitled, "Where Bats Hide." Among the more unusual roosting places he mentions are old birds' nests, used by many species; the hollow joint of a bamboo stem, entered by a mere crack in the culm, used by one of the smallest of bats, *Tyloycteris pachypus* of the East Indies; and the alteration of leaves, particularly of palms, "by biting partly through large leaves or fronds so that the terminal half bends to form a peak, beneath which they may rest secure from the sun and rain," practiced by several species.

CARNIVORA

Bears

Dens of the black bear *(Ursus americanus)* have been described by many workers, especially because they are of interest in relation to dormancy studies. Black bears den alone, except for females with cubs, or cubs which have recently left their mother. A variety of sites are used by bears for winter dens, including caves, windfalls, holes in hillsides, excavations under logs or roots and hollow trees, old hot springs and geyser openings (in Yellowstone), and drainage culverts. Some use little or no shelter; hunters have stumbled over snow-covered bears that have just curled up on the ground. Many individuals, especially pregnant females, line their dens with grass, leaves, and clubmoss raked in from an area up to 26 m (82 ft) from the den entrance. Occasionally the den entrances are plugged with the same materials used for nest lining. The cubs, weighing less than 0.5 kg (1 lb), are born in January; there are usually two, but may be from one to four (Erickson, 1964; Morse, 1937; and others).

The denning habits of the polar bear *(Ursus maritimus),* spending its entire life within the Arctic Circle and out of the range of human habitation, have long been a matter of mystery and conjecture. Recent research techniques and perseverance, especially the work of Harington (1968), have shed light on this matter. Both males and females use shelters as a temporary escape from bad weather (in the Arctic?), or to return to after hunting and feeding. These natural shelters may be dried-up stream beds beneath protective river ice, snow bridges, in caves or icebergs. Dens may vary from small shallow pockets in the snow, to room structures with long entryways. Pregnant females and females with young occupy more permanent shelters, or maternity dens, which are also more elaborate, sometimes with several rooms, each at a different level, and a ventilation hole leading to the exterior (Fig. 11–2). Female polar bears also become dormant or lethargic during the winter, but the males apparently remain active throughout the winter, except for a few which may be dormant for short periods (Harington, 1968).

Raccoons

Raccoons *(Procyon lotor)* are most often associated, in our minds, with water and hollow trees. Hollow trees serve as dens, and often as nests for the young (Cabalka, 1952; Steuwer, 1943; and Whitney and Underwood, 1952). In an

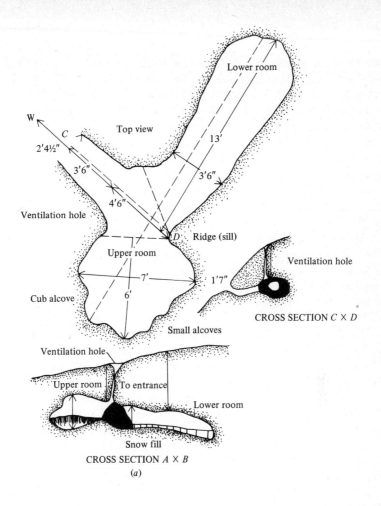

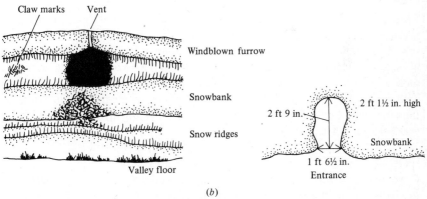

Figure 11-2 A maternity den *(a)* and temporary den *(b)* of the polar bear, *Ursus maritimus.* *(From Harington, 1961. Reproduced by permission of Information Canada.)*

early study in Michigan a lack of tree dens was cited as one of the most important factors limiting raccoon numbers (Steuwer, 1943). More recent studies have shown that raccoons avail themselves of a variety of sites, at least for resting. In Kansas, raccoons have been reported using ground holes (Stains, 1956). In Iowa, raccoons used ground beds in grassy fields (Cabalka, 1952). In east-central Illinois, four raccoons carrying radio transmitters were always located on the ground in herbaceous growth beneath tree canopies (Ellis, 1964). In a forested area in east-central Minnesota, the fall resting sites of four intensively studied individuals, carrying transmitters, were ground beds, in a cattail marsh, in elder or cedar swamps; only a few were in trees (Mech, Tester, and Warner, 1966).

With this informational background, the raccoons' well-known, but poorly documented, movement out onto the prairies was not so surprising. At Sand Lake National Wildlife Refuge, Columbia, South Dakota, raccoon were reported as uncommon in 1936. In 1954, 433 individuals were taken on the refuge. Apparently this was the peak population. Since then the population has fluctuated but never reached the level of 1954 (Lyle Schoonover, personal communication). The raccoons made use of old muskrat houses and culverts for both nesting and resting. At the lower Souris Refuge, Upham, North Dakota, no raccoons were seen until 1935, when tracks of one were seen. In November 1938 four raccoons were taken in traps set for mink. In 1956–1957, 109 were taken, and in 1957–1958, 186 were caught, a peak, at least through 1960 (Merrill Hammond, personal communication). On the prairies they use old fox and badger dens and just let the snow drift over them in phragmites and cattail marshes. In colder winters there is practically no emergence from the dens. Here, then, is a species, seemingly once restricted to the use of tree dens for its survival, which has become adapted to use a variety of dens.

Weasels

There are apparently no marked differences between nests of the least *(Mustela nivalis),* short-tailed *(Mustela erminea),* and long-tailed *(Mustela frenata)* weasels. Their homes can be under tree roots, stumps, logs, woodpiles, or rockpiles. Sometimes the nest is a shallow, reworked burrow of a ground squirrel, mole, or pocket gopher. The breeding nests consist of leaves and grasses mixed with fur of rabbits, meadow mice, and other prey. Near the entrance and in the burrows are remains of partly eaten prey. The young are born blind, naked, and helpless (altricial).

Canidae

Coyotes *(Canis latrans),* wolves *(Canis lupus),* and foxes (*Vulpes* and *Urocyon*) use dens for rearing their young and may return to the same den year after year. Coyotes are believed to be monogamous, and if one of the pair is killed, a new mate is brought to the den. On occasion there may be more than one litter in a den; this is believed to be the result of polyandry, i.e., the male is polygamous (Young and Jackson, 1951). Coyotes sometimes enlarge the abandoned burrows of other animals. Coyotes begin the preparation of the burrows several weeks prior to parturition, as do foxes and wolves. Several dens are cleaned out, and if one den is disturbed, the young are moved to another den.

The dens of the wolf *(Canis lupus)* have been described by Murie (1944), Mech (1966, 1970), and others. In this species, as in other members of the family Canidae, the dens are used for rearing the young. Murie (1944) discussed in detail the life around the den for four wolf groups in Mt. McKinley National Park in Alaska. Two dens were located on an island, one-third of a mile wide and a mile in length, in the Toklat River. One of these dens was a renovated red fox *(Vulpes vulpes)* den. One entrance out of eight or nine was enlarged by the wolves; it passed under the roots of a spruce to a chamber in the center, about 1 m (3 ft) below the surface. In 1937 and 1940 seven pups were present in this chamber. The entrance to this den was 35 cm high by 52.5 cm wide. A trail of 300 m connected this den with a second den, which contained four pups in 1941. Another family, whose den Murie did not find, consisted of at least six pups, cared for by three adults. The sex and age of the extra adult could not be determined. A third den was also a renovated red fox den. Since fox dens, many unoccupied, were common in the region, it was not surprising that timber wolves preempted them. Murie wiggled into this burrow, whose entrance was 35 cm high and 53 cm wide. There was a right angle turn 5 m from the burrow entrance. At the turn was a rounded and worn hollow, used as a bed by an adult. From this turn the burrow slanted slightly upward and at a distance of 5 m was a chamber containing six pups. Two adults were seen at this den, but even after all this disturbance they did not move the pups.

A trapper at Grand Marais, Minnesota, along the north shore of Lake Superior told this writer of finding a den in a hollow, downed red pine, which the female could not enter, but when she appeared near the end of the log the young would come out to nurse, seemingly in response to some kind of a signal. Wolf pups leave their dens when 8 to 10 weeks old but remain all summer and even into September at "rendezvous" sites, until they eventually join the packs. These sites are characterized by a system of trails, beds, and activity area, and a "view" over the surrounding territory (Murie, 1944; Joslin, 1967).

Foxes

Red fox *(Vulpes vulpes)* territoriality, home range, and homes have been discussed by Komarek (1939), Murie (1936 and 1944), Livezey and Evenden (1943), Sheldon (1950), Scott (1943), and others. Red foxes, along with other members of the canids, are considered to be monogamous. Scott (1943) described the life history events of red foxes in Iowa. Their activity centered on the family as a social unit, progressing through a winter breeding and predenning period, a spring and early summer denning period, a late summer and early fall period when the family is not as closely tied to the den or to one another as previously, and then a period of dispersal. During the first part of March pairs of foxes were moving together or bedded down not far apart. In late February they began enlarging and cleaning out several dens, one of which would be used for the young (Scott, 1943) (Fig. 11–3). Murie (1944) found that dens in Mt. McKinley National Park, Alaska, were often in sandy loam, but some were on open flats, others were in the woods. He described a typical den as having six or seven entrances. Livezey and Evenden (1943) described an Oregon fox den which extended for 40 m. Komarek (1939) recorded a burrow which went to a depth of 12 m. Sheldon (1950), in New York, found that foxes were gregarious during the denning season, and that two litters occupying the same den was not unusual. He further found that vixens occasionally moved litters into dens already occupied by other litters.

The den site of an arctic fox *(Alopex lagopus)* has been described by Gunderson, Breckenridge, and Jarosz (1955). The burrows occupied the greater part of a pear-shaped gravelly knoll which varied from 1 to 3 m in

Figure 11–3 Den of a red fox *(Vulpes vulpes)* at the base of an elm tree.

height. The mound, about 23 m in diameter at the large end and 54 m in length, was covered with a luxurious stand of vegetation, easily delineated from the air. This was probably the result of added fertilizer, including nitrogen from the urea and from decomposing, partly consumed prey such as fish from the nearby Back River. In the center of the mound was a bare gravel area about 1.5 (5 ft) across, which served as a "watchpost." From here an adult could scan the surrounding countryside for potential prey or approaching predators. At this den John Jarosz and I observed an interesting behavior. After watching the den for several hours from a distant hill we began to think the young had been moved. Then an adult appeared on top of the mound near the area of the "watchpost," walked into the area and sat down. It carefully scrutinized the surrounding countryside for a minute or two. As if on a signal the three kits popped out of the burrow almost simultaneously, their tails waving vigorously. There was a great deal of frolicking around the adult by the three kits. As in the timber wolves, discussed previously, the young must have appeared in response to some signal. Conversely, something also kept them underground until an adult appeared.

Lagomorpha

The Lagomorpha contain two families, the Ochotonidae and the Leporidae. In the family Leporidae (seven genera) the newborn rabbits are altricial, naked, and blind, and are placed in a nest, while the young of hares (one genus) are precocial, not naked or blind, and able to move shortly after birth. Bailey (1926) gave this description of "nests" of the white-tailed jackrabbit *(Lepus townsendi)*, "The young are usually found in some shallow burrow or concealing cavity in the ground. . . ." These very shallow depressions are given the name *forms*. In my youth in northwestern Minnesota, I found many young jackrabbits in the spring, but they were either completely exposed or at best protected by only a very shallow depression. The young were most difficult to see, being surprisingly well camouflaged. Bear and Hansen (undated), in their study of the white-tailed jackrabbit in southern Colorado, made this terse statement, "Newborn young jackrabbits are rarely obtained for study." Bailey (1926) observed that even "on the short-grass prairie they absolutely disappear from view when squatted flat with ears laid low and ears tucked in."

The nests of the cottontail *(Sylvilagus floridanus)* are well prepared to receive the blind and naked young. They are often found in lawns, gardens, and golf courses. The female scoops out a hollow in the ground and lines it, first with grass and then with hair plucked from the venter. The young are concealed by arranging dead grass over the nest cavity (Lord, 1963). A frequent statement heard by a museum worker in spring, along with the report of finding a cottontail nest, is this, "But the young must be abandoned for we have never seen a female near the nest." Lord (1963) constructed an ingenious device to record the female's presence at the nest and obtained information

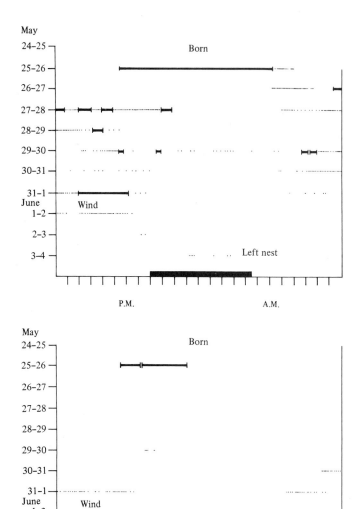

Figure 11-4 Nest attendance by female cottontail *(Sylvilagus floridanus). (From Lord, 1963.)*

on two nests. He learned that nest attendance was continuous for a long time the first full night after the birth of the young (Fig. 11–4). For one nest this was a 12-h period. Then the females were absent, one for a period of 55 h, the other for 22 h. Thereafter the female visited the nest for short, variable intervals. With such a routine it would be difficult to catch sight of the female at the nest. The young remained in one of the nests for 10 days and in the other for 11 days.

RODENTIA

Woodchucks

The woodchuck *(Marmota monax),* originally a forest animal, has now extended its range into borderline areas which support trees, such as gallery forests extending out onto the prairies. They are excellent diggers, with four strong claws on each front foot, and strong incisors to tear away roots. Hamilton (1934) did an extensive life history study of the rufescent woodchuck *(Marmota monax rufescens)* and described the burrows and nests in detail. The following descriptions are from his work.

Woodchuck burrows differ from those of skunk, rabbit, and fox because fresh dirt is constantly being added to the entrance mound of the woodchuck's burrow. Evidence of occupancy of the burrow is furnished by foraging trails leading from one burrow to another and from the main burrow into surrounding fields.

Although burrows go down 5 m, the nest is not located at the deepest part of the burrow. Most often it is located within 65 cm of the surface. The nests seldom measure more than 38 cm in diameter and half that in height, and are constructed of the nearest material. In the woods the fallen leaves of deciduous trees are used; in the fields, dead leaves of goldenrod and asters. Some nests are constructed wholly of grasses. Often the animal travels long distances to gather nesting materials.

Woodchuck dens resemble those of other mammals in serving as temporary havens and, in the case of abandoned dens, as permanent homes for other animals. For instance the den might serve as a temporary refuge for cottontails *(Sylvilagus floridanus),* but red foxes *(Vulpes vulpes)* use as homes the dens made by woodchucks they conquer. Skunks *(Mephitis),* weasels *(Mustela),* chipmunks *(Tamias),* and even house cats have been found to use woodchuck dens.

Black-tailed Prairie Dogs

The burrows of prairie dogs *(Cynomys ludovicianus)* are of great interest, for they not only serve as homes in the usual ways but also help shape the social organization of a prairie dog colony (King, 1955) (Fig. 11–5).

The external appearance of the burrows, based on the excavated soil,

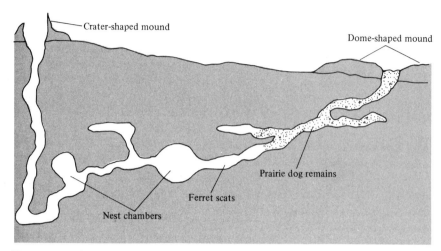

Figure 11–5   Composite drawing of 15 prairie dog *(Cynomys ludovicianus)* burrows. *(From Sheets, Linder, and Dahlgren, 1971.)*

varies from practically no soil to a craterlike mound, 1 m high and 2.5 m across. A dome crater is usually the larger type, shaped like a rounded dome with the burrow opening in the center. The dome crater is made from soil excavated from the burrow. The rim crater is a smaller, crested mound made by scraping the top soil into a mound about the entrance. Thus a dome crater is present where the burrow had its beginning, a rim crater where the burrow came up to the surface from below. The burrows within a coterie (a socially organized subunit in the town) are connected, and King (1955) believed the burrow system to be the basis of the coterie organization.

The nest chamber was 45 cm in diameter, 30 cm in height, 3.5 m from the burrow entrance, and 1 m below the surface. The nest was composed of fine dried grass. The density of burrows in King's study was 57 per acre in 1948 and 52.5 per acre in 1950. Sheets, Linder and Dahlgren (1971) measured 18 burrow systems and found they varied from 1 to 12 m (3 to 14 ft) deep and from 4 to 44 m (13 to 109 ft) long.

The black-footed ferret *(Mustela nigripes)* was reputed to live entirely within prairie dog towns. Its dwindling population was blamed on the extensive poisoning campaigns against the prairie dogs. But black-footed ferrets were rare long before poison campaigns became extensive, and it may be that it was never a common animal. Poisoning has further reduced their numbers. Recent methods of searching and intensive search have led to a belief that there may be more black-footed ferrets than were thought to exist a decade ago.

Thirteen-lined Ground Squirrels

One of the most detailed studies of the burrow system of the thirteen-lined ground squirrel *(Spermophilus tridecemlineatus)* was that of G. E. Johnson (1917). Near South Chicago he dug out more than a hundred burrows. Most of them (67 percent) were less than 3.5 m in length and 32 cm in depth.

Nests were most often found in the longer, deeper burrows; sometimes there were several nests in the same burrow. These spherical nests, from 10 to 25 cm in diameter, were made of dry grass, "well woven" together. Nests were also used for the storage of food, as well as for care of the young and resting (Fig. 11–6).

The habit of plugging the holes to the burrow during dormancy (estivation or hibernation) must have great survival value or it would not be so prevalent. It is a protection against predators and, perhaps more important, against unfavorable weather conditions.

Wade (1930) studied the behavior of several species of ground squirrels as they were entering into and during estivation and hibernation near Lincoln, Nebraska. He found that ground squirrels constructed cells to one side of the

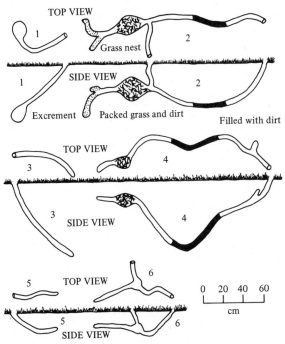

Figure 11–6 Structures of thirteen-lined ground squirrel *(Spermophilus tridecemlineatus)* burrows. *(From Rognstad, 1965.)*

burrow to serve as hibernating nests. The cells were just large enough to receive the animal and enough nest material for its protection. The burrow extended below the nest, forming a drain.

Eastern Chipmunks

Nests and 30 burrow systems of the eastern chipmunk *(Tamias striatus)* were studied in Wisconsin by Panuska and Wade (1956), who made a clay model of a system which extended for 12 m and included many chambers and passageways. They differentiated the systems into two basic types: (1) simple systems ranging in depth from 30 to 70 cm and (2) extensive systems which descended to at least 50 cm. The floors of large chambers were regularly from 50 to 75 cm beneath the surface. They wrote, "Considering the previously mentioned report on frost penetration it appears that the chipmunk's burrow system does not penetrate beneath the frost line." D. L. Allen (1938) described a single burrow that extended 70 cm below the surface and had two entrances. Eastern chipmunks *(Tamias striatus)* living under the patio of the author's home in Minneapolis, Minnesota, plugged their burrow entrances during both short and long periods of dormancy; such plugging probably serves the same purpose for chipmunks as for ground squirrels.

Tree Squirrels

Both the gray squirrel *(Sciurus carolinensis)* and the fox squirrel *(Sciurus niger)* use two well-defined types of nests. Descriptions of gray squirrel nests are taken from Fitzwater and Frank (1944), and those of fox squirrel nests from Allen (1943). One type is found in tree cavities. These are used primarily for wintering and rearing the young, in other words, as permanent residences. The other kind are leaf nests, which may be transitory structures but may also be the usual shelter where the trees are not large enough for dens.

Pocket Gophers

Anyone who in his youth has trapped pocket gophers for "pocket money" has some idea of the architecture of a pocket gopher burrow system. For one thing the trapper usually found that there is a "Y" just below the surface where the mound has been formed. He also found trapping most successful in the spring, and learned that rarely is a pocket gopher found completely above ground.

Pocket gophers are admirably adapted for a fossorial life (Hill, 1937). On each front foot they have three very long claws, which Howard (1953) has calculated grow at a rate of 0.23 mm per day, or 84 mm per year. These adaptations of the front feet are useful to a majority of fossorial mammals.

Extensive ecologic work done on several species of pocket gophers has included descriptions of their burrows. These burrows follow the general pattern of a subsurface system plus a deep system of runways which are complex and dynamic. The subsurface systems are more extensive and are more prevalent in the spring. During the hot, dry months of late summer gophers are active at deeper levels, presumably because moisture and temperature are more favorable there. This may help to explain why my youthful trapping was most successful in spring. Pocket gophers live individually, and there is a size difference between the male and the female, the female being the smaller. This sexual dimorphism is reflected in the diameter of their burrow systems. Both Wilks (1963) and Kennerly (1959) measured these differences in the plains pocket gopher *(Geomys bursarius)*. Wilks (1963) found the average diameter of females' burrows to be 7.6 cm and of males' burrows to be 8.3 cm. Kennerly's measurements showed a smaller difference—7 and 7.2 cm.

The nest of a female plains pocket gopher was $18 \times 9 \times 16$ cm ($7 \times 3\frac{1}{2} \times 6\frac{1}{2}$ in.), constructed of three kinds of grasses—purpletop *(Tridens flavus)*, yellow bristlegrass *(Seteria lutescens)*, and smoothbrome *(Bromus inermis)* (Smith, 1948). No fecal material was found in the nest or main burrow, but such material was present in the short plugged-up sections. The total length of the main burrow was 63 m (206 ft), and the diameter was 7.5 cm (3 in.) wide and 10 cm (4 in.) high. Smith (1948) calculated that more than 59 m^3 (70 ft^3) of earth had been moved to make this main burrow. Downhower and Hall (1966) excavated three burrow systems and also found the subsurface burrow systems, some of them several hundred feet long, to be more extensive than the deep ones; the subsurface burrows were at a depth of 13 to 17 cm (6 to 10 in.). The deep runways attained a depth of $1\frac{1}{2}$ m ($4\frac{1}{2}$ ft). Only one nest was found, and it was about $3\frac{1}{2}$ cm (1 ft) above the deepest part of the burrows. They believed that deep runways and nests serve as a refuge from the extreme temperatures of both summer and winter.

Davis, Ramsey, and Arendale (1938) believed that the pocket gophers of the southern plains (called *Geomys breviceps* at that time) required at least 10 cm (4 in.) of sandy topsoil for burrowing and that this requirement restricted their range. Undoubtedly soil plays some part in the distribution of various species of pocket gophers. In Minnesota the northern pocket gopher *(Thomomys talpoides)* is restricted to the heavy clay soils of the very northwestern corner of the state, and the plains pocket gopher *(Geomys bursarius)* is found over most of the rest of the state (Gunderson and Beer, 1953). Quimby (1942) believed that *Geomys* encounters difficulty in digging its larger burrows in the heavy clay soil, and that *Geomys* prevents *Thomomys* from invading the lighter soils.

The environment of fossorial mammals, such as pocket gophers, certainly differs from that of mammals living above the ground surface. For one thing, the temperatures are more stable underground (Fig. 11–7). For another, the humidities are high in the burrows. Kennerly (1964) showed that relative humidities in the burrows of the plains pocket gopher *(Geomys bursarius)*

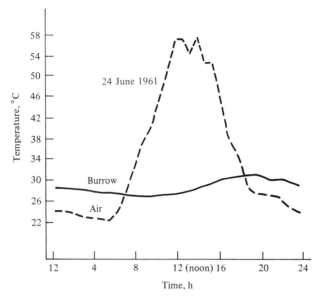

Figure 11-7 The burrow does much to moderate the environmental temperature of pocket gophers *(Geomys bursarius). (From Kennerly, 1964.)*

ranged from 86 to 95 percent even in July, when the soil of the burrow floor contained 1 percent water. Such high humidities reduce the water balance problem. Most striking were the concentrations in the burrow atmosphere of CO_2, which ranged from 10 to 60 times that contained in atmospheric air.

Beavers

The beaver *(Castor canadensis)* and the muskrat build both burrows and above-surface houses. The habitation of beaver includes the houses, dams, reservoirs, and canals, and it would be difficult to leave any of them out of a consideration of the beaver's home. This description has been drawn from several sources (Bailey, 1926; Bradt, 1938; Henderson, 1960; Knudsen, 1962; Longley and Moyle, 1963; and Rutherford, 1964).

The beaver usually inhabits a stream of moving water of moderate depth. A source of rather highly specialized food, the bark of aspen, cottonwood, or willows, is also necessary. To have water in sufficient quantity near a suitable source of food usually dictates that a dam be built. The dam is built of sticks and mud, and one dam provides the impoundment for the main lodge. Secondary dams are built which function as transportation aids. To serve as shelter and for protection a lodge or house is commonly constructed by the animals, and studies have shown how a beaver lodge can moderate extreme climatic conditions (Fig. 11–8). Sometimes a house is in water 1.5 to 1.9 m (5 or 6 ft) deep, sometimes it is on the bank of a lake or stream, sometimes on a floating marsh, but always an exit leads from the house into water deep enough to remain open all winter under the ice (Bailey, 1926). Most new beaver houses begin where a feeding spot has been in use, and so there are remains of sticks

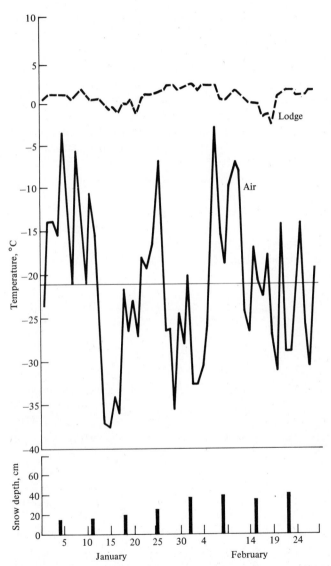

Figure 11–8 A beaver *(Castor canadensis)* lodge's interior temperature is greatly moderated by insulation of logs and snow. *(From Stephenson, 1969.)*

and saplings from which the bark has been peeled. The first step is a water hole or entrance. Old, peeled sticks are added, and bushes are cut and added to the pile, as well as limbs and sections of small trunks of trees. Finally mud and sod are pushed up over the sides and top. A house is never "finished" as long as it is occupied. Sometimes the top of a new house is so porous that on a cold day one can see the water vapors coming from the inside (Fig. 11–9).

There is usually only one room. The floor of the room, or platform, strewn with grass, roots, and bits of bark, is a few inches above water level. The altricial young are usually born on a softer bed of grass and rootlets, which may also serve as food before the young are able to leave the nest. The beavers bring food into the house to be eaten, especially in the winter. Often one plunge hole is at the edge of the platform; thus the entrance is sealed with water. This keeps out many but not all predators. When the beaver leaves the house it does not surface nearby, but may go up to a half mile away before coming to the surface (Bailey, 1926). Beaver houses which have been in use for many years may reach sizes up to 4 m high and 12 m wide, with chambers large enough to hold a person.

Beavers may also use burrows, and where no lodge is present, these beaver receive the colloquial name of "bank beavers," which is not a different kind of beaver. Some burrows are small and used temporarily during floods. Others have several entrances and one or more chambers. The burrows have their openings under water. In Kansas, tunnels vary from 3.5 to 16 m (12 to 30 in.) in diameter; underground chambers may be anywhere from 1.8 to 17.5

Figure 11–9 A beaver *(Castor canadensis)* lodge showing extensive snow cover.

m (6 to 50 ft) back from the bank and large enough "to accommodate two grown men sitting up" (Henderson, 1960). Sometimes beaver use a combination of bank dens and lodges.

A colony is basically a family unit. In Michigan 57 colonies averaged 5.1 individuals (Bradt, 1938); in Minnesota a colony contained from 5 to 13 beaver (Longley and Moyle, 1963). Most workers agree that the 2-year-olds are driven from colonies prior to the appearance of a third litter (one litter per year).

Wood Rats

It would be inappropriate in any discussion of mammal homes not to include the legendary activities of rats of the genus *Neotoma,* especially the western kinds or pack rats, which have the unusual habit of putting curious objects in their shelters. Pack rats have a variety of common names—among them, brush rat, cave rat, and trade rat. Tales of the early Western travelers, who slept in tents, covered wagons, or even more permanent shelters, were laced with stories of things that pack rats stole or traded. Such objects were a bizarre assortment, including money, watches, false teeth, and in mining camps, dynamite! Even without these embellishments a wood rat's home is remarkable. The nest itself is a globular mass about 25 cm in diameter, woven of shredded bark of juniper or sagebrush, yucca fibers, leaves, soft grasses, and other materials. The nests may be in burrows, as for instance, an old armadillo burrow, root crevices, or overhanging rocks. Around the nest a protective house of twigs, sticks, and other material is usually, but not always, built. Some houses contain several nests. The complexity of the structure varies with species and with the availability of material. Construction of new houses begins at sites used for the deposit of feces and urine.

That complexity of houses varies with species was shown in a study of the ecology of the wood rats of Colorado by Finley (1958). The Mexican wood rat *(Neotoma mexicana)* had the weakest collecting instinct of any species of wood rat in Colorado, and built stick houses only where a natural rock shelter was present. One unusual nest was in a vertical cleft between two blocks of sandstone. The cleft was crammed to a height of 1.5 m with sticks. Among the sticks were a cervical vertebra and the pubis of a 10- to 12-year-old Ute Indian who had been buried 40 m from this den many years before. There were also a partially chewed deer scapula and two cottontail bones. The bushy-tailed wood rat *(Neotoma cinerea)* had what Finley (1958) called a "moderately strong collecting instinct."

Muskrats

Muskrats *(Ondatra zibethica)* are familiar inhabitants of most of North America where suitable habitat exists. They are not found in Florida or parts of adjoining states. No other fur-bearing mammal (economically speaking) has been more thoroughly studied than the muskrat, which derives its common name from the odor produced by glands situated beneath the ventral skin near the external genitalia. The most comprehensive publication on muskrats is that of Errington (1963), and his descriptions of their burrows, houses, and runways have been freely used here.

Muskrat habitations may be divided into two types, lodges and burrows, each with numerous variations. Errington (1963) considered lodge building as a behavioristic advance over burrowing.

The simplest burrow occurs when a muskrat crawls into or under loose vegetation. During the spring dispersal, transients sometimes dig short, shallow burrows with underwater entrances, which they occupy for a short time, from several hours to several days. These burrows sometimes have enlarged chambers above the water in the banks. The other burrow system, which may be decades old, contains mazes of caved-in and renovated diggings, with old and new chambers at different levels. The burrows into the bank may be short, with many lateral ramifications following the water's edge, or they may lead inland from the bank 90 m or more.

Lodge building begins in late summer and reaches a peak just after the first frosts. It often begins with a sitting place, just as with pack rats and beaver. The sitting place may be on solid ground or on floating vegetation. After a preliminary heap has been made, a passageway from beneath and a chamber are hollowed out. The lodges, up to 2 m above the water surface, may have multiple chambers, not necessarily connected. In lodges with bases of 2.5 m or more in diameter, rings of chambers connected by tunnels encircle a solid center. But in the typical lodge, the dwelling chamber is centrally located, with its bed a few inches above water and with two or three plunge holes leading downward into the water. The nests are lined with shredded vegetation. These are the nests in which the naked, helpless young are born. The outer sides of the lodge may contain temporary nests for transients, or even a nest for young, or on occasion for the male consorts of the female whose young are inside the lodge. Errington (1963) wrote, "A concept of a loose monogamy would seem most consistent with reality," for the muskrat.

The Smaller Rodents: Mice of the Families Cricetidae and Zapodidae

One would expect to find an abundance of descriptive material written about the nests of small mammals. There is an abundance of material, but much of it is concerned with nest building by caged individuals; the information on nest building in the wild is both fragmented and scattered. After reading many descriptions, one finds a rather generalized pattern emerging for many of the species. The globular nests are made of plant material, the coarsest on the outside, the finest (sometimes mixed with bits of fur, feathers, and milkweed or thistledown) on the inside. The nests are constructed in a variety of places—crevices in rocks, holes in trees, under pieces of bark (whether on a log or on the ground), in logs, under roots, and under matted grass. Sometimes small mammals place their nests in the burrows of other mammals and sometimes in burrows dug by themselves.

The prairie white-footed mouse *(Peromyscus maniculatus bairdii)* and northern white-footed mouse *(Peromyscus leucopus)* are capable of excavating burrows but are reluctant to do so (Howard, 1949). Both species use the burrows, presumably abandoned, of ground squirrels and red foxes. Howard (1949) found several burrows of prairie white-footed mice *(Peromyscus maniculatus bairdii)* which were shallow and believed that they constructed only shallow burrows. These mice lived in small social groups (in artificial nest boxes placed in their habitat), composed of one or both parents and their young. In winter larger groups of up to 12 members were found. There was apparently a continuous association of the males with the females and offspring.

The northern white-footed mouse *(Peromyscus leucopus)* is semiarboreal. Unlike the prairie white-footed mouse, it seemingly never digs its own burrow, but utilizes burrows of other mammals. It seems to prefer tree nests up to 7 m (20 ft) above the ground. It may use abandoned bird's or squirrel's nests, or sometimes it may construct the entire nest itself. Hollow limbs and old stumps are often used as nest sites.

The nest of the smallest North American rodent, the pygmy mouse *(Baiomys taylori)*, of the Southwest, is characterized by a network of runways which lead away from the nest. The nests which Raun and Wilks (1964) found were most often under fallen logs or prostrate cactus pads, or in thick clumps of grass. Nesting material included finely shredded grass or cactus fibers and sometimes fur, in addition to the usual finely shredded grass. Just as the pygmy mouse uses materials available to it in the arid Southwest, so does the bog lemming *(Synaptomys cooperi)*, which generally lives in the mesic areas of the temperate regions. Connor (1959), working in New Jersey, found a nest of dry, shredded sedge leaves under the surface in a sphagnum bog.

Another species of small mammal which often inhabits bogs is the red-backed vole *(Clethrionomys gapperi)*. A high-pitched squeaking sound emanating from a hummock in a bog in Minnesota led me to a nest of this species. The nest was near the top of a sphagnum hummock which projected about 10 cm (4 in.) above the immediate vicinity. The diameter of the hummock was between 10 and 13 cm (4 and 5 in.). The nest itself was simple, just a hollow in the sphagnum without any other nesting material. There was an indistinct entrance of looser sphagnum.

One of the commonest and most widespread of voles is the meadow vole *(Microtus pennsylvanicus)*. Bailey (1924) indicated that new nests with soft linings are usually prepared a few days in advance for each litter of young. Unlike the white-footed mice, the male of which seems to help with nest building and care of the young, the meadow mouse seems to illustrate the other extreme. Bailey (1924) wrote, "There seems to be no moral necessity of life with them other than the most-rapid increase possible of the individuals of the species." He was impressed with their productivity, for it was he who gave the production figure of one pair multiplying to 1 million individuals in one year, if all lived. This could happen with an average litter of five, 17 litters per year, an equal sex ratio, and attainment of breeding age at 46 days.

The prairie vole *(Microtus ochrogaster),* which occupies more arid areas than the meadow vole, makes a tortuous network of paths through the grass and honeycombs the topsoil (Jameson, 1947). The burrows are about 50 to 100 mm below the surface, but those leading to food chambers go deeper. Nests were anywhere from 6 to 18 in. below the surface and had two entrances. Food was stored in underground chambers. One of these, 30 mm below the surface, was 250 mm wide, 400 mm long, and 200 mm high, and contained 8 qt of seeds of the Kentucky coffee tree *(Gymnocladus dioica).*

The jumping mouse *(Zapus hudsonius)* is a hibernating species, whose nest requirements seem greater than those of other small mammals discussed here. Summer homes were globular nests, about 10 cm in diameter, above ground but usually well concealed in tussocks of grass, under vegetation, or under logs. Maternity nests were located in situations offering more protection, such as a rotten willow log, the inside of an upright rotten tree with the nest at ground level, and the exposed and hollow root of a willow tree. Several other nests containing young were not in trees or roots but were well below the surface. Hibernating nests were quite simple (only a few elm leaves in one case), but up to 91 cm (3 ft) below the surface.

Two species of small mammals with rather unusual nests are the harvest mouse *(Reithrodontomys megalotis)* and the golden mouse *(Ochrotomys nut- talli).* Fisler (1963) believed that none of three species of harvest mice he studied (which included *Reithrodontomys megalotis*) made their own burrows, but they made use of other burrows. The western harvest mouse *(Reithrodon- tomys megalotis)* builds a grassy ball nest of about 3 or 4 in. outside diameter, with two or three entrances. The nest may be placed on the ground or in shallow burrows, but many are placed above the ground. In grasslands, nests have been described as delicately built and supported by prairie grasses or sedges.

The nestlike structures of the golden mouse *(Ochrotomys nuttalli)* have been described by Goodpaster and Hoffmeister (1954). There are two types of nestlike structures. One is a well-constructed globular mass, about 8 in. in diameter, of leaves, inner bark, and grass, with an inner lining of fur, feathers, and milkweed "down." Sometimes the outer part is an abandoned bird's nest which may be capped over. These structures are the *nests,* where a single individual or an entire family may live. Nests are usually 60 to 525 cm (2 to 15 ft) above the ground in trees or vines. The other nestlike structure is the *feeding platform,* which is similar in appearance to the nest but is not as bulky and may be found as high as 17.5 m (50 ft) up in a tree. Feeding platforms were reported to be six times as numerous as nests. Sometimes feeding plat- forms were converted to nests and vice versa.

Behavior

The day has come when the field of behavior is being recognized as one of the major branches of endeavor in the life sciences. The function of behavior is to enable an animal to adjust to external and internal conditions, i.e., to maintain homeostasis (Nissen, 1958).

Behavior is difficult to define. One definition is that it is an animal doing something. It is a stimulus-response action.

Ethology is the study of behavior, and Dilger (1962) defined it in this way: "Ethology has now come to mean the study of function, biological significance, creation and evolution of species typical behavior."

Johnsgard (1967) wrote:

"Although the study of animal behavior is sometimes described as the newest branch of the zoological sciences, it is in a broader sense the oldest. There can be little doubt that early man's survival depended upon an intimate knowledge of animals and their behavior; doubtless the teaching of successful methods of stalking game was one of the pre-eminent concerns of early attempts at human communication.

Morphology, physiology, and behavior are not independent, but are simply different ways of looking at living organisms. It is not just a whim that causes a bison *(Bison bison)* to get up front feet first, and the zebra (*Zebra* spp.) hind feet first. These behaviors are dependent on the form and function of the *astragalus,* one of the bones in the hind legs. Its form is such that it allows the hind legs of the bison to bend a great deal, the hind legs of the zebra to bend only a little.

The interest in behavioral studies has probably been stimulated by a change in the philosophic outlook of life scientists. It would have been unthinkable several decades ago for a biologist to accept a statement such as De Vore's (1965), "The effects of early social deprivation on the behavior of the adult monkey or ape have clear implications for the understanding of human psychiatric problems." Insight gained from studying the behavior of nonhuman mammals could never have been applied to human behavior. Our attitude has mellowed, at least enough to accept sensible interpretations. The work of Harlow and his associates (Harlow, 1958; Harlow and Harlow, 1961, 1962) at Wisconsin on primates practically revolutionized the philosophy of child rearing and has pointed the way to research on the "battered-child syndrome."

Ethology is of vital practical importance to the wildlife biologist (Stokes and Balph, 1965). Pack, Mosby, and Siegel (1967), studying gray squirrels *(Sciurus carolinensis)* in Virginia, became interested "in the influence of behavior on population ecology, particularly the relationship of squirrel behavior and age, breeding attainment, size of home range, feeding activity, and the correlation, if any, between hierarchial rank and the consistent occurrence of 'shock' losses observed during livetrapping operations." They came to agree with Wynne-Edwards's (1965) statement that lack of privilege, or low social rank, may result in the individual's suffering "such active persecution as to bring on pathological consequences." These surplus individuals can thus be identified when the population needs to be thinned out.

Good general accounts of animal behavior include Johnsgard (1967); Scott (1958, 1969); Hafez, editor (1969), whose first 233 pages of *The Behavior of Domestic Animals* is an excellent general discussion of behavior; Eibl-Eibesfeldt (1970), who deals with animals in general; and Ewer (1968), whose book is devoted entirely to mammalian behavior.

THE GENETICS OF BEHAVIOR

Behavioral characters are inherited just as physiologic and morphologic characters are; behavior responds to selection as readily as the other characters do, and genetic changes in behavior of adaptive significance survive by selection just as do other genetic changes.

Two rather interesting examples, one spanning evolutionary time, the other a short-term effect of selection (by humans, in this example) on evolutionary adaptation, will be discussed here.

The first example, from Colbert (1961), is also another example of the correlation between behavior and morphology in one species, which separated its members from those of another species, in their social behavior while hunting.

Cats and dogs are members of the order Carnivora. Both are common pets. Dogs are sociable, what we would call "teamworkers." They can be trained to do many things and are very companionable. Cats might be called "loners," and though friendly, never become the close companions that dogs do. Cats and dogs both had their origin in the Eocene, and the rift separating them had appeared by the beginning of the Oligocene.

In the canids the emphasis, evolutionarily, has been on the pursuit of prey, with long legs, compact feet, and long jaws. They are relatively unspecialized carnivores, physically, and therefore more flexible. They have learned to work together. Wolves *(Canis lupus)* hunt aggressively in packs (Murie, 1944; Mech, 1966, 1970; and others). Coyotes *(Canis latrans)* use ruses for chasing deer *(Odocoileus* spp.) and "spell" each other in the chase (Cahalane, 1947).

Cats depend on sneaking up on their prey, catching it in ambush, or catching it by making a sudden short dash at high speeds. The skull of the cat is truncated and the rostrum shortened in comparison with canids. Rollings (1945) described the hunting habits of the bobcat *(Lynx rufus)* thus, "Trail sign indicated that a bobcat obtained prey either by creeping behind cover, or by patient waiting, crouched atop a log, stump, or other vantage point, until the prey chanced to pass on a nearby trail that could be reached by a few short leaps." Wright (1960) described the stalking by lionesses *(Panthera leo)* in East Africa. Although the stalking began in midafternoon, they had not reached their prey before dark. Sometime during the night they had killed one of the Thomson's gazelles *(Gazella thomsoni).* Over the past 40 million years cats have become morphologically highly specialized, with a dentition which is purely for stabbing and shearing. The dentition of canids, with their specialized flattened cheek teeth, has fitted them for a varied diet. Some of them, such as the jackals *(Canis)* and hyenas *(Canis),* are to a great extent carrion eaters. Kleiman and Eisenberg (1973) have summarized comparisons of canid and felid social systems from an evolutionary perspective, including distribution, habitat, preference, morphologic characteristics, and behavior. They believed that provisioning the female and young by the male and the existence of a pair bond are old traits in the Canidae which evolutionarily favored selection for the formation of larger social groups. In felids the absence of a pair bond or close contact between a pair in care of the young would indicate that when selection for the formation of groups was favored in the felids, selection acted upon the female and her offspring as the basic social unit. Carnivores provide a good example of how behavior reflects structure.

The other example shows how behavior in the red fox *(Vulpes vulpes)* can be modified by the actions of the same genes which modify coat color. An intriguing article titled, "Melanin, adrenalin, and the legacy of fear," by

Keeler, Mellinger, Fromm, and Wade (1970), gave the rationale which forms the background for the following discussion.

The red fox *(Vulpes vulpes)* has been called a bundle of jangled nerves. When an adult red fox is brought into captivity it displays many symptoms of schizophrenic psychosis. Such foxes exhibit panic, anxiety, fear, a distrust of white objects, eyes, lenses, man, and open spaces. In the wild this behavior obviously has survival value. In captivity these psychotic symptoms can be effectively treated with psychomimetic drugs such as oxazepam.

Any youth who has tried to tame wild red fox cubs has usually found them to become more vicious as they mature, and they make every attempt to escape. There is no affection in a fox similar to that displayed by a dog *(Canis familiaris),* a tame coyote *(Canis latrans),* or a tame timber wolf *(Canis lupus).* Only those familiar with fox behavior can appreciate the way in which fox behavior differs from that of other canids.

A standardized test, "the startle distance," was used for measuring fear in foxes. It was based on the distance at which an approaching observer would cause the fox to start running away. Measurements on the priming range of adult red foxes showed that the startle distance was greatest for red foxes, but the distance decreased through the black and amber color phases. Thus a color phase with no mutations or one mutation had significantly greater startle distances than a color phase with two mutations.

The study also showed with statistical certainty that the size of the adrenal medulla per unit of body weight follows the same descending order.

There is apparently a pleiotropic (a gene that has more than one effect) association of pigment gene mutations with variation of epinephrine output as well as with relative quantities of fear. The "fight or flight" reactions are caused by norepinephrine and epinephrine secreted by the adrenal medulla. These two hormones are in turn derived from tyrosine, which through a different chemical pathway produces eumelanin or phaeomelanin. Eumelanin imparts the brown or black color to hair, while phaeomelanin imparts the yellowish or reddish color. Keeler, Mellinger, Fromm, and Wade (1970) wrote, "It appears certain from the present studies that adrenal and thyroid functions are altered by the presence of coat genes, and that through these altered functions fear is modified." The chemical mechanisms involved have not yet been definitely determined.

INSTINCT AND LEARNED BEHAVIOR

The problem of instinct and learned behavior has not yet been resolved. Instinct, or innate behavior, is usually thought of as genetic, while the other type of behavior is learned. Hinde (1966) points out that these dichotomies "are not only false but sterile." The ethologists' insistence that instinctive behavior is a discrete category has been a heated issue. Reflexes and innate behavior are probably genetically fixed but are sometimes changed by environmental factors.

Johnsgard (1967) has suggested one way in which the gap between innate and learned behavior may be narrowed. He wrote, "Furthermore, if synaptic transmission efficiency is affected by frequency of use, a basis for behavioral changes through experience, or learning, can be visualized."

Apparently there is no sharp line between instinct and learned behavior. We speak of learning to talk, but we must have the basic mechanisms (muscles, voice box, nervous mechanisms, and affectors) to do so. Most mammals have the basic mechanisms but are unable to talk.

PHYSIOLOGIC BASIS OF BEHAVIOR

The study of behavior differs from other branches of the life sciences only in that the level of organization is the whole animal rather than its parts, or systems. Behavior can be thought of as the integration of all the processes that take place in the animal.

Behavior involves the reception of stimuli or information through receptors or sense organs from the environment, and transforming these stimuli into neural activity by relaying them through neurons and synapses to an effector in, for instance, muscles. Along with the neural mechanisms are endocrine mechanisms for processing and programming both activity and information.

The influence of hormones, for instance, on reproductive behavior, is vast and complex; this subject has been discussed in Chap. 10.

BEHAVIOR SYSTEMS

J. P. Scott (1958) has combined a group of behavior patterns with a common function into a behavioral system. He defined a behavior pattern as an organized segment of behavior having a special function. Since Scott has identified nine general functions, it follows that there are at least nine behavior systems. There are of course great variations in these systems, and their general organization varies from species to species.

1 *Ingestive behavior* is behavior connected with taking in any sort of nourishing substance, both solids and liquids. The artiodactyls have a common behavior pattern of ruminating. Bison *(Bison bison)*, after grazing for a period of time, will lie down, regurgitate a mouthful of food (the bolus), and chew it a second time. Along with this behavior pattern is the morphologic characteristic of a stomach divided into many compartments. Bison *(Bison bison)*, elk *(Cervus canadensis)*, and moose *(Alces americana)*, which lack upper incisors, wrap a mouthful of vegetation with their tongues, and jerk the vegetation so that it is cut by the lower incisors. The perissodactyls (horses, zebras) are grazers but have upper incisors. They bite their food off, rather than tearing it off, and they do not ruminate. Rodents bite off their food also. Carnivores swallow their food in large chunks and chew it very little.

2   *Eliminative behavior.* The process of elimination is not a very complicated behavior pattern in most mammals. The canids and felids are exceptions. The felids cover their excreta, probably in an effort to keep their lairs from being discovered. The canids tend to deposit theirs at specific spots and especially to use their urine to mark scent posts.

3   *Sexual behavior.* The relationships between males and females may be very weak and tenuous, as in herd animals, or very strong, as in some carnivores. In bison *(Bison bison),* males will mate with any female, and the female is receptive to any male. A temporary "tending bond" is established as a female comes into heat. This bond may last from a few hours to no longer than a few days. McHugh (1958) calls this a "temporary monogamous mateship." In carnivores the bond lasts much longer, at least for the season. Schenkel (1947) reported that for wolves *(Canis lupus),* pairing begins in early winter and the bonds strengthen as winter progresses. Mech (1966) wrote, "by mating season pairs are well established." Buechner (1961) and Buechner, Morrison, and Leuthold (1966) have described a mating arena for the male Uganda kob *(Adenota kob)* much like the leks of grouse, where the males gather together, each with a small territory, and the female comes to the male's territory.

4   *Care-giving (epimeletic) behavior,* attentive behavior, maternal behavior. In mammals, the care of the mother for its young is the most common example of this behavior. All mammals allow their offspring to suckle. In *Peromyscus,* the female may stay with the young a great deal (up to 20 h, King, 1963), while in cottontails *(Sylvilagus floridanus),* the female, after the first few days, has been known to stay away from the young for 55 h (Lord, 1963).

After the birth of her calf the bison *(Bison bison)* cow licks it for at least 10 to 25 min, and "for the first few days the calves remained particularly close to their mothers" (McHugh, 1958).

5   *Care-soliciting (et-epimeletic) behavior,* means calling or signaling for care and attention. Distress signals by the young may result from hunger or from pain. A bison *(Bison bison)* calf after straying too far or waking up after a rest will call in a grunt. Usually the mother responds with a grunt. After a series of interchanges, cow and calf will come together, and the cow may lick her calf and the calf may nurse. The calf has solicited care. Adults may also solicit care or attention. *Imprinting* is a simple learned behavior which is primarily concerned with mother-young behavior, and I have chosen to include it here, although not everyone would agree with me. Lorenz (1937) introduced the term "as the process of acquiring the biologically 'right' object of social relations by conditioning them not to one individual fellow-member of the species, but to the species as such." Lorenz (1937) imprinted newly hatched goslings to follow himself, a practice which Chinese gooseherds have practiced for centuries. Goslings follow the first object they see after hatching. Lorenz (1937) showed that this type of imprinting, sometimes called "following behavior," occurs only during a very short period early in the individual's life, and the effect is irreversible. This type of imprinting should be distin-

guished from sexual imprinting, a result of early experience or mate selection. This is sometimes referred to as "preference behavior," as opposed to "following behavior."

The major function of imprinting appears to be that of establishing a social relationship between the young and others of its own species. Most of the research has been done on birds, but enough has been done to show that it also exists in mammals. Hess (1959) reported a "following" response in guinea pigs *(Cavia)* and in sheep *(Ovis),* and Altman (1958) has reported it for moose *(Alces americana),* although in moose it lacks irreversibility and the period of sensitization is much longer. Early explorers of the American West frequently mentioned being adopted by orphaned bison *(Bison bison)* calves. Catlin's (1965) account represents a deliberate case of imprinting. He reported holding his hands over the eyes of a calf and blowing into its nostrils. He "rode several miles into camp with the little prisoner busily following the heels of my horse the whole way as closely and as affectionately as its instinct would attach it to the company of its dam." To an extent, imprinting appears to occur in timber wolves *(Canis lupus)*. Woolpy and Ginsburg (1967) found that captive pups formed ties with humans, but lost this attachment if kept from humans for 6 months or more. The process of attachment in the wild begins when they start following each other at 3 to 4 weeks old. Mech (1970) called this "allelomimetic" behavior.

Imprinting is usually concerned with the behavior of the young toward the mother and other conspecifics. Among mammals, the reverse happens and it is the behavior of the mother toward the young that is the critical factor. A well-known example of this is the rejection of lambs by any but their own mother. Collias (1956) has reported the period from birth up to 4 h as the period of sensitizing. Bartholomew (1952) has reported that the female Alaskan fur seal *(Callorhinus ursinus)* will nurse only her own pup.

6   *Allelomimetic behavior,* also called contagious behavior and imitative behavior. Allelomimetic behavior consists of two or more animals doing the same thing at the same time, with a degree of mutual stimulation. Bison *(Bison bison)* do this when grazing. If you try to change their direction of grazing by simply changing the direction of a few, you will not meet with success. The few will momentarily move out of the way then continue in the general direction of the rest of the herd. It is obvious that the animals are reacting to one another, rather than to a single individual. At times when there is a single leader, it always pays attention to the followers.

Darling (1937) has recorded an interesting effect of individual red deer *(Cervus elaphus)* stimulating one another. He wrote, "I have noticed repeatedly how a number of stags will frighten each other into movement from a source of disturbance which would not have moved them individually or in two and threes."

7   *Shelter-seeking behavior.* All mammals try to find an environment in which they are most comfortable. Quoting Darling again, he wrote (1937), "Deer in their behavior, usually defer to the changing environment by movement, but the lemming of the tundra, in its small way, defies it by creating its own world of higher temperature under the snow." This is an important

part of maintaining homeostasis. For many species of animals the bodies of their fellows protect them to some extent against environmental changes. There has been some speculation that the higher types of social behavior had their origin in this adaptation. Whatever the means, the function is to reduce the rate of thermal conductance from the individual to the environment or vice versa.

8 *Agonistic behavior.* This includes conflict behavior, such as fighting, flight, or submissive behavior. These reactions stem from conflicts. The term is often used to cover all types of response seen not only in fighting but also in territorial behavior. Fighting is most pronounced in the males of mammals. As an example of agonistic behavior, groups of bison *(Bison bison)* bulls which have lived together quite amiably (although a hierarchy exists, but does not show) become antagonistic during the rutting season. McHugh (1958) described their agonistic behavior during this period. He wrote, "While tending cows, many bulls showed intolerance toward surrounding group members of either sex which ventured close. Most intolerance was displayed toward nearby bulls, which were repelled by threat, aggressive charges or battle (Fig. 12–1). The tending bull also kept away even the calf or yearling of the cow he was tending."

In the large ungulates, horns and antlers are of use in fighting. In elk *(Cervus canadensis)* the branching of antlers is such that when they interlock during a fight there is little chance of injury from the tines (McCullough, 1969). In oryx *(Oryx* sp.), with scimitarlike horns, fighting is done with the heads in contact and the horns are not brought into play (Walther, 1958). In bison *(Bison bison)* the horns seem to serve to keep the heads locked together, preventing flank attacks.

In bison *(Bison bison)* bulls, turning the head away from an aggressor is an *appeasement* gesture, which can serve to avoid a fight. During the rutting

Figure 12–1 Agonistic behavior between adult male bison *(Bison bison).*

season, there is a pronounced increase in wallowing by the bulls. Wallowing in this situation is a *displacement activity*. When two bulls are in a situation of potential conflict between a fight or a flight, one may pursue a third activity, that of wallowing. His aggression is turned to attacking the ground and rolling in it. "Displacement activity" is a term applied to a diverse range of behavior patterns, the common characteristic of which is their apparent irrelevance to the situation in which they appear.

9 *Investigative behavior.* This is also called exploratory behavior. It can be seen whenever an animal is placed in new surroundings. If a mouse is placed in a new cage it will spend a great deal of time exploring the cage with its nose and whiskers. Raccoons *(Procyon lotor)* will dip a new object in water when water is available, and then feel the object all over. This is not a washing process; it is a sensory activity. A trapper I once knew told me he set weasel *(Mustela)* traps in shoeboxes, with a small hole cut in the end. The weasel, "overcome with curiosity," would enter the shoebox and get caught, knocking the lid off the shoebox. Thus a long line of traps could be quickly examined at some distance. I cannot vouch for the success of this method, but if it is successful, it shows how investigative behavior can be put to extremely practical use. Bison *(Bison bison)* calves will investigate a bison wallow if they are in its vicinity when it has been used by an adult. Morris (1964) has discussed the importance of an understanding of investigative behavior in caring for zoo animals. He concluded, "They are medically cared for, protected from the elements, well-fed and well-housed. They lack nothing except variability, novelty and stimuli to maintain a high activity level."

MATING SYSTEMS

The behavior of mammals in establishing mating systems is extremely variable. Perhaps the most common kind is *promiscuity,* which occurs in many small mammals such as rodents, insectivores, and bats. In promiscuity the receptive female is mounted by the first male that comes along, and after copulation no relation exists between the partners; the male may go on to copulate with other females, and the female to copulate with other males. In the other kinds of sexual activity a relationship between sexual partners remains for a longer period of time. Although the term *pair bond* is widely used by ornithologists, its use has not, as yet, become common practice among mammalogists. Johnsgard (1967) has called a pair bond a sexual association which may be established some time prior to fertilization and may persist through the reproductive cycle. A characteristic of promiscuity is that no pair bond is formed.

In *monogamy,* a pair bond is established between a male and a female which persists through one or more breeding seasons. Monogamy is not common among mammals but does exist. Beaver *(Castor canadensis),* gibbons *(Hylobates lar),* and some species of white-footed mice and deer mice *(Peromyscus)* supposedly have permanent monogamous bonds. Among carnivorous terrestrial mammals, wolves *(Canis lupus)* and coyotes *(Canis latrans)* are believed to have pair bonds which last for more than one season, as do those of the red fox *(Vulpes vulpes)* and gray fox *(Urocyon cinereoargenteus)*. Murie (1944) published some interesting observations on the family life of timber wolves *(Canis lupus)* in Mt. McKinley National Park in Alaska. In the summer of 1940 he found a den where a pair of timber wolves *(Canis lupus)* raised five young. Associated with this pair were two adult males and an adult female, and two more males joined the group in the late summer. This entire group of seven adults and five young hunted together the following winter. The following spring the original pair returned to the den of the preceding year and raised four young. This evidence shows that this pair remained together at least 2 years. The unmated (or at least unproductive) female of the previous year mated with one of the extra males and raised six young in a nearby den.

More recently courtship and mate choosing among timber wolves *(Canis lupus)* have been shown to be rather complicated. Observations of a pack kept in a large outdoor area at the Brookfield Zoo near Chicago have given researchers some insight into the complexities. Courtship and mating take place in the bitter weather of late winter. Females begin claiming males, but as the season of heat approaches, males begin to court. Being the top (alpha) male of the males in a pack does not necessarily mean having a choice of partners, for there may be strong mate preferences which bear no relationship to the existing dominance hierarchy. Some of these preferences last for years, supporting the idea that the wolf is monogamous. Indeed the responsibilities of leadership may be such that the male leader devotes less effort to mating activities than do his subordinates. Because of the interplay of dominance relationships, mate preference, and leadership responsibilities, very few courtships result in mating (Rabb, Woolpy, and Ginsburg, 1967). This complicated interplay may also be a mechanism for population control.

Polygamy is a term sometimes used to designate a mating system in which one male sets up a pair bond between two or more females, or a female sets up a pair bond between two or more males. But since each of these situations has its own definition, the term polygamy is sometimes discarded (Lack, 1968). The term *polygyny* has been applied to the situation in which one male has formed a pair bond with two or more females, either simultaneously, called *harem polygyny,* or successively, called *successive polygyny.* The term *polyandry* has been used when one female has formed a pair bond with two or more males. Examples of polyandry are difficult to find among mammals. C. R. Carpenter (1934) believed, but was not certain, that the estrous female howler *(Allouata palliata)* monkey may satiate one male and then move on to another. In the rhesus monkey, *Macaca mulatta,* this does definitely occur (Carpenter, 1942). Apparently no pair bond is established, so that these examples may not fulfill the definition of polyandry and may in reality be examples of promiscuity.

Polygyny is probably the most common type of mating system in mammals, particularly among the larger ones. There are, of course, many variations of this arrangement, and harem formations are, at least popularly, the best-known one. It is very common among the Cervidae, and is also found among Bovidae and among the Pinnipedia. It has recently been reported for a bat, a sac-winged bat, *Saccopteryx bilineata,* of Trinidad. A harem is a group of females loosely held together by an aggressive male who tries to keep other males away during the rutting season. This does not necessarily mean that the male dominates the females. Darling (1937) has described this activity for the red deer *(Cervus elaphus)* as "similar to that of a collie keeping together a knot of sheep." He considered the social organization of the red deer to be a matriarchal society, and added, "We must realize then, the dual nature of the harem—the continuation of normal care and responsibility for the group by the leading hind and the more spectacular but extremely circumscribed, ego-centric and casual dominance of the stag." The red deer furnishes a well-documented example of harem polygyny, and the matriarchal society seems the usual system in this pattern.

Examples among mammals of what Lack (1968), writing about birds, called successive polygyny are exhibited by the North American bison (McHugh, 1958) and by the caribou, *Rangifer tarandus* (Lent, 1965). In the bison *(Bison bison)* a temporary bond was established between the cow and the bull during the rut (McHugh, 1958). The tending bond lasted only a few days, usually less (Fig. 12–2). McHugh (1958) called this a *temporarary*

Figure 12–2 A tending bond in bison *(Bison bison).*

monogamous mateship, but the bulls successively served several cows in any one season. The tending bond was essentially matriarchal, in spite of the efforts of the bull to isolate the cow he was tending. The two terms used for the bison mating system point up the lack of uniformity in terminology, even though both terms are descriptive.

The Uganda kob *(Adenota kob)* exhibits still another type of polygny. The kob is a medium-sized antelope found in parts of Africa. Buechner (1961) and Buechner and Schloeth (1965) have described an arena system of territorial behavior during the rutting season for this species. Clusters of about 35 somewhat fixed territories are concentrated in one arena or territorial breeding ground. The central area consists of 10 or 15 territories, each occupied by a single mature male. Estrous females leave the large herds and enter the arena for a day or two until inseminated. Almost all mating occurs on the territorial ground of the arenas. Buechner, Morrison, and Leuthold (1966) wrote, "Prolonged, intense fights occur during interchanges when a male enters an arena to claim or reclaim a territory. Competition among males is keenest for territories preferred by females."

TERRITORY AND SOCIAL HIERARCHIES

It can safely be said that territory most often functions in some way to perpetuate the race, i.e., to ensure reproductive efficiency and to enhance the survival of the young.

Agonistic behavior includes all kinds of hostile reactions between individuals, ranging from vigorous attack to abject response. Obviously a great deal of time and energy expended in fighting would be deleterious to a population, so various substitutes for fighting have evolved in many species.

Quite often there has evolved a system of social ranking, the familiar "peck order," which avoids excessive stress and fighting. Another and perhaps more common method which has evolved for avoiding energy waste by fighting is the establishment of territoriality. This has been more thoroughly investigated in birds than in mammals. Although a territory has been defined in many ways, the simplest definition is that it is *any defended area.* In most instances it is restricted to defense against others of the same species. A more complete definition is an *area defended by its occupant(s) against other members of the same species.* There are extremely few exceptions, as, for example, that of a red squirrel defending its food cache against species other than its own. Good general discussions, not restricted to mammals, may be found in Lack (1968), Lorenz (1966), and Wynne-Edwards (1962); for discussion of this subject in mammals only, see Kaufman (1962). Sanderson (1966) has reviewed the literature on mammalian movement in regard to territory and home range, and Carpenter (1958) listed 32 functions of territoriality in animals, with examples.

He wrote, "The list is impressive and suggests the apparently very great biological significance of territory in animal life."

The best-known examples of territoriality naturally occur among the more easily observed larger mammals, and among the economically important species.

A territory may be fixed, such as that of the male Uganda kob *(Adenota kob)* during the rutting season, discussed earlier, or it may be mobile, such as that of the bison *(Bison bison)* where a male may defend an area around an estrous cow (McHugh, 1958). More recently this type of territory during the rutting season has been described for the barren ground caribou, *Rangifer tarandus* (Lent, 1965). This is also the case in a much smaller mammal, the swamp rabbit *(Sylvilagus aquaticus),* among which territorialism was also restricted to defense by the dominant male of an area surrounding the female (Marsden and Holler, 1964). Territoriality has recently been demonstrated for the white-lined bat *(Saccopteryx bilineata)* of Trinidad.

In the once economically important Alaskan, Northern, or Pribilof fur seal *(Callorhinus ursinus),* the male comes out on land during the breeding season, defends a chosen breeding area against other bulls, and collects a harem within this area (Kenyon, 1960). At this season—June and early July—about 100,000 individuals gather on the Pribilof Islands. This is the time of the commercial fur harvest. July 5 or 6 is about the midpoint between the beginning of territorial stability and territorial abandonment by the males. In August temperatures climb over 12°C and the seals show signs of discomfort at temperatures above 12°C. The northern fur seal has carried to the extreme the combination of polygamy with territorialism, although it exists in many of the carnivorous aquatic mammals. An unusually short period of time, 5 days, elapses between parturition and copulation. Two features of the female's reproductive physiology make this workable: (1) her uterus is bicornuate (i.e., has two horns), and one horn lies dormant while the fetus develops in the other; (2) the egg lies dormant for 4 or 5 months after the first divisions (delayed implantation). Thus the female can mate and bear her young just a year later, at the only time of the year when climatic circumstances are optimal.

In the California sea lion *(Zalophus californicus)* territories are less well defined than those of most other pinnipeds. Reasons for this appear to be that (1) there is no inland boundary which the bull defends, (2) the seaward edge of the territory may be under water, (3) a bull rarely patrols the entire perimeter of his territory, and (4) intruding bulls are sometimes successful for short periods, probably because the owner is unaware of intruders. Also, in contrast with other pinnipeds, the bulls of this species are relatively inactive in courtship behavior; instead, the female solicits the male (Peterson and Bartholomew, 1967).

Although it has been generally believed that pronghorn *(Antilocapra americana)* bucks defend harems, Bromley (1969) showed that mature prong-horn bucks are territorial during the rut and defend recognizable, static areas. The bucks were believed to mark territories in three ways: (1) the visual mark of the male's presence, (2) marking vegetation with scent from the postman-dibular scent glands, and (3) special vocalization. Similar types of territories have been described for many other kinds of ungulates.

In a smaller species of animal, the black-tailed prairie dog *(Cynomys ludovicianus)*, the territory of a "coterie" (an adult male, several adult females, and some young) is kept intact by an adult, dominant male. This behavior is not restricted to the mating season. Within this coterie all burrows are commu-nal, except when there are young in the burrows (King, 1955). At this time the female defends an area around the burrow entrance, a good example of a territory presumably serving to enhance survival of the young.

In some cases dominant males may defend a core area and satellite males may defend fringe areas or areas of less intensive use. These are the gathering grounds of many males which are visited individually by the females in estrus. This is the situation which exists in the Uganda kob *(Adenota kob)* (Buechner and Schloeth, 1965); in the southern elephant seal *(Mirounga leonina),* among which a "beachmaster" shares his control with one or more "assistants" (Corrick, Csordas, Ingham, and Keith, 1962); and in the European wild rabbit *(Oryctolagus cuniculus),* among which secondary males maintain territories within that of the dominant male (Lockley, 1954).

In the last-mentioned examples, two types of social system are obviously in operation at the same time, territory and social hierarchies. Quite often both systems are present, but not always are they so obvious. Usually territoriality is associated with reproductive behavior, while social hierarchies are more permanent and are often associated with other activities, such as feeding. For example, a male bison *(Bison bison)* may defend the territory surrounding an estrous cow, but there is apparent a type of social dominance between males which may have been established over a long period of time; it is quite obvious during the rutting and present but not as obvious during the nonrutting season.

In the red deer *(Cervus elaphus),* a mature female leads the herd and has therefore established some type of social dominance. During the rutting season males try to dominate other males, but must fit into the basically matriarchal social system (Darling, 1937). In my work on the striped ground squirrel *(Spermophilus tridecemlineatus),* I found evidence for a linear dominance among at least three of the most active males. Among the carnivores, some may defend hunting areas, as Schaller (1972) has mentioned for lions *(Pan-thera leo),* but others, such as the red fox *(Vulpes vulpes),* apparently do not defend a territory (Scott, 1943), although more recently Ables (1969) noted a suggestion of territorial behavior between two females on his study area in south-central Wisconsin.

In the cottontail *(Sylvilagus floridanus)* and swamp rabbit *(Sylvilagus aquaticus),* a linear dominance hierarchy exists during sexual interaction, and this male hierarchy is fundamental in the prevention of reproductive period fighting (Marsden and Holler, 1964).

This brief discussion has served to emphasize to the reader that a great variety of behavior patterns serves to reduce stress from agonistic behavior during reproductive periods.

HOME RANGE

The concept of territory seems to be most often associated with birds, while the concept of home range seems to be almost restricted to mammals. This may be more a characteristic of man's expediency in research than a characteristic of the classes of animals. Man must have been long aware that not all mammals are nomads, and this knowledge may have eased his existence when man was primarily a hunter. Seton long ago (1909) wrote, "No wild animal roams at random over the country: each has a home region, even if it is not an actual home." Burt (1940) defined home range as "that area about its established home which is traversed by the animal in its normal activities of food gathering, mating, and caring for young." Familiarity with an area also decreases an animal's vulnerability to predation. Anyone who has live-trapped and released animals within or outside their home range must have been impressed with how quickly a small mammal disappears when released in its home range. Metzger (1967), studying owl predation on white-footed mice *(Peromyscus leucopus),* wrote that "possession of a familiar home range confers a considerable advantage since transient mice are subject to greater danger from predation." Bovet (1968) has summarized some of the work on homing in deer mice. A more comprehensive definition of home range might be, the *area used by an individual or group of individuals which provides all the requirements for survival of the species.* Home ranges are not defended, while territories are. Home ranges of individuals of the same species can and often do overlap.

Obviously, home ranges must vary greatly to be suitable for mammals ranging in size from the very small pigmy mouse *(Baiomys)* to a 10,400-kg (2300-lb) bison *(Bison bison).* They also vary according to circumstances and exhibit great elasticity, not only for a species but also for an individual. When food is plentiful, for instance, home ranges may shrink, although there is a lower limit of "compressibility" (D. E. Davis, 1953).

McHugh (1958) suggested that the home range for the herd of bison *(Bison bison)* in the Hayden Valley of Yellowstone Park in winter at the time of his study was 94 km^2 (36 mi^2) (Fig. 12–3). Murie (1944) thought the family home range of wolves *(Canis lupus)* in Mt. McKinley National Park, Alaska, may have a diameter of 80 km (50 mi). Home ranges of gibbons *(Hylobates lar)* vary from 12 to 40 hectares (30 to 100 acres) (J. R. Carpenter, 1940). A

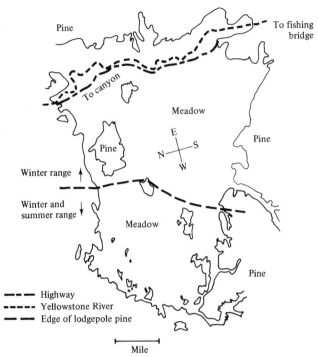

Figure 12-3 Home range of a bison *(Bison bison)* herd in Yellowstone Park. *(From McHugh, 1958.)*

group of 65 to 75 chimpanzees *(Pan troglodytes)* may range over 15.6 to 78 km (6 to 30 mi) (Reynolds and Reynolds, 1965). On the other hand mountain gorilla *(Gorilla gorilla beringei)* groups, which are smaller (2 to 30 animals), range over an area of 26 to 39 km^2 (Schaller, 1965).

Lechleitner (1958) gave the home range of the male black-tailed jackrabbit *(Lepus californicus)* as about 17.6 hectares (35 acres). Schwartz (1941) gave the average home range for the cottontail in Missouri (which he believed to be an intergrade between *Sylvilagus floridanus mearnsi* and *Sylvilagus floridanus alacer*) as 0.6 hectare (1.4 acres) for the male and 0.5 hectare (1.2 acres) for the females. Haugen (1942), working on the cottontail *(Sylvilagus floridanus mearnsi)*, at Swan Creek Wildlife Experiment Station near Allegan, Michigan, calculated a winter home range of 6 hectares (14 acres) and breeding home ranges of 90 hectares (22.5 acres) for the female; he believed the home range of the males to be over 40 hectares (100 acres). He found that the areas occupied by most females were separate and distinct from those of other breeding females.

Territoriality in pocket gophers serves a different function than in most other animals. In the case of pocket gophers, territory and home range may be synonymous, for these mammals appear actively to defend their home range from all others of the same species, regardless of sex (Howard and Childs, 1959). These workers calculated territories by connecting, for the most part, the outer points of capture. Territories of male *Thomomys bottae* averaged 240 m² (2700 ft²); those of females averaged 117 m² (1300 ft²). Wilks (1963) worked on another species of pocket gopher, *Geomys bursarius,* in southern Texas, and wrote, "The home range of the pocket gopher is, in the strictest sense, a tube several hundred feet long and two or three inches in diameter." He also found that males had a significantly larger home range, 474 ± 55 m² (5040 ± 596 ft²), than females, 150 ± 35 m² (1560 ± 307 ft²). These figures were estimated for the entire year. Howard and Childs called the area territory, and Wilks (1963) called it home range. Wilks (1963) wrote, "The home range and territory are probably synonymous with respect to the pocket gopher."

Allen (1943) found so many variables when trying to calculate the home range of the fox squirrel *(Sciurus niger)* on his study area in Michigan that he felt he could not give the average home range as so many acres. He believed that a fox squirrel used an area of at least 4 hectares (10 acres) in any one season, but that on a yearly basis this could be over 10 hectares (40 acres). In his studies of the gray squirrel *(Sciurus carolinensis),* Flyger (1960) found that the average home range for the males was 1.9 (0.2 to 7.2), and for the females 1.2 (0.2 to 2.6). Their home range is three-dimensional. Robinson and Cowan (1953) recorded the home ranges for an introduced population of gray squirrel in British Columbia as 2 to 6 hectares (5 to 15 acres) for females, and up to 20 to 22 hectares (50 to 55 acres) for the males.

The home range of striped ground squirrels *(Spermophilus tridecem-lineatus)* studied by the author in Minnesota was much smaller than that of those studied by McCarley (1966) in Texas. Live traps were placed at random in the Minnesota study; the average home range for females was 0.18 hectare (0.44 acre), and for males 0.35 hectare (0.84), although males would at times range great distances. In Texas, traps were spaced 100 ft apart. Home ranges for the females averaged 1.5 hectares (3.4 acres), and for males, 4.7 hectares (11.7 acres).

Home ranges have also been calculated for many of the smaller rodents. For the prairie white-footed mouse *(Peromyscus maniculatus bairdi)* in Michigan, Blair (1940a) calculated the home range as being a little over 0.2 hectare (½ acre) in prairie. The woodland deer mouse *(Peromyscus maniculatus gracilis)* had a slightly larger home range. For this species in northern Michigan the average for adult males was 2.31 ± 0.27 acres and for adult females 0.92 ± 0.31 acre (Blair, 1942).

Much work has been done on the meadow vole *(Microtus pennsylvanicus),* and home-range calculations vary. Hamilton, in 1937, wrote, "The home range of an individual vole seldom encompasses an area in excess of one-fifth of an acre," in New York. Blair (1940) worked on the meadow vole in southern Michigan and found the home range of females to be 0.19 ± 0.02 acre, and that of males to be 0.31 ± 0.02 in moist grassland. In drier habitat it was 0.28 ± 0.03 acre for females and 0.50 ± 0.07 for males. His traps were spaced 50 m (60 ft) apart. Gunderson (1950) found the average home range for 11 meadow voles *(Microtus pennsylvanicus)* in Minnesota to average 0.11 hectare (0.28 acre) [0.06 hectare (0.14 acre) to 0.22 hectare (0.55 acre)]. Getz (1961) found that home-range sizes varied between males and females, as well as between marshes and old fields. In addition they vary throughout the seasons. One cannot help but wonder what happens to home ranges and territories under extreme conditions. Linduska (1950) took over 300 individuals of meadow voles *(Microtus pennsylvanicus)* from 315 corn shocks, on 11 hectares (28 acres) of land in the Rose Lake Wildlife Experiment Station in Michigan in late January. Seventy percent of the mature females were pregnant or lactating.

I (1962) found that the home range for the red-backed vole *(Clethrio-*

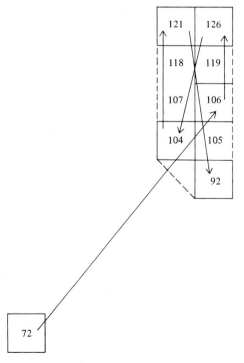

Figure 12–4 Home range of red-backed mouse *(Clethrionomys gapperi)*. Numbers represent trap stations and surrounding area. Traps were set in a grid and would be in the middle of each square. *(From Gunderson, 1950.)*

nomys gapperi) in southeastern Minnesota was 0.30 ± 0.02 (2 S.E.—two standard errors) acre for 131 adults. It was slightly larger for the males (71 individuals), with an average of 0.33 ± 0.008 (2 S.E.) acre, than for the females (60 individuals), 0.28 ± 0.01 (2 S.E.) acre (Fig. 12–4).

From this discussion it has probably become apparent that territory and home-range figures are subject to a great deal of variation. Certainly these figures must be used with a great deal of caution. Methods of gathering data include such basic techniques as simple observation for the larger mammals, capture-mark-recapture methods for small animals (L. E. Brown, 1962; Burt, 1940; Sanderson 1966), and, more recently, marking with radioactive materials (Godfrey, 1954; Kaye, 1961) and radiotelemetry (Adams, 1965; Sanderson, 1966). The calculation of home-range data is also extremely subjective. Though it is possible to gain an approximate idea of the size of the home range of a mammal, such statistics should not be accepted as absolute.

SOCIAL BEHAVIOR

The simplest social systems are those of mammals considered to be solitary. An extreme example would be any member of the family Geomyidae. Its home range and territory, except during the breeding season, would be its burrow system. Similarly, some ground squirrels are also somewhat asocial (Edwards, 1946). In my work with the thirteen-lined ground squirrel *(Spermophilus tridecemlineatus),* I began to wonder whether they are asocial or whether they are a very loosely organized, widely scattered social unit. I concluded that they are a widely scattered social unit. Whales tens of miles apart have now been recognized as part of a social unit (Payne and McVay, 1971). I should like here to present some examples of group or social structure among mammals to show the variety and complexity of such groups.

The Primates

For the primates the discussion will proceed from a species with the smallest social unit, the gibbon *(Hylobates lar),* to a species with a very large social unit, the baboon *(Papio).* For information on more species of primates, DeVore (1965) is a good reference.

The gibbon *(Hylobates lar)* society is based on a monogamous family unit consisting of an adult male and female and their young. These groups are relatively cohesive, and there is apparently little dominance between adults. There were 21 groups in DeVore's study area, ranging in size from two to six. Eighteen groups lived in an area of 10.9 km² (4 mi²). Young left the family group as they became sexually competitive with either adult. Carpenter (1940) suggested that gibbons "have a low degree of sexual drive." They seemed to acquire and defend sections of a forest, and territories were defended by vocalizations and actual fighting.

Of the other apes, little is known of the orangutan *(Pongo pygmaeus)*, although incidental observations (Schaller, 1961; Harrison, 1962; and others) indicate that the groups are small and unstable. Units of the African apes such as chimpanzee *(Pan troglodytes)* and gorilla *(Gorilla)* tend to be larger, numbering 25 or more animals at times. The orangutan *(Pongo pygmaeus)* and chimpanzee *(Pan troglodytes)* have unstable social units. The gibbon *(Hylobates lar)* and gorilla *(Gorilla)* have relatively stable groups.

Schaller (1963) described the central core of a mountain gorilla *(Gorilla g. beringei)* group as composed of a dominant male and adult females and young. The other males tended to be peripheral. The gorilla groups (*Gorilla* sp.) varied from 2 to 30 individuals, with an average for 10 groups in Kabara of 16.9 individuals. Apparently the oldest male was dominant. The home ranges of groups ranged from 10 to 15 mi^2 each.

Another primate studied intensively by Carpenter (1934) was a New World monkey, the howler *(Alouatta palliata)* of Barro Colorado Island, an island in Lake Gatun in the Isthmus of Panama. Since Carpenter's first study, both he and others have studied this isolated group of monkeys (Carpenter, 1962, 1965; Collias and Southwick, 1952; and Altmann, 1959). Groups of howler monkeys range in size from 4 to 35, with a median of 18, according to Carpenter (1934). These groups vary in constitution, and there are subgroups. In censuses conducted from 1932 to 1959, the median was 15.8 individuals.

There apparently was no strong individual leadership in a howler group. Females were conspicuously compatible but might order themselves along a dominance-subordinance gradient of low slope, with older females having more control and rights than younger females. This same arrangement seemed to hold for the males. The male components of a group exercised some control over the females, but the females of a group constituted a kind of majority control. Males were seen to roar together and together approached an infant which fell to the ground, thus possibly giving it protection until the female recovered the infant. Males mated communally with a female, with no competition or dominance.

Although females never seemed to leave a group, extragroup males of all ages were apparent in the howler populations. "These are not permanently isolated" (Carpenter, 1965). This acceptance of extragroup males may provide a mechanism for interbreeding among other closed groups and thus keep the gene pool more plastic.

The usual definition of territory does not describe accurately the territory of howler *(Alouatta palliata)* groups. Carpenter (1965) wrote, "They defend the area where they are, and since they are most frequently in the familiar parts of their total ranges, these areas are most frequently defended." The howls are a kind of spacing behavior.

Among the largest socially integrated groups of primates that have been studied are those of baboons which belong to the genus *Papio* (Hall and DeVore, 1965). In contrast to the primates already discussed, which are arboreal, the baboons are primarily terrestrial. Hall and DeVore (1965) reported wide variations in group size, from 12 to 87 members in Nairobi Park, with an average number of 80. Although no observations of territorial defense were made, baboon groups spent most of their time in circumscribed areas or home ranges varying from 5.2 to 40.3 km^2 (2.0 to 15.5 mi^2). The 5.2-km^2 area was judged as not a complete home range. The social cohesiveness of a baboon group is based on the dominance hierarchy of adult males. Hall and DeVore (1965) wrote, "The simplest form of organization is probably that of groups in which one, and only one, adult male is conspicuously dominant." The dominant male was most successful sexually, he was the first to threaten or attack when a disturbing situation occurred, and the females with newborn infants tended to cluster near him.

The importance of social groupings and all their ramifications is fast becoming apparent in primate studies.

The Prairie Dog

Among the rodents, the most detailed behavioral study of an indigenous species is that of the black-tailed prairie dog *(Cynomys ludovicianus)* by King (1955) (Fig. 12–5). Black-tailed prairie dogs, "were once incredibly numerous—perhaps as numerous as the extinct passenger pigeon," according to Dobie (1947). Merriam (1901) surveyed a dog town north of San Angelo, Texas, which he estimated to be 169 km (100 mi) in width and 419 km (250 mi) in length. If the population averaged 25 prairie dogs to the acre, then the expanse of 6,400,000 hectares, or 42,000,000 km^2 (16,000,000 acres), contained 400,000,000 animals, according to Merriam's calculations. As late as 1931, Bailey estimated that there were 6,400,000 prairie dogs in Grant County, New Mexico. Today few "dog towns" of any size remain outside of refuges, parks, or sanctuaries, although the existence of the species is not endangered.

The black-tailed prairie dog was a characteristic inhabitant of the shortgrass prairie; its geographic range extended from Texas, with its extremes of heat in summer, north into the Canadian provinces of Manitoba and Saskatchewan, with their extreme cold in winter. The east-west range occupied the belt of 500-cm (20-in.) rainfall, which is the prairie-producing rainfall. Within this geographic range, suitable habitat for the prairie dog was maintained by the grazing buffalo *(Bison bison)* and prairie fires.

Distinctive features of prairie dog towns are the burrows. The town in Shirttail Canyon, Wind Cave National Park, of the Black Hills of South Dakota had from 20 to 50 burrow entrances per 0.4 hectare (acre), according to King (1953). The burrow entrances range from no mound at all to craterlike mounds, which are the largest and most characteristic. Such a mound may be nearly 1 m (3 ft) high and 0.28 hectare (7 ft) across. A burrow entrance

Figure 12-5 The prairie dog *(Cynomys ludovicianus)* is an extremely social mammal.

with little or no soil about the entrance may be the opening of a new entrance or the beginning of an exploratory burrow; the digging is usually undertaken after a rain.

The burrow system is an extremely important part of a prairie dog's social organization. Aside from a refuge and a place to rear the young, the burrow system "contributes to the distinct spatial patterns of their social organizations. Since the burrows remain essentially the same from generation to generation, they have a stabilizing effect on the organization of the colony" (King, 1955). The mounds themselves serve several other purposes. They prevent the burrows from being flooded when a flash downpour strikes the hardpan of the prairie, and they serve as lookout stations, where the prairie dogs can watch for predators or competitors. Koford (1958) has also recorded that insects are attracted to these areas and thus the mounds serve as a feeding ground for burrowing owls and prairie horned larks, as well as for other mammals and reptiles. Many early accounts of prairie dog towns mention the presence of the prairie rattlesnake *(Crotalus)*.

The feeding habits of prairie dogs eliminate the taller plants. This in turn deprives predators of cover and encourages fast-growing weeds without abundant seeds and fruits. The earth around the burrows may become occupied by plants not growing abundantly elsewhere. Tall vegetation in the middle of the town is removed entirely.

Prairie dog towns may be divided by topographic features into what King (1955) called wards, but the most significant divisions are the territories of the prairie dog clans which are imposed by the animals themselves. These self-imposed divisions, called *coteries* by King (1953), are the basic units of a town's social organization. Each member of the coterie is known to the other, and the burrow systems are shared by all the members of the coterie except when the young are still in the nest. A coterie covered about 0.2 hectare (seven-tenths acre) in Shirttail Canyon, and the average composition was one adult male, three adult females, and about six offspring. Though numbers may change the size of the territory, it remains relatively stable from season to season. The stability of the territory is based on the network of burrows on which the prairie dogs depend for survival. Members of the territory are recognized by "the kiss." When two individuals come together, if they are members they bare their teeth and "kiss"; they may groom each other or feed together. The bared teeth are a threat rather than a show of affection, but each may recognize that the other is defending the same area, and therefore is a member of the same coterie. An intruder, facing bared teeth, will leave.

Another means of maintaining coteries, wards, and towns was vocalization. The prairie dog's bark and also its gait are responsible for its name. This short, nasal yip is a warning signal and evidently has many nuances and many interpretations. Another kind of vocalization is the two-syllable territorial call. This seemed to be an "all's-well" call given by a foraging dog in the home territory, or after the departure of danger. If the others join in "an all-clear signal becomes a cacophony of togetherness" (King, 1955).

Prairie dogs have developed a significant pattern of social behavior which is central to their survival. These patterns include modification of the environment to make it more suitable to their own needs, and a social organization which protects them from predators and the risk of famine. These patterns are transmitted to each generation by the preceding one, a type of education which necessitates close association over a relatively long period of time. The adult male seems to determine the size of the territory; thus the prairie dog's social system is essentially patriarchal.

The Lion

The lion *(Panthera leo)* is unique among cats, which tend to be solitary, in the extent of its social life. Schaller (1972) has described in great detail the social organization of the lions living on the Serengeti Plain of Africa. There were two basic types of lion: (1) *resident* ones, which remain in a limited area a year or more, sometimes their entire lives, and (2) *nomads,* which wander widely, often following the migrating herds of ungulates. Nomads may become residents and vice versa. A *pride* was considered to be any resident lionesses with their cubs, and any attending males, all of which interacted peacefully. Members, or groups of members, may be scattered widely even though belonging to the same pride.

In the 14 prides whose identity was certain, the average number was 15 individuals, with a range of from 4 to 37. All pride lionesses were directly related and consisted of daughters, mothers, grandmothers, "and perhaps another generation." Pride males may move alone, with other males, or with the females. Females tend to be permanent members of the pride while males seem to be rather transitory.

The hunting efforts of lions *(Panthera leo)* have been reported upon by both Schaller (1972) and Darling (1960). Darling discussed the interesting hunting habits of the lions on the Mara Plains of Africa. The lionesses have developed a sociality and loose structure in the "pride" (group of more than two lions) which is of high survival value. The cubs are born blind and helpless and quite accessible to jackals *(Canis)* and hyenas *(Hyaena)*. Since lion cubs therefore cannot be left alone, two or more lionesses with cubs of the same age cooperate. One stays with the cubs (baby-sits!) while the others hunt. Even nursing may not be restricted to the mother. When the cubs become mobile they join grown-up members of an earlier litter, and thus a pride is formed. These prides feed without jealousy, "a habit which makes for efficient utilization of the kill and less active killing of the game by separate individuals" (Darling, 1960). This behavior has its survival value, but, "If lions are so reduced in an area that lionesses cannot combine, they may be assumed to be finished as a successful breeding species" (Darling, 1960). Darling also noted that lions would watch the vultures for signs of meat. He told of shooting a severely lame zebra *(Zebra)* for humane reasons. Soon vultures appeared, then lions, and finally hyenas *(Hyaena)* and jackals *(Canis)*. A typical lion hunt was described by Wright (1960), who also discussed the results when he wrote, "When the lionesses have made the kill, the males come forward and appropriate it. Only when the males have finished eating do the females and cubs make their meal."

The Red Deer

In Darling's (1937) study of the red deer *(Cervus elaphus)* in Scotland, he reported the red deer's social groups to be matriarchal. He wrote, "Landseer showed himself a faulty observer when he entitled his painting 'Monarch of the Glen,' yet this phrase has continued in the popular imagination. The stag never attains to leadership, and the first stag in rut which may be running round a group of fifty hinds still has no power of leadership."

Three types of territory were described for the red deer:

1 Winter territories for red deer females include calving grounds in June and the place where they gather in October for the rut. The hinds (female red deer) are on the winter territory from October to June. These areas were on the lower slopes of the hills. The winter territory of stags is not the same as that occupied by the hinds.

2 Summer territories extend from the winter territories to the tops of the hills. Summer grounds, near the summit, are necessarily smaller in area, and the stags and hinds may graze communally near the summit, but retain their own social groupings.

3   The breeding or rutting territories are always on the territories of the hinds and are formed toward the end of September. The rutting territory of a stag may be 5 acres or 25, depending on many factors. Rutting territory of a stag (a group of hinds may range from 5 to 50) is arbitrary and psychologically determined. For one thing, the rutting territory must be an area which the stag can scan and run over easily. The hinds are usually very unconcerned and tend to drift to rougher ground, and the stag is kept busy herding them back. At the beginning of November the rutting territories begin to disintegrate. The density was estimated to be one per 16 hectares (40 acres) on Darling's study area in Scotland.

In red deer, territory and social systems are related. The sexes are separate for the greater part of the year. The female ("hind") group is a cohesive social unit, derived from the stability of the family. Maternal care may extend to the third year of life, and a hind may be followed by two or three offspring. This high degree of sociality extends the educational period of the young. A hind group has as its leader a mature hind with a calf or an old hind.

At calving time (June 7 to July 15 in Scottish Highland deer), the hind group breaks up and each pregnant hind goes by herself. She drives off the older youngsters, at least temporarily, but it is at this time that the young stags join the company of older stags. The young stags leaving a hind group flock together. Once the calf is able it follows its mother, and together with the yearling and the 2-year-old, the group moves off to the high ground of the summer territory, joining with the older hinds and their young into a well-knit group.

The stag group is much looser than the hind group. Darling describes stag company as a "number of egocentric males. . . . There is no apparent leader, though one animal may be in a position to bully the rest, which is quite a different thing." Stags do much more wandering than hinds, and may leave their territory at rutting time and travel many miles.

Paths, wallows, and rubbing trees are important to the sociality of red deer and "provide evidence of the sub-human presence of tradition" (Darling, 1937).

Chapter 13

Communication

Behavior is greatly influenced by communication. We think of communication as the transfer of information from one individual or group to another individual or group. Such transfer may be divided into that within species, *intraspecific,* and that between species, *interspecific,* e.g., warning signals. For this there must be a sender, a receiver, and some method of transmitting the signal. A broader definition of communication might be, "The reception of information by any stimulus from the external world." This would then include exchanges of information with the inanimate environment (exteroreception), e.g., echolocation, which Griffin (1968) has termed "solipsistic" communication, a sort of self-communication. It has an indirect effect on physiologic and behavioral phenomena.

A common, but artificial, way of classifying systems of communication is by the sense organ which is the receptor. It is artificial because there is an assumption that only one kind of communication is received at one time. There are many functions of signaling behavior. Some of the major ones are individual, species, and sex recognition; warning, including intimidating; and social integration, to which Marler (1965) believed the greatest part of the whole system of communications is devoted. The mechanisms of communication discussed in this section will include (1) visual, (2) chemical, (3) tactile, (4) acoustic, and (5) echolocation (exteroreception).

VISUAL COMMUNICATION

Visual communication occurs widely among diurnal and crepuscular mammals. Some visual signals—such as those used by dogs *(Canis familiaris),* cats *(Felis domesticus),* and horses *(Equus)*—are familiar to everyone. Dogs, wolves *(Canis lupus),* and horses have a highly developed facial and ear musculature, which can signal intensities of threat, submission, or greeting. Warning colors and postures, such as those of the striped *(Mephitis mephitis)* and spotted *(Spilogale putorius)* skunks, are other well-known examples of visual communication. Piloerection of hair on the neck or tail and flash patches also serve as signals of warning, fear, or anger. Flash patches on the rump are very prominent in pronghorn *(Antilocapra americana)* and in elk *(Cervus canadensis).* When pronghorns are threatening or threatened (nervous) their neck hair (as well as flash patch) is erected. Grant's gazelle *(Gazella granti)* and Thomson's gazelle *(Gazella thomsonii)* have an unusual repertory of similar warning signals (Estes, 1967; Walther, 1965). These include snorting, stamping with the foreleg, twitching of the flank skin, "stotting" or stiff-legged bounding, and erection and flaring of the hair on the rump patch, are all used as warning signals. North American deer characteristically hold their tails, "flags," in the air in flight.

In the bison *(Bison bison)* bull a high-intensity threat is indicated by bellowing with head turned toward the antagonist and tail standing straight up.

A great deal of work has been done on visual expression in the timber wolf *(Canis lupus),* primarily by Schenkel, who included 19 pages of line drawings of body postures and facial expressions in his 1947 publication.

In the wolf *(Canis lupus)* two extremes of agonistic behavior indicate social rank: (1) the threat or dominance pattern of dominant individuals; (2) the defensive or submissive pattern of subordinate animals. Between these extremes are several moderate patterns. There is even an "active submission" pattern, in which a subordinate individual holds its head and tail down when a dominant animal approaches. Schenkel (1947) has called the head of the wolf *(Canis lupus)* the most important center of visual and acoustic expression. The tail of the wolf is also an important indicator of a wolf's mood, just as is the tail of a dog *(Canis familiaris).* Mech (1970) has given a detailed discussion of social behavior in wolves. It is not surprising that the rich and varied social life of dogs and wolves should produce a wide range of facial and body expressions.

In tree squirrels and ground squirrels agonistic behavior is shown by piloerection of tail and body hair, lashing back and forth of the tail or standing stiff-legged, as in the African ground squirrel, *Xerus erythropus.*

In smaller mammals visual communication quite often involves whole-body postures, as exemplified in shrews and rodents. Olsen (1969) has described five postures in the agonistic behavior repertoire of the short-tailed shrew *(Blarina brevicauda)* which serve to induce retreat in the opponent (Fig. 13–1). The adaptive significance of threat postures in shrews (or any animal) is open to debate. It could be a substitute for fighting, or, as Crowcroft (1957) and Eisenberg (1964) have suggested, it could function to disperse individuals, reducing the competition for food.

Postures during an agonistic encounter between kangaroo rats *(Dipodomys)* have been described by Eisenberg (1963). In this species the extreme upright posture (class I in Fig. 13–2) was often sufficient to cause an approaching animal to leave.

Figure 13–1 Visual communication in the short-tailed shrew, *Blarina brevicauda. (From Olsen, 1969.)*

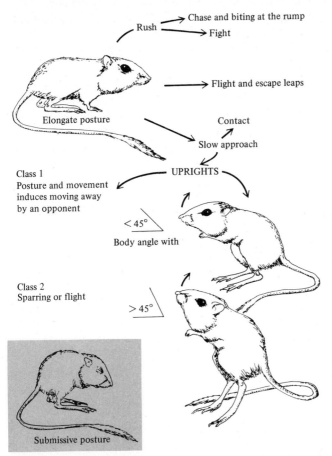

Figure 13-2 Visual communication in *Dipodomys. (From Eisenberg, 1963. Originally published by the University of California Press; reprinted by permission of The Regents of the University of California.)*

OLFACTORY (CHEMICAL) COMMUNICATION

The sense of smell is the most widespread communication mechanism among mammals (Hediger, 1950), occurring in all but some of the anthropoid apes, seals, walruses, and cetaceans. Moulton (1967) wrote, "The ability to detect, analyze and exploit odors appears to reach its highest degree of development among mammals."

To this end there is a marked increase in the number of olfactory receptors, as well as modification of supporting structures in the mammalian nose. The sensory or olfactory receptors are true nerve cells, located in the olfactory epithelium and, in some mammals, in an accessory olfactory bulb. Still another organ is present in most mammals, the vomeronasal or Jacobson's organ, which also has a sensory epithelium similar to that of the olfactory organ. The function of Jacobson's organ is still a matter of debate, although it is presumed to be an extension of the olfactory epithelium surface. The organ is an elongated dead-end tube, opening either into the nasal cavity or into the buccal cavity by way of the nasopalatine canal. Among ungulates a characteristic behavior of males is to "scent" cows to determine individuals entering estrus, or breeding condition (Fig. 13–3). Scenting is by smelling the female or her urine. This is often followed by a curling of the upper lip so as to expose the upper gum region, while the head is held horizontally, forming a straight line with the extended neck. The nares are extended. Schneider (1931) has named this posture *Flehmen*. Knappe (1964) believed this act serves to make Jacobson's organ more effective. Although Dagg and Taub (1970) do not agree entirely with Knappe's hypothesis, his general thesis is probably correct. Other words used for the German word *Flehmen* are grimacing and lip curling. Although it is a typical reaction in males after intraspecific scenting, it may occur in females when they are examining an unfamiliar odor. The gray meerkat *(Suricata suricatta)*, a civet of Southern Africa, is said to "flehmen" in response to sherry wine (Ewer, 1968). The twitching of rabbits' noses probably retracts the flaps of the cleft lip, thus exposing Jacobson's organ, and other olfactory tissue, to odors. For a thorough discussion of the functioning of *Flehmen* and Jacobson's organ, see Eisenberg and Kleiman (1972).

Figure 13–3 Olfactory communication in the bison *(Bison bison).*

Chemicals used for social communication by mammals originate in urine, feces, or in cutaneous scent glands. Scent glands are discussed in Chap. 6. The information transmitted by chemical cues includes trail substances; alarm substances; individual, race, species, and age, sex, and reproductive state recognition and parameters of territories.

The word *pheromone* has come into general usage as the chemical agent, but in the narrowest definition, not all chemical cues fit. The familiar concept of hormones as internal secretions regulating physiologic processes led to the use of the word ectohormones for materials secreted externally to regulate physiologic processes. Karlson and Lüscher (1959) introduced the term *pheromone*. In its strictest sense a pheromone is defined as "a chemical signal which triggers a response in a conspecific receiver" (Wilson, 1963). In contrast, when a signal is used to communicate with a member of another species it is called an *allomone*. The release of an odor by the skunk when frightened certainly communicates effectively with all species of mammals!

The chemical composition of very few mammalian pheromones is known. The odor of skunk is produced by *n*-butyl mercaptan, and resembles other odors, e.g., odors around a red fox *(Vulpes vulpes)* den which has been disturbed by other predators or a dog *(Canis familiaris)*. In the civet cat *(Viverra civetta)* a ketone called civettone has been isolated, and in musk deer *(Moschus moschiferus)* another ketone, muscone, is the chemical stimulant. In the black-tailed deer *(Odocoileus hemionus),* one of the more active agents in the tarsal gland secretion has been identified as a gamma lactose (Müller-Schwarze, 1971). Androgens may provide the musky odor in swine, according to Sink (1967). Kingston (1964) has reviewed the chemistry of musk, civet, and castoreum.

A distinction between *releaser* and *primer* effects of pheromones has come into common usage. Releasers elicit an immediate and sometimes reversible response in the recipient, while primers trigger a chain of physiologic events and the effect is delayed and not behaviorally visible. But the same substance can have both effects. A chemical stimulus which causes the release of a specific behavior pattern can also act on the reproductive state. Thus some olfactory stimuli can have multiple effects.

Another division which has been made for pheromones is into physiologic regulatory functions and communicative functions.

Physiologic Regulation

The greatest stimulus to research on mammalian pheromones was provided by the pioneering work of Bruce (1960), Lee and Boot (1955), Whitten (1959), and others on the reproductive processes of mice *(Mus musculus)*. The increased incidence of spontaneous pseudopregnancy among grouped females in contrast to singly housed females was first reported by Lee and Boot (1955). These pseudopregnancies were believed to result from an olfactory-mediated stimulus (Lee and Boot, 1955). Bruce (1960) demonstrated that the introduction of a strange male to a recently inseminated female blocked implantation. Later Parkes and Bruce (1962) and Dominic (1964) showed that this effect was produced by odor. Housing of female mice in large groups of 30 produced highly significant incidences of anestrus (Whitten, 1959), contrasting with the Lee-Boot effect (pseudopregnancy). Female mice *(Mus)* or deer mice *(Peromyscus)* caged together and showing irregular estrous cycling or high incidence of pseudopregnancy can be rapidly brought into synchronizing estrous cycles by the presence of a male, the urine of a male, or the odor of an adult male (Whitten, Bronson, and Greenstein, 1968). They have shown the mediating substance to be a pheromone transported in the air. Recently Vandenbergh (1969) has shown that female mice reach puberty earlier when housed with adult males or exposed to their odor. The reproductive-priming pheromones which inhibit pregnancy, induce anestrus, induce estrus, or advance puberty in mice could be the same substance present in the urine of all adult males. Breeding synchrony for a number of other mammalian species, including humans (McClintock, 1971), has been described as a possible function of olfactory signals. For a good résumé of the physiologic regulation of the reproductive process in mammals, see Gleason and Reynierse (1969).

Communicative Functions

Olfactory stimuli as communicating devices occur in great diversity in a great variety of mammal species. Whether or not they are pheromones depends on how narrowly the term is defined.

Sex Attractants; Discrimination of Sex and of Reproductive State   Sex attraction and sex recognition are not identical. Sex attractants actively recruit approach behavior, while sex recognition requires only discrimination. It has long been known that the odor of a female in estrus is a prime stimulus indicating her condition to the male. This stimulus serves all three purposes given above, and additionally, species recognition. How else could one explain the behavior of a bison *(Bison bison)* bull, from a small herd (probably with no estrous female), pacing the fence line closest to a large herd of rutting buffalo *(Bison bison),* contained in a pasture whose nearest fence was 2 miles away?

Discrimination between estrous and anestrous females by olfactory cues has been documented for many mammals, including dogs *(Canis familiaris)* (Beach and Gilmore, 1949), among which urine is the source of scent; in marsupials (Sharman, Calaby, and Poole, 1966), where epocrine glands in the pouch may be the source of the chemical stimulant; in heteromyid rodents (Eisenberg, 1963); and in many ungulates. Hafez, Schein, and Ewbank (1969) provide a word of caution concerning domestic cattle. They wrote, "It has often been assumed that the sense of smell plays an important, if not primary, role in sexual stimulation of the bull." The studies of Hart, Mead, and Reagen (1946) indicated that olfactory cues were "all important." Hale (1966) reported that, "Attempts to elicit male sexual behavior from bulls with olfactory stimuli while holding visual stimuli constant have proven unsuccessful. It may be that the association of smell with sexual excitation is a secondary conditioning." Not enough is known about these chemical stimulants to call them pheromones, but with our present knowledge they seem to fit the definition.

In many species, male pheromones also stimulate sexual behavior. In pigs *(Sus)* the preputial glands have been implicated in the male sex odor which elicits the "mating stance" from females (Hafez and Signoret, 1969). In the European rabbit, *Oryctolagus cuniculus* (Mykytowycz, 1965), and in the ring-tailed lemur, *Lemur catta* (Evans and Goy, 1968), skin glands are better developed in males than in females and may permit discrimination not only of sex but also of sexual state.

The discrimination of the reproductive state can also be dependent on differential excretion of steroids affecting the reproductive system, as well as changes in metabolism. These can be detected in the waste products. Such chemical communication raises several interesting points. Wynne-Edwards (1962) proposed that all functional odors have been derived from natural selection of metabolites. So in this chemical system we seem to have an example of an evolutionarily significant mode. Secondly, some of these odors obviously are not glandular secretions, and again may not fit into the precise classification as pheromones, but they certainly are chemical stimulants.

McCullough (1969) has given a most graphic description of this type of pheromone and olfactory communication and its significance in tule elk *(Cervus canadensis)*. In elk a harem is the social organization during the reproductive season. The male regularly smells the rump of each female so as to recognize those that are entering estrus. McCullough (1969) recognized four categories of bulls during the breeding season. Primary bulls are mature individuals which shed their velvet earliest and establish harems first. Secondary bulls are large individuals which take over the harem by defeating the primary bulls when they become exhausted from lack of food and rest. The same cycle is repeated for tertiary bulls. Opportunist bulls, the fourth category, meet cows by chance. Mature bull elk during the breeding season do not urinate in the usual way. Instead the urine is squirted on the belly and the thick hair on the chest. The bull frequently thrashes the low vegetation, wallows, and engages in other agonistic activities. By "self-marking" he also extends his scent onto vegetation and wallows. This scent reflects the deteriorating physical condition to other males, and may postpone useless disruption by fighting, until an appropriate time, when a fresh bull can replace an exhausted one. Most importantly, this increases the probability that a fertile mating will occur.

Species and Racial Identity Godfrey (1958) has demonstrated that European bank voles *(Clethrionomys)* could distinguish between odors of their own and another subspecies and also that the males preferred the odors of closely related females. In this case olfactory cues are effective reproductive isolating mechanisms. Others have shown that *Peromyscus, Mus,* and *Meriones* can distinguish the odor of conspecifics from that of others. The existence of a community odor in social animals has been demonstrated in the common marmoset, *Callithrix jacchus* (Epple, 1970), and in the Australian sugar glider, *Petaurus breviceps* (Schultze-Westrum, 1969).

Individual Recognition   Individual recognition on the basis of pheromones appears to be a more general phenomenon than sex recognition. Identification of individual prairie dogs of any sex or age belonging to the basic social group, or "coterie," is accomplished through a "kiss" which has an olfactory component (King, 1955). In black-tailed deer *(Odocoileus hemionus),* the tarsal scent is most important in individual recognition (Müller-Schwarze, 1971). Mother-young recognition by odor has been documented for Alaskan fur seals, *Callorhinus ursinus* (Bartholomew, 1959); goats, *Ovis* (Blauvelt, 1955); bats (Kleiman, 1969); and other species.

Alarm Substances   The release of scent during stress situations has been reported for rats *(Rattus)* (Valenta and Rigby, 1968), mice, woodchucks *(Marmota monax)* (Hamilton, 1934), metatarsal scent in black-tailed deer *(Odocoileus hemionus)* (Müller-Schwarze, 1971), the house shrew *Suncus murinus*

Figure 13-4   The recognition "kiss" between coterie members of the prairie dog *(Cynomys ludovicianus)* may be both visual and olfactory.

(Dryden and Conaway, 1967), and members of the weasel family, the best known of which are the skunks.

Trail Substances   Trail-marking substances in mammals probably serve not only one but several functions, including alarm, individual and sex recognition, and territorial marking.

Scent Marking and Territory   Territory has been described as any defended area. Its size and function are discussed elsewhere. A very common kind of scent marking is that which has been described as "territorial marking," which implies the marking of a fixed area that the individual will defend. Ralls (1971) has pointed out that not all scent marking is territorial but that mammals tend to mark most often in a situation where they are dominant to, and intolerant of, other members of the same species, and the marking of the territory is coincidental. It should also be pointed out that activities related to marking behavior render an area conspicuous through disturbance and the release of odor.

Scent-marking postures, such as leg lifting and piloerection of neck fur in a dog, suggest that scent marking may be an extension of a visual display which lasts only momentarily. Scent marking has the advantage of advertising ownership even when the owner is absent (Hediger, 1955).

Territorial marking may help an individual attain or maintain dominance by serving as a threat. An interesting point is that the frequent copulations by dominant males may lead to significant increase in plasma testosterone, as shown for rabbits by Haltmeyer and Eik-nes (1969). This would tend to reinforce their dominance. In addition to glandular scents, urination and defecation are frequently used for territorial demarcation.

Scent marking is practiced by the social rabbits (Mykytowycz, 1965), the short-tailed shrew *(Blarina brevicauda)* (Pearson, 1946), the muskrat *(Ondatra zibethicus)* (Errington, 1963), and the *Mustelidae,* which leave deposits of oily, strong-smelling substances. Members of the family *Felidae* mark their territory with urine. They often share a large, moving hunting territory, with temporal rather than spatial borders. Fresh markings provide temporary boundaries which deter conspecifics, but which, when not actively used, may be noticed but not respected (Leyhausen and Wolff, 1959). Dogs and wolves use scent marking (Mech, 1970). The home range of the black-tailed deer *(Odocoileus hemionus)* is marked by rubbing the forehead glands against twigs (Müller-Schwarze, 1971).

Ralls (1971) suggested that species which tend to mark frequently are both intolerant of and dominant to other members of the same species. She wrote:

An animal which marks frequently may be the dominant individual in a group, the dominant individual in a fixed area or territory, both of these, the dominant individual only when close to certain other animals (for example a male near females) . . . or a solitary individual which does not defend a territory but which habitually wins agonistic encounters with other animals of the same species.

The sugar glider *(Petaurus breviceps),* a small Australian marsupial, lives in a community consisting of up to six adults and their young. Each community has a territory. Male sugar gliders produce odors in the cloacal region and in the frontal and sternal glands. They use the products of the frontal gland to mark other members of the community. They mark their territory in several ways. Females rarely mark. All marking of the territory and individuals is done by one or two dominant males (Schultze-Westrum, 1969).

Research in olfactory communication is a newly emerging field of research and I have devoted much space to it. Those who would like more information would profit from reading review articles by Tembrock (1968), Wilson (1963), Gleason and Reynierse (1969), Ralls (1971), and Poulter (1968).

TACTILE COMMUNICATION

Tactile stimulation is more restricted in its use than are other forms of communication, but it is highly important, especially in sexual behavior and the care of young. Both moles *(Talpa)* and pigs *(Sus)* have well-developed olfactory and tactile functions in the snout. In the pig the cortical area for the snout is proportionally larger than in any other ungulate. In rooting, tactile and olfactory stimuli must be used simultaneously for exteroreception (Moulton, 1967). In some parts of France, pigs are trained to search for truffles (underground mushrooms).

In cattle *(Bos)* (Hafez, Schein, and Ewbank, 1969) and in bison *(Bison bison)* (McHugh, 1972), chin resting by the male conveys information to both the male and the female. Before mounting, the male rests his chin on the female's rump, and may exert a mild forward pressure. Nonreceptive individuals escape, but estrous cows respond by "standing" or even exerting some back pressure. The male then mounts. Tactile stimulations are perhaps of greatest importance in the sexual activities of mammals.

The phenomenon of milk let-down is a well-known response of females to the suckling of young. In nursing whales the female forcibly ejects milk when the young touches the nipple or its vicinity (Slijper, 1962).

Mutual grooming in primates not only cleans a partner's fur but probably has an important function in promoting social contact.

AUDITORY COMMUNICATION

Auditory or vocal communication in mammals is widespread and well known. The "bugling" of the elk *(Cervus canadensis)* and "bellowing" of the bison

(Bison bison) are known to many, especially hunters and Western history buffs. The howling of coyotes *(Canis latrans)* in the West and of the wolf *(Canis lupus)* in limited areas of the northern part of North America have also been represented in many ways in American literature. Mech (1970) has described the uses of wolf vocalization. Surprisingly, wolves do not vocalize when nearing their prey, as popular myths would have us believe. There is now a record available from the American Museum of Natural History (New York) called, "The Language and Music of the Wolves." Another record makes available to human ears "the songs of the humpback whales" (published by CRM Records, Del Mar, California). Among social ungulates, vocal communication is nearly constant. This author has spent much time among bison *(Bison bison)* herds; during the summer there is an almost continuous sound of grunting or bellowing bulls (Fig. 13–5). McCullough (1969) reported "a continuous array of sounds—foot bone creaking, stomach rumbling, teeth grinding and others," for the tule elk *(Cervus).* Kelsall (1968) reported that creaking and snapping of foot bones allows the members of a herd of caribou *(Rangifer tarandus)* to remain in auditory contact.

In prairie dogs *(Cynomys ludovicianus)* vocalization has already been mentioned as an important and well-known method of communication. The well-known "bark" is a short, nasal yip, used as a warning signal. They also use a two-syllable call, which has been interpreted as a territorial call. It seems to be an "all's-well" call given by a foraging dog in the home territory, or after the departure of danger. If others join in "an all-clear signal becomes a cacophony of togetherness" (King, 1955).

Recently, research has shown that ultrasonic sounds are uttered by all young myomorph rodents. Parents respond by bringing them back to the nest. With the increase in homeothermy (decreasing cold stress) as the infant matures, these calls decrease (Noirot, 1969; Sewell, 1970; and Hart and King, 1966). Hart and King (1966) and Noirot (1969) have suggested that audible and ultrasonic sounds might elicit search behavior in the female and also reduce aggression toward the pups. J. C. Smith (1972) proposed that audible sounds of deer mouse *(Peromyscus* spp.) pups were in response to cold stress by younger pups, while pure ultrasonic calls were characteristic of older pups when subjected to tactile stimulation. Smith (1972) believed that while both types of calls induced some maternal searching, the audible calls induced maternal search behavior and ultrasonic calls reduced maternal aggression.

ECHOLOCATION

Although not in the strictest sense communication (information transmittal between individuals), echolocation is the perception of information within an individual's environment.

At present, echolocation is known to occur in four orders of mammals —Insectivora, Chiroptera, Pinnipedia, and Cetacea (suborder Odontoceti). Its

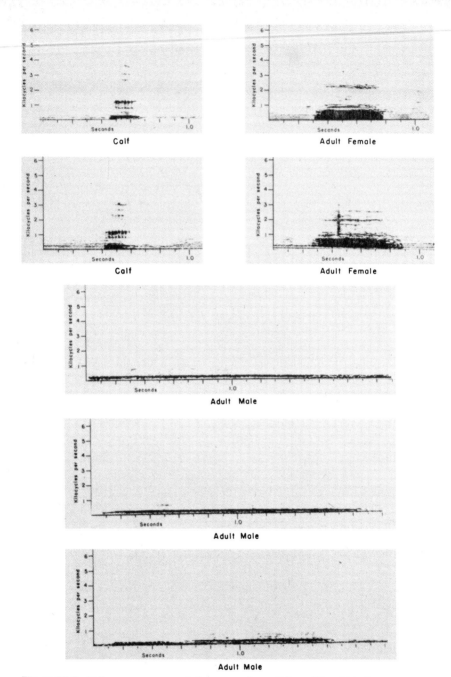

Figure 13–5 Using sonograms made from recordings of bison *(Bison bison),* the frequency and duration of vocalizations can be determined. For both calves the duration was 0.2 s and the fundamental frequency was 231 Hz. For the females the duration was 0.6 and 0.5 s and the fundamental frequency was 192 and 347 Hz. For the adult males (bellowing) the duration was from 0.18 to 0.9 s and the fundamental frequencies were from 146 to 239 Hz.

occurrence in the orders Dermoptera, Marsupialia, and Rodentia is suspected but not yet definitely proved.

Bats Except for a very few, such as the large fruit bats, bats are nocturnal and insectivorous. For this unusual mode of life, entailing rapid maneuverability in the dark, a sophisticated navigational system has evolved. Echolocation is sending out sound (energy) and sensing the time it takes for the sound to return from an object. It is used both to avoid objects and to capture prey. Radar involves the same principle, but the form of energy is different. Naturalists long ago realized that bats could fly about in a room at night without hitting walls, furniture, or even threads stretched across a room. As early as 1794 the Italian Lazaro Spallanzani suggested that bats use acoustical perception to avoid objects. In 1908, Hahn strung wires from a ceiling and released bats to fly through them. When nothing was done to the bats, 25 percent hit the wires; when their eyes were covered 22 percent hit the wires; but when their ears were plugged 66 percent hit the wires. In 1926 a British physiologist (Hartridge, 1920) suggested that bats use a system of sending out vibrations whose echoes are reflected back to their ears. Some years later Griffin and Galambos (1940, 1941) and their colleagues sparked the research taken up by many to unravel in great detail the "sonar" which nature has been using for at least 15 million years. Echolocation has allowed bats to exploit at night the zone which is exploited by birds during the day.

In bats, the cerebral hemispheres of the brain are small and the olfactory portions are reduced, but the cerebellum, which is of importance in the coordination and regulation of motor activities and maintenance of posture, is large. The ears are highly specialized, with a large cochlea and a lightly stretched basilar membrane. The external pinnae are often huge, e.g., in *Plecotus,* and within each pinna is a *tragus,* a fleshy projection of the anterior border of the pinna. The auditory bulla is loosely attached to the skull, allowing the ear to act somewhat independently of skull vibrations. In those bats which apparently do not use echolocation (e.g., most of the large fruit bats, *Pteropus),* these specializations are lacking. In the suborder Megachiroptera, the large fruit bats (whose food is immobile), echolocation occurs only in members of the genus *Rousettus.* Their pulses are audible tongue-clicking noises (Kulzer, 1956 and 1958).

Nearly all members of the suborder Microchiroptera use echolocation. In Vespertilionidae, a family of the suborder Microchiroptera, the sound or cry is emitted through the mouth. Griffin (1958) and his associates found that bats such as the big brown *(Eptesicus fuscus),* little brown *(Myotis lucifugus),* and Keen's bats *(Myotis keeni),* among others of the vespertilionids, gave out different classes of sounds. These varied from those with a frequency of 7 kHz and audible to the human ear up to ultrasonic sounds of frequencies between 30 and 70 kHz, not audible to the human ear. A resting bat may give out only

5 to 10 of these ultrasonic "cries" per second. When a flying bat is approaching an object these cries increase, sometimes to over 100 per second.

The basic unit of the ultrasonic "cry" is an individual sound of short duration and high frequency, which may contain from 40 to 3500 individual sound waves. What is still more interesting is that these pulses are frequency modulated. The longer wavelengths reach farther out and inform the bat about distant objects, the shorter waves inform the bat about both closer and smaller objects. Bats are able to control both the duration and the repetition of the pulses, as well as the frequency modulation.

In the horseshoe-nosed bats (Rhinolophidae), the sound is emitted through the nose instead of the mouth, and is beamed by interference of the nostrils, which are set a half-wave length apart (Mohres, 1953). In these bats the pulse duration is longer, 50 to 65 ms (as compared to 1 to 4 ms in vespertilionids), and of a constant and high frequency. (A millisecond is $\frac{1}{1000}$ of a second.) This system is believed to be even more effective than that of many other bats; horseshoe-nosed bats can detect insects as distant as 6 m, while for the vespertilionids an insect must be within a meter of the bat for detection. Mohres (1953) has shown that the horseshoelike membrane surrounding the nostrils serves as a horn, beaming the pulses forward. Horseshoe-nosed bats, even when stationary, seem to search about by movements of the ears, which suggest that their system is effective enough to take advantage of the Doppler effect, i.e., of change of pitch due to motion. Fish-eating bats are even able to locate their underwater prey by echolocation (Bloedel, 1955). For a detailed discussion of the anatomy of bats' ears and their use in echolocation, the reader is referred to Henson (1970).

Although echolocation is very efficient, there is a minimum size beyond which objects cannot be detected. Constantine (1958) found that *Lasiurus cinereus, Tadarida brasiliensis,* and *Antrozous pallidus* are unable to avoid wires 0.006 in thick. This has led to development of a method of capturing bats as they emerge from caves at dusk, by extending a screen of thin nylon "string" hanging down over the opening. Bats also apparently can become conditioned to the presence of objects and so "learn" to avoid them (Van Gelder, 1956).

As already mentioned, bats are not the only mammals, or even animals, to use a system of echolocation. One of the most interesting examples is that of a moth, which is sometimes the prey of echolocating bats, using echolocation to escape (Roeder, 1964, 1965).

The pulses used by most bats during echolocation are produced by the larynx, and are of frequencies from 30 to 120 kHz, but frequencies differ widely among species. We can detect frequencies up to 20,000 Hz (20 kHz), which is above our hearing range. Pulse duration and emission rate also vary widely. Pulse rate during "searching" flight is slow, 10 per second, and increases to 200 per second during the final stage in capture of a flying insect.

Cetaceans   A whole gamut of sounds is made by cetaceans. Schevill and Lawrence (1949) were the first to listen to underwater sounds of whales. Their subject was the white whale *(Delphinapterus)*. Payne (1970) has made the "songs" of the humpback whale *(Megaptera)* available to the public through a record. The profits from its sale go to the New York Zoological Society whale fund, which is devoted to the study and preservation of whales. The humpback whale has a varied song repertory. Payne and McVay (1971) reported that "the function of the songs is unknown," but it is generally assumed that whale "songs" serve a communication purpose. A great variety of sounds are made by both the baleen whales (suborder Mysticeti) and the toothed whales (suborder Odontoceti), but only the odontocetes are believed at least at present, to echolocate. The common dolphin *(Tursiops truncatus)* has been intensively studied; it uses short pulses, as bats do, to avoid obstacles and locate food. This dolphin emits whistles and clicks, some audible to man and some ultrasonic. For a discussion and literature review of work on *Tursiops* the reader is referred to a paper by Lilly (1966). The pulse rate rises as the bottle-nosed dolphin approaches a target, just as it does in bats. This dolphin can even distinguish between two different thicknesses of the same kind of metal (Evans and Powell, 1967).

One of the most unusual of whales presumably using echolocation is the blind river dolphin *(Platanista gangetica)*, which inhabits the muddy waters of the Ganges, Indus, and Brahmaputra Rivers in India and Pakistan. It is likely that its atrophied eyes can detect light, but the absence of a lens probably does not allow it to focus an image. This dolphin swims on its side, and the lower flipper is usually touching, or near, the bottom. Captive specimens continuously produce pulse rates of from 20 to 50 per second, which are reflected from the concave front of the skull and are focused by the "melon," the lens-shaped fatty structure present in many odontocetes. Large flanges from the maxillary bone, with an intricate radial pattern of latticework on the inside, serve to concentrate the pulses into a narrow beam (Herald, Brownell, Frye, Morrisse, Evans, and Scott, 1969).

How whales produce sounds is still open to speculation, since the larynx of the odontocetes lacks vocal cords. Despite the lack of vocal cords, there are well-developed muscles, and the larynx has a complicated structure. Norris and Prescott (1961) suggested that the muscular valves at the blowhole that close the nares, and the membranes against which they rest, as well as the lips of the blowhole, might produce sound. Purves (1966) believed the larynx to be the site of sound production. The bulk of the most recent research seems to indicate that, at least in delphinid cataceans, sound is produced in the nasal region. For a good review the reader is referred to Mead (1972) and Evans and Maderson (1973).

As would be expected of an organ used for echolocation, the ear of whales have become very highly specialized. Reysenback de Haan (1966) had described the whales as "hearing animals par excellence." Norris (1969) reported anecdotes of whales, one of which was that "the rubbing of oarlocks could send a sperm whale into flight." Whales have evolved from land mammals, and their marine life is a secondary development. Reysenback de Haan (1966) interestingly pointed out that if we accept the lateral line organ of the fish as a "hearing organ," then we have the primitive form of hearing, as well as the most highly developed form of hearing, in animals living in the water.

In general, whales hear with the same structures as land mammals, but with major modifications in the hearing apparatus. The discussion that follows is based on the reports of Fraser and Purves (1954, 1960, 1960a), Purves (1966), and Norris (1964, 1968, 1969). In whales the external ear opening (external auditory meatus) may be very small or may be covered over completely. The cavity of the external auditory meatus is filled with wax. The conductance property of the wax approximates that of wood. Three routes have been proposed for sounds to reach the cochlea of the ear. One theory is that sound reaches the middle ear via the region of the external ear and then passes through the tympanic ligament to the ossicular chain (Fraser and Purves, 1960; Purves, 1966). Another suggestion is that sound reaches the tympanic ligament from tissues of the side of the head, passing through the ossicular chain, the oval windows, and finally the cochlea (Reysenback de Haan, 1957). Still a third view (Norris, 1964, 1968) is that echolocation sounds of delphinids are received through the thin back part of the lower jaw, through the inframandibular fat body which leads directly to the wall of the bulla.

The most useful departure from the generalized mammalian ear is the insulation, "suspension," of the middle and inner ear so that they are not rigidly attached to the skull. Each ear must function in such a manner as to localize the sound. The middle ear is surrounded by cavities containing an albuminous foam. The cavities, in turn, are surrounded by fatty tissue in the toothed whales and by connective tissue in the baleen whales, which serve as acoustic insulators. In this way sound vibrations can reach the cochlea only by the way of the eardrum. If the bone that houses the middle ear were attached rigidly to the skull, as in most land mammals, vibrations from the water could be transmitted through the bones of the skull to both ears and would reach the ears from various directions, making it impossible to echolocate.

We have noted that in bats and whales the tympanic bulla is "insulated" from rigid connections with the bones of the skull. This is an adaptation to localize the direction of sound origin. In the watery environment of whales, this adaptation has reached its greatest development.

Other Mammals   The potential to echolocate may be present in many kinds of mammals, even in people who are blind, as Griffin (1958) has pointed out. Among the insectivores, Gould, Negus, and Norvick (1964) first reported echolocation in shrews. The common short-tailed shrew *(Blarina brevicauda)* of the Eastern United States produced pulses of about 5 to 33+ ms duration at between 30 and 60 kHz. Later Gould (1965) revealed that the primitive shrews of the family Tenrecidae from Madagascar produced clicking noises, audible to man, with their tongues, in the 5- to 17-kHz frequency range.

Echolocation also occurs in seals. The Antarctic Weddell seal, *Leptonychotes weddelli,* which spends much time under ice, emits chirps with frequencies up to 30 kHz (Watkins and Schevill, 1968).

Other mammals presumed to use echolocation include several species of rodents, order Rodentia (Kahmann and Ostermann, 1951; Rosenzweig, Riley, and Krech, 1955).

Auditory Adaptations of Kangaroo Rats   Sound is absorbed in different degrees under different conditions of temperature and humidity (Knudsen, 1931). The smaller mammals living in hot, arid conditions have relatively large auditory bullae, along with varying degrees of saltatory (jumping) locomotion. Each arid region of the world has its own ecologic equivalent of this type of mammal. These include the Heteromyidae, or kangaroo rats of the New World; the Dipodidae, jerboas of Africa, Arabia, and southern Russian, and southeastern China; the subfamily Gerbillinae, gerbils, also in North Africa, the Near East, and parts of Asia; the Muridae, the Australian hopping mouse, *Notonys alexis;* the Gliridae, the fat dormouse *(Glis);* the Cricetidae, the golden hamster *(Mesocricetus);* and the Macroscelididae, the elephant shrews of North Africa.

Among the small mammals, the kangaroo rats *(Dipodomys)* of the Southwestern United States have received the greatest amount of attention, with a view to finding out how a mammal's adaptation has fitted it to its specific environment. One of the more interesting of these adaptations in the kangaroo rat is that of the ear. The auditory bullae of the kangaroo rat have a greater total volume than that of the braincase. The bones of the skull are very thin, and kangaroo rats cannot gnaw their way through cardboard or wood the way many rats and mice can. Webster (1961, 1962, 1966) and Webster, Ackermann, and Longa (1968) have elucidated the auditory specializations; the following discussion is based on their work.

The arid air of the desert has poor sound-carrying qualities in comparison to cooler, humid air. The inflated bullae of the kangaroo rat increase the volume of the air-filled chambers surrounding the middle ear, which in turn reduces the resistance to the inward movement of the tympanic membrane.

Anatomically the malleus is allowed to rotate more freely, and the manubrium of the malleus is greatly lengthened, giving increased leverage. This transforms relatively weak vibrations of the greatly enlarged tympanic membrane into more powerful movements which are transmitted to the incus and the small footplate of the stapes, which rests against the oval window of the inner ear. The result of all this is to transform a rather weak pressure on the large tympanic membrane into a relatively great pressure on the fluid within the inner ear. This ratio is called the transformer ratio, which in man is 18:1 but in the kangaroo rat is 97:1.

Further experimental work showed the kangaroo rats to be most sensitive to sounds of low intensity with a frequency between 1 and 3 kHz.

The reactions of rats to two nocturnal predators, the sidewinder rattlesnake *(Crotalus cerastes)* and the barn owl *(Tyto alba),* were tested. The wings of the owl and the short burst of sound the rattlesnake makes prior to a strike, when the scutes scrape the substrate, are both of low frequency and low intensity. In the laboratory, kangaroo rats made a sudden vertical leap when they heard these sounds! Obviously these adaptations of the ear and hind legs are highly advantageous. A graphic representation of this behavior is shown in the Disney film "Vanishing Prairie," as well as in a more recent film presented by the National Geographic Magazine called "The Mojave Desert."

Chapter 14

Migration

Where winter climates are severe, cold and lack of food can place undue stress on an organism. Some of the adaptations which allow mammals to maintain an internal homeostasis and remain active under these conditions have already been discussed under thermoregulation. Some mammals escape these stresses by hibernation, which has also been discussed. Still another way of escaping these stresses and retaining a fairly constant internal environment is to move between external environments which do not differ greatly—i.e., migrate.

The word migration has been used to denote many kinds of movements of living organisms, but because of its widespread application to the regular seasonal movements of birds, its most common usage is in terms of Heape's (1931) definition. He wrote, "Regarding migration, that is to say, a movement which involves a journey to a definite area, and a return journey to the area from which the movement arose. . . ." He gave three causes for migration: (1) alimental (food and water), (2) climatic, and (3) gametic. Welty (1962), in his well-known ornithology textbook, wrote, "These more or less regular, extensive movements of birds between their breeding regions and their 'wintering' regions are known as migration." Urquhart (1958) has discussed, as has Heape (1931), the terminology of animal movement, including migration, immigration, emigration, nomadism, and dispersal. He has suggested use of the word "remigration" for the north-south seasonal movement of birds. Heape (1931), an Englishman, believed the annual movements of North

American ungulates to be not migrations but rather examples of nomadism, since there is not always the alleged precise regularity associated with birds (i.e., the swallows at Capistrano). Thomson's (1926) definition of migration probably best describes the movements of North American ungulates. He wrote, "True migrations are changes of habitat, periodically recurring and alternating in direction, which tends to secure optimal conditions at all times." It is in the light of this definition that migration will be discussed in this chapter. Nomadism is a random type of movement controlled by the organism. Emigration means literally "movement out of," and is a movement out of a given area, with no return. Immigration is a movement into an area with no return. For a further discussion of these terms, as well as other movement terms not mentioned here, the reader is referred to Heape (1931), Milne and Milne (1958), and Urquhart (1958).

Among the mammals, migrations occur among bats, pinnipeds, ungulates, and whales.

BATS (CHIROPTERA)

Bats are mostly tropical in distribution, but some species have evolved in or invaded the temperate regions. These are primarily insectivorous and so must cope with marked seasonal changes in food supply as well as in climate. They do this by either migrating or hibernating, as well as migrating to hibernation sites.

The successful banding of birds for migratory studies led the well-known ornithologist A. A. Allen (1921) to try his luck with banding bats. He banded four pipistrelles *(Pipistrellus subflavus)* with leg bands and 3 years later recovered three of them at the same roosting place.

Bat banding increased gradually until the early 1950s, when the discovery of rabies in the Brazilian free-tailed bat *(Tadarida brasiliensis)* gave banding a tremendous impetus. Over 100,000 were banded in Oklahoma, Texas, New Mexico, and Mexico. Cockrum (1969) and his coworkers banded 162,892 individuals between September 1952 and September 1967. These figures give some indication of the tremendous effort required to map the migration of one species, especially when one remembers that the percentage of bats recovered at a distance is less than 1 percent. A good review of bat migration in North America is provided by Griffin (1970).

Merriam, as early as 1887, pointed out that three species of North American bats, the red bat *(Lasiurus borealis),* the hoary bat *(Lasiurus cinereus),* and the silver-haired bat *(Lasionycteris noctivagans)* seldom hibernate. Although all have been found in a torpid condition in the southern part of their range, the marked seasonal changes in abundance confirm the hypothesis that at least the majority migrate, at least for several hundred kilometers. These three species usually roost individually in trees and are for this reason called tree bats. The more gregarious bats of the genera *Myotis, Eptesicus, Pipistrellus,* and *Tadarida* are grouped together as cave bats, for they roost in caves, crevices, and old houses. Griffin (1940 and 1945) banded over 13,000 individuals belonging to the genera *Myotis* and *Eptesicus,* and found that the summer colonies were widespread over New England but that a very high percentage of them wintered in caves in western Massachusetts and Connecticut, and in Vermont. The greatest distance covered was 270 km (168 mi).

More recently Davis and Hitchcock (1965) have banded more than 70,000 little brown bats *(Myotis lucifugus)* in eastern New York and the New England states (except Maine). Most interesting was the fact that the 300 or more recoveries, made of the thousands that were banded at Mt. Aeolus cave in Vermont, showed a migratory route of southeast in the spring and northwest in the fall. Distances up to 320 km (200 mi) were recorded.

The most interesting story of bat migration and biology is that unfolded by banding studies of the Brazilian free-tailed bat *(Tadarida brasiliensis).* An excellent review has been provided by one of the participants, E. L. Cockrum (1969). These bats occur by the millions in caves of the Southwestern United States and Mexico, among them the well-known Carlsbad Caverns of New Mexico. First studies produced conflicting reports on the movements of these bats. Benson (1947), working in California, believed that they made only local seasonal movements. Christensen (1947) concluded, on the basis of his observation at Carlsbad Caverns, that they performed long migrations, a conclusion later confirmed by Constantine (1967). It turns out that both were right. Cockrum (1969), adding all other information, has hypothesized that four or more behaviorally (and possibly genetically) separate populations occur in the Western United States during the summer months. These four he specified as:

Group A, inhabiting southern Oregon and western California. They appear to perform only local seasonal movements, as reported by Benson (1947).

Group B, inhabiting western Arizona, southern Nevada, and southeastern California. They evidently migrate, because they are not found there in winter. Their winter range is as yet unknown, but it may be that they migrate southward into Baja California or west into California.

Group C, of central and eastern Arizona and western New Mexico. They appear to have a well-developed flyway through Sonora and Sinaloa, Mexico.

Group D, including the populations of eastern New Mexico, western Kansas, Oklahoma, and Texas. They migrate into Mexico. These bats are evidently excellent fliers, for Villa (1956) reported one as traveling at least 1280 km (800 mi) in 68 days, and R. B. Davis, Herreid, and Short (1962) reported them to fly at least 80 km (50 mi) to their foraging area each night.

Another interesting aspect of the biology of this bat is its reproductive behavior, which includes separation of the sexes and the use of maternity colonies. These are also discussed in Cockrum's (1969) review. On the basis of the season and population composition, five types of roosts were defined for Arizona. *Maternity Colonies* contain primarily adult females and young of the year that appear to have been born there. These caves are occupied during the warmer part of the year. At Eagle Creek Cave an estimated 25 to 50 million bats were present by late June. On the first of June the population was estimated to be less than 100,000. The adult females do not carry their young with them when they go out to feed, but leave them clinging to the walls. *Summer Male Roosts* consisted of small groups of 10 to 300 individuals in Arizona. *Transient Roosts* are occupied early in the year and again in late summer and early fall. Populations vary widely in sex ratios and numbers. *Winter Roosts.* No major roosts of this kind were found in the study area. *Multi-use Roosts* are those localities which serve, in season, as transient roosts or as maternity colonies.

The migration sequence for the free-tailed bat populations of Arizona has also been described by Cockrum (1969). Most of these bats spend December, January, February, and early March in Mexico. Breeding occurs in late February and early March, before the northward migration, although some breeding occurs during migrations. Beginning in March subpopulations move northward, occupying transient roosts. Males appear to migrate more rapidly than the females. By mid-June to early July, when parturition occurs, most females have congregated in relatively few maternity colonies such as the previously mentioned Eagle Creek Cave. Large colonies of males remain in Mexico during the summer months and apparently do not migrate—at least not northward. The maternity colonies break up in late August, September, and early October, when the young are grown, and again the transient colonies are occupied during the southward movement.

SEALS, SEA LIONS, AND WALRUS (PINNIPEDIA)

The most extensive mammalian migrations are made by some of those mammals which live in the ocean. Whales make the longest migrations, but some pinnipeds, e.g., the northern (Pribilof, Alaskan) fur seal *(Callorhinus ursinus),* run them a close second. This discussion will deal primarily with pinnipeds living in the Northern Hemisphere. As a general rule, pinnipeds do not live in coastal waters in which the summer temperature exceeds the 20°C isotherm.

Their movements and migrations are primarily for reproductive purposes (gametic). Some seals' movements may consist only of rather short movements, such as drifting on ice, or the males may move away from the coast during the herding season. The Steller's sea lion *(Eumetopias jubatus)* and the California sea lion *(Zalophus californianus)* may have some seasonal movements. For instance, in the California sea lion, on the west coast of North America, the adult and subadult males generally move northward after the summer breeding season and return to the rookeries in the spring. The females and young either remain in the area of the rookeries or some may move southward in winter (Peterson and Bartholomew, 1967). Little is known concerning the Steller's sea lion.

Walrus *(Odobenus rosmarus)* are found in fairly shallow water, close to land or ice, around the arctic coasts. They are circumpolar in distribution, with an Atlantic and a Pacific population. The Atlantic walrus migrates, but not all individuals do so each year. The Pacific walruses apparently make more extensive migrations, moving north in spring and summer, mostly on ice floes, which melt during the summer. In November there is a general movement south again. These movements take them through Bering Straits, which forms a funnel. Through this funnel occurs the most spectacular migration of marine mammals and birds to be seen anywhere in the world. In addition to the thousands of walruses, four species of seals, and five species of whales, and millions of northern waterfowl move through this watery funnel (Perry, 1967). Neither of the other two circumpolar species, the bearded seal *(Erignathus barbatus)* or the ringed seal *(Pusa hispida),* make long seasonal migrations, although they may float some distance on ice.

Some pinnipeds have developed a social structure in which there is a movement to rookery areas, where birth and breeding occur. The best example of this would be the northern (Alaskan Pribilof) fur seal *(Callorhinus ursinus)*. Between 1 and 2 million of these animals gather annually on the Pribilof Islands in the Bering Sea. The sea in the immediate area of the rookery could possibly not even support this population. The males may go without food and water for over 2 months during the breeding season (Bartholomew and Hoel, 1953). The breeding rookeries were discovered in 1786 by Gerassim Pribilof, whose name was bestowed upon these islands. Kenyon and Wilke (1953) gave a detailed description of the migration. According to them "fur seals migrate loosely." They begin leaving the islands in October; most of them are gone by December, although a few remain near the breeding islands throughout the year. Others may travel southward up to 4800 km (3000 mi) to near the United States–Mexican border on the east side of the Pacific and just off Japan on the west side of the Pacific. Adult females and young of both sexes move the farthest. During migration fur seals are generally found from 8 to 80 km at sea in water several thousand kilometers deep (Kenyon and Wilke, 1953). At the time of the discovery of their breeding grounds, fur seals numbered 2½ million, but extravagant sealing operations brought their numbers to a low of 200,000 in 1911, when the United States took over their control. Since then the herd has increased. The commercial sealing season starts June 20 and lasts for 4 to 5 weeks. Only bachelors 2 to 5 years old and about a meter (41 to 45 in.) in length are taken. After the Japanese and Canadian skins are dispatched, those belonging to the United States are sent to a firm in Missouri for processing. With the new ecologic awareness, the demand for fur has decreased greatly. If this decrease continues the fur seals may again attain a population of 2½ million.

WHALES (CETACEA)

The longest migrations undertaken by mammals are accomplished by some of the whales. Whales are found in all the oceans and in some of the smaller bodies of water (e.g., Hudson Bay). Some kinds are restricted in distribution, while other kinds occur throughout the world, although populations may be geographically restricted. For instance, among the great whales there are, in the broadest sense, two populations, of each species, one in the Northern Hemisphere and one in the Southern Hemisphere.

The greatest amount of information has been gathered on the large baleen whales, because they are the ones that have been, and in many cases still are, pursued by the whaling industry. These baleen whales (as opposed to toothed whales) belong to three families, the Balaenidae (right whales), the Balaenopteridae (rorquals), and the family Eschrichtidae, which contains only one species, the gray whale. At present the blue *(Balaenoptera musculus),* humpback *(Megaptera novaeangliae),* gray (*Eschrichtius robustus,* according to Rice and Wolman, 1971), bowhead *(Balaena mysticetus),* and right *(Eubalaena glacialis)* whales are protected. At the 1973 meeting of the International Whaling Commission a proposal to stop all whaling for 10 years lost by a few votes.

These large whales must consume tons of food, mainly plankton, and the areas of greatest plankton production are near the poles. But apparently this water is too cold to allow the newborn young to survive. The whales migrate to the warmer waters near the equator for the birth of their young, whose very thin blubber coat and relatively large skin surface allow much heat loss. But the tropical waters are relatively barren of plankton, and so these huge whales may go as long as 4 months without food.

Such migratory whales as the blue, fin, and humpback may travel up to 4800 km (3000 mi) in a season. Although these whales are found in both hemispheres, the northern and southern populations are isolated by the seasons. Since gestation is nearly a year, pairing and parturition occur in warm water near the equator. As the seasons are opposite, the southern and northern populations do not meet in tropical waters, although a few individuals may cross from one population to the other (Dawbin, 1966; Mackintosh, 1966; and Jonsgard, 1966).

Most species of whales are social animals, moving about the ocean in groups, sometimes called herds. A group of females is sometimes called a "pod." Sperm whales are organized into either a harem group or a bachelor group of younger, smaller males. How whales maintain contact in the vast oceans is a problem engaging many researchers. The answer appears to be in some kind of vocalization, whether "singing" or echolocation. Singing in whales has been investigated by Payne and McVay (1971) and others. Payne

(1970) has recorded the singing of humpback whales and made these recordings available to the public. The sounds made by whales are believed to allow travel in coherent groups even "when individuals are separated by tens or perhaps hundreds of miles" (Payne, 1970). Echolocation, used by toothed whales (Odontoceti), has not yet been found in baleen whales (Mysticeti).

The longest and most highly publicized of whale migrations is that of the gray whale *(Eschrichtius robustus),* whose seasonal migratory cycle has been thoroughly discussed by Rice and Wolman (1971). Their appearance along the coast of California near San Diego in mid-December is viewed by thousands of people every weekend. Cabrillo National Monument on a promontory near San Diego has a glassed-in observation point and exhibits. Here millions of people a year view the migration. Visitors are taken in excursion boats and to islands for an even closer look. The gray whales swim slowly and stay close to shorelines. Thus they became easy prey for whalers and were almost exterminated by the turn of the century. In 1946 they were given complete protection. It is this same behavior (staying near the shore) which now makes their migration "one of the world's outstanding wildlife spectacles" (Rice and Wolman, 1971) (Fig. 14–1).

Although not the largest of whales, the gray whale is still a magnificent

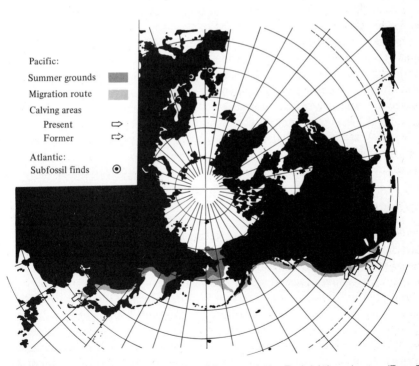

Figure 14–1 Distribution and migration of the gray whale, *Eschrichtius robustus. (From Rice and Wolman, 1971.)*

animal, reaching a length of 16 m (50 ft) and a weight of 27,200 kg (30 tons). Until several centuries ago it was found in the North Atlantic and in the North Pacific. Only the North Pacific population remains, and that is divided into two geographically isolated stocks. The western Pacific stock migrates between the Okhotsk Sea and South Korea. The eastern Pacific stock migrates between Bering and Chukchi Seas and Baja California, a distance (round trip) of 9600 km (6000 mi).

During the summer, May to October, the California stock of gray whales is found in shallow waters of the Chukchi Sea and parts of the Bering and Beaufort Seas, and a very few remain along the west coast of North America. Their feeding areas are apparently only in shallow waters.

In October, after 4 months of gluttonous feeding, they begin their southward movement, keeping near the coast and moving at a rate of 185 km/day on this southward migration and only half that fast on their northward migration (Pike, 1962). There seems to be a temporary segregation of the whales according to sex, age, and reproductive status. In both the north and south movement, the females are first, those closest to parturition being the very first, then the males. Adults migrate earlier than the young. Thus those arriving off the coast of California during the last half of December are predominantly females near parturition time.

During January and February the eastern Pacific stock of gray whales is in the warm waters on the west coast of Baja California and the southern Gulf of California, where parturition and pairing occur. Six calving areas are known (Gilmore, Brownell, Mills, and Harrison, 1967).

In northward migration, the first individuals are the newly pregnant females, which in the southward movement were the females that had recently ovulated. Female gray whales normally have one estrous cycle every 2 years. On the northward migration, starting in mid-February, all the newly pregnant females pass within a limited period of 15 days. The peak passage of adult males was 2 weeks later (Rice and Wolman, 1971).

Migrating whales move singly or in groups of up to 16 individuals.

UNGULATES

Of the North American ungulates, the caribou *(Rangifer tarandus)* make the longest migrations (Fig. 14–2). They may travel between 200 to 400 km on their seasonal journeys. Caribou migrations, always known by the Eskimos, have been particularly well documented in the last 50 years, especially by Banfield (1954), Clarke (1940), Murie (1935), and Pruitt (1960). During the summer, caribou live on the tundra north of the timber line, perhaps partly to get away from insects, although Murie (1935) does not believe the insect-pest theory. In July a southward movement begins and breeding takes place. They remain in the winter area, usually the taiga and the edge of the tundra, where there are a few scattered trees. In the spring, the northward movement begins, and the young are born on this journey. The caribou migration is relentless, and mass drownings occur when the animals cross swollen streams. Clarke (1940) mentioned an occasion when over 500 drowned.

The bison *(Bison bison)* made extensive migrations, according to Hornaday (1889) and Sandoz (1954), among others, but Roe (1951) does not agree with this, as discussed later in this chapter.

In the mountainous regions of western North American vertical migrations are made by all the ungulates, the most pronounced by elk *(Cervus canadensis)* and the deer *(Odocoileus),* the least pronounced by the moose *(Alces americana).*

For elk *(Cervus canadensis),* Altman (1956) has described altitudinal migration in Wyoming, where more than 10,000 elk gather in the National Elk Refuge at an altitude of 6000 m for winter feeding. With the beginning of spring this group breaks up into subgroups of smaller and smaller numbers, which begin to follow the receding snow line. The pregnant cows, together with their yearling calves, seek secluded calving areas, leaving the now predominantly male groups following the melting snow. These bull groups, ranging in size from 2 to 16 individuals, remain separate until the beginning of the rutting season in September. The cows remain on the calving grounds in groups of 12 to 30 until the calves are about 3 weeks of age, when the cow and calf herds begin moving up the mountains and arrive on the summer range. For a time after their arrival the bull groups and nursery groups retain separate identities, although both are in the same area.

In early August spike bulls in the nursery group become restless, the calves are being weaned, and adult bulls attempt to join a band of cows without success. Now the colder nights and absence of food drives them to lower areas. The characteristic unrest of August and early September develops into the "rutting season." A bull will gather a harem of 10 to 15 cows and their calves, and these groups now drift down to 2700 to 3000 m (9000 to 10,000 ft). The rutting season wanes as the elk continue to the valley floor, where they arrive as the snow begins. Near the upper reaches of the Gallatin River in Montana the elk also have an altitudinal migration. Brazda (1953) wrote, "In summer most elk move to higher elevations adjacent to the river, while in winter they are generally concentrated in areas much closer to the river." The Sun River elk herd in the Bob Marshall Wilderness had a winter range at an elevation of about 1.7 km (5000 ft) "in the warm chinook belt along the eastern edge of the mountains where they rise abruptly from the Great Plains" (Picton, 1960). He summarized their movements in these words:

> During the calving season, in June, the majority of elk utilized the grassland areas at lower elevations. . . . In late June and early July, elk moved through the forest types until by mid-July many were found in the subalpine barrens and along the

upper edges of the subalpine forest, where they favored the forb subtype. In late August, there was a downward movement into the forested areas, and by September these types were heavily utilized. All of the harems and sexually active males were seen in the forest types; nevertheless, a few scattered elk had been present in all negative types throughout the summer.

Craighead, Atwell, and O'Gara (1972) studied the elk migrations in Yellowstone National Park and found that five of the six discrete herds using the Park in summer were migratory.

Murie (1951) wrote, "On the whole, the elk is a migratory animal, but not much can now be known in that respect about the eastern elk. It is probable, however, that in many parts of the eastern United States there was little drift of elk from summer to winter range, although undoubtedly there were local wanderings."

Of the moose *(Alces americana)* Peterson (1955) wrote, "Moose show no well marked migrations as do wapiti and caribou. The only population shifts which might be distinguished as seasonal movements are found in the mountain regions of the west."

Denniston (1956) wrote of the Rocky Mountain moose, which he studied in the Jackson Hole region, "Moose spend their winters where food in the form of willow or aspen browse or hay is available and where the snow is not too deep." In the late spring the bulls and some barren cows follow the receding snow lines to higher and higher elevations. Cows carrying calves do not go so high but move into secluded areas in May to have their calves. The yearling calves remain with them. This group may remain in a small area of several acres through the summer. The bulls and dry cows may move as much as 60 km (30 to 40 mi), "often moving across passes from one drainage to another" (Denniston, 1956). In the Gravelly Mountains of Montana, Knowlton (1960) found, "The moose, in this area migrated to willow bottoms below 7,000 feet elevation during the winter. During summer and fall, most of the moose were found at elevations above 7,500 feet."

The mule deer *(Odocoileus hemionus)* resides mostly in mountainous areas. Where there is a distinct difference between summer and winter ranges, mule deer develop migratory habits. Some migrations may involve a horizontal shift of only 6½ to 8 km (four to five miles) and a vertical shift of only 305 m (1000 ft), while others may be 80 km (50 mi) in length and 2.5 km (8000 ft) in altitude. Lovaas (1958) found the mule deer in the Little Belt Mountains moved higher in summer and lower in September but gave no figures.

Evidently man's activities have broken up the older and longer migrations of the mule deer and the white-tailed deer *(Odocoileus virginianus)*. One of the longest migrations of white-tailed deer started from the shore of Lake Superior at its eastern end and ended in the region of north central Wisconsin, a distance of over 80 km (50 mi). In Minnesota (Olson, 1938; Morse, 1942), there is a movement from summer range to winter "deer yards" of white cedar *(Thuja occidentalis).*

The best-known migrations of any North American ungulates are those of the caribou *(Rangifer tarandus).* Murie (1935) wrote, "The seasonal movements of caribou have caught the attention of man from the earliest times, and naturally an explanation of this most interesting habit has been sought." One of the first to record in detail the migrations of caribou was Samuel Hearne, who was governor at Fort Prince of Wales (across the Churchill River from Churchill, on Hudson Bay) for 7 years after 1775, and who led an expedition to the Coppermine River. Hearne (1795) reported that the southern limit of the barren-ground caribou was the Churchill River. He wrote of "woodland caribou" which spent the winter in the vicinty of the Nelson River. They traveled east in the spring, crossing the Nelson and Severn Rivers, returning in the autumn.

Banfield (1954) has discussed in detail the caribou movements in Canada and the Northwest Territories. He warned, "Migration is here used to describe wanderings or movements of caribou bands over long distances. It is not meant to imply that these movements are necessarily comparable to bird migration, for instance." The herds travel from summer to winter ranges over distinct migration routes. Banfield (1954) wrote, "The residents [Eskimo] of the caribou range are familiar with the routes generally used by caribou during migra-

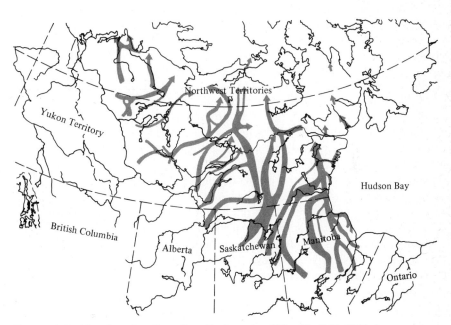

Figure 14-2 Migration of caribou, *Rangifer tarandus. (After Banfield, 1954a.)*

tion. Since they are unable to forecast with certainty the appearance of the herds, if the caribou fail to use the favorite route, the hunters may suffer, as Back did in 1833–4."

In addition to the spring and fall migration, caribou also have a regular midsummer migration, first described by Clarke (1940) as a "backwash" from the spring migration. Murie (1935) also noted this July and early August migration but probably believed it to be a part of the main fall migration. By late July and early August the herds from Eskimo Lake to Hudson Bay retraced their routes toward timber line. In September the herds move away from timber line again, but not as far as in the spring migration. The first severe snowstorms of October or early November turn the caribou south again. Murie (1935), in his careful and extensive study of the Alaska-Yukon caribou, described the migration, winter ranges, and summer ranges of five separate but not completely isolated herds. He made this comment, "The principal feature of caribou migration appears to be its uncertainty."

Murie (1935) discards "the need-for-shelter theory," "the fly-pest theory," "the response to hormone secretions," but is convinced that "the prime cause of migration is the search for suitable food." He summarized the reasons for caribou migration in these words:

Late in summer there is a general searching for better food, necessitated by local failure or seasonal changes of the vegetation. Local wanderings then take on the nature of a migration, probably at first to reach the lichen areas, and later they are augmented by the general unrest of the rutting activities. By that time the migration has a definite form, and the animals retrace their ancestral routes.

Another North American ungulate whose migrations have been the subject of debate is the bison *(Bison bison)*. Even with the voluminous reports of the observers of early herds, it is impossible to determine whether the herds responded to a seasonal urge, were prompted by temperature variations, or wandered at random. No single herd seemed to move over 563 km (300 mi). McHugh (1958) reported that a herd in Yellowstone Park had a predictable seasonal migration of 16 to 40 km (10 to 25 mi) one way. Soper (1941) determined that most herds in Wood Buffalo National Park migrated from 40 to 48 km (25 to 30 mi), while one herd migrated 241 km (150 mi), but 40 percent of the animals in the Park did not migrate at all. Roe (1970), in his voluminous historical compilation of the buffalo, came to the conclusion "that these wanderings were utterly erratic and unpredictable and might occur regardless of time, place, or season, with any number, in any direction, in any manner, under any conditions, and for any reason, which is to say, for no 'reason' at all."

Chapter 15

Historical Zoogeography

Mammals have been dispersed to all parts of the world from their places of origin. The historical account of how they became so distributed may be defined as historical zoogeography, generally shortened to just zoogeography, and partly helps to explain their present distribution. The fields of ecology and ecologic zoogeography are concerned primarily with the present distribution of animals. This chapter will be concerned mostly with historical zoogeography, but no sharp line will be drawn between this and ecologic zoogeography.

One cannot explain the present distributions of tapirs found in South America and in Southeast Asia, and nowhere in between, in terms of available habitat (Fig. 15–1). The camelids of South America are a great distance from their closest relatives in Asia and Africa. These distributions require a historical explanation. Zoogeography is a study of past movements of animals over geologic time spans.

Before Darwin and Wallace announced their theory of natural selection, naturalists believed that each species lived in a region best suited to its needs, because it had been specially created for that place. This was not a very satisfactory explanation. The rabbit did alarmingly well when introduced into Australia, and the mongoose has become a pest wherever it has been introduced. Obviously they did not occur in all suitable places.

As in any beginning science, the first studies in zoogeography were de-

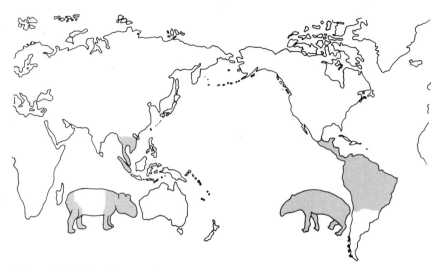

Figure 15-1 Discontinuous distribution as shown by the family Tapiridae. *(After George, 1962.)*

scriptive. Wallace (1876) was among the first to describe the geographic distribution of animals and was concerned with existing distributions. Darwin's (1859) two chapters on zoogeography in *On the Origin of Species* kept to the evolutionary point of view and are still among the best discussions of the evolutionary principles of animal distribution.

There were of course zoogeographers prior to the time of Wallace and Darwin. Some of them were concerned with finding "centers of creation." One of these was Sclater, who, in a publication in 1858, divided the world into six regions, based on the distribution of birds. Although his hypothesis ("center of creation") was shaky, his regions have remained, with but slight alterations. Lydekker (1896) understood that where animals were and what they did there required a historical approach, but he had scanty data to work with. Land bridges were built and continents were moved with great abandon in these times. Gadow (1913) was another who moved continents to fit his theories, but his book was a good summary of ideas up to that time. The feverish activity of paleontologists at this time helped provide some of the factual data needed to support the historical approach to animal distribution. Osborn's (1910) *The Age of Mammals* was convincing proof of the progress that was being made, as was W. B. Scott's (1913) *History of the Land Mammals in the Western Hemisphere.* Mathew (1915, reprinted 1939), a few years later, did much to present evidence for his reconstruction of the history of animal dispersal. Myers (1938) wrote, "Generally speaking, zoogeographers, like Gaul, are divided into three parts—those who build bridges, those who do not, and the proponents of continental drift." More modern interpretations of animal dispersal and distribution can be found in discussions by Allee, Emerson, Park, Park, and Schmidt (1949); Burt (1958); Darlington (1957); George (1962); Hesse, Allee, and Schmidt (1937); Scott (1937); Simpson (1947, 1947a, and 1953); and Udvardy (1969).

PALEOGEOGRAPHY

Before an organism can live in any region a suitable environment must be present. Suitable environments have been a long time building, approximately 4½ billion years, if we go back to the earth's beginnings as a whirling mass of matter. The oldest rocks, 3.2 billion years old, are known from almost every continent. In North America, these Precambrian rocks are exposed primarily in the northeastern quarter of the continent.

In succeeding years, mountains were thrown up and were worn away by wind and water, oceans invaded new areas, leaving others high and dry. Sediments were left containing plants and animals, shells, and skeletons. Thus, the shifting of the seas left a record of the succession of plants and animals as fossil remains. Much of the history of the earth has been unraveled by paleontologists studying these fossils. In 1868 Sir Thomas Huxley delivered a lecture to the working men of Norwich, in Great Britain. The title of his lecture was "On a Piece of Chalk." I know of no better illustration of how fossils have been used to unravel the past history of the earth. This lecture has become a literary work of enduring excellence.

Sir Charles Lyell (1797–1815) was the first to convince the public of the enormous span of geologic time. Up until that time the earth had generally been considered to be not over 6000 years old. Lyell believed that the great continents had always existed in about their present form and place. He also argued that the continents have changed in detail, through the same geologic forces that exist today. His ideas have come to be known as the theory of the permanence of continents. Although paleogeographic maps stretching back into Cambrian days have been drawn, they do not take on a very definite shape until the Mesozoic, when the continents are illustrated much as they are today. The most obvious difference between then and now was the Tethys Sea, a vast sea separating the southern and northern land masses.

By Cretaceous times South America and Africa had both briefly gained land connections with the northern continents. For most of the Paleocene and the Eocene neither had any land connections with continents to the north. Africa regained its connection with the north in Oligocene times, and South America became connected with North America in Pliocene times. During the same time the British Isles became a part of Europe. Several times during the Cenozoic there was land across what is now the Bering Straits. At times some of the southerly islands may have been included in this land bridge. Australia has been isolated most of the time during the last 180 million years.

Mountain ranges have appeared during periods of great movement in the earth's history. The Caledonian Mountains of Great Britain were formed during the Silurian, the Appalachians during the Permian, the Andes and Rockies in the late Cretaceous, and the Himalayas, the Alps, and the Pyrenees during the Cenozoic. Volcanic activity was probably prominent all through the time of mountain upheavals, and is present today in Melanesia, the Malay Archipelago, through Japan and the Aleutian Islands, along the west coasts of Central and South America, and around the Mediterranean.

Another physical movement that changed the details of some northern continents, producing the contours that we see today, took place during the Pleistocene and comprised the several advances and retreats of great sheets of ice over vast areas of the Northern Hemisphere.

According to the theory of permanence, the main changes in the last 180 million years have been the shrinking of the Tethys Sea, the making and breaking of land connections between the northern continents with Africa and South America, the formation of mountains, and changes caused by glaciation in the Northern Hemisphere.

Land bridges of continental size during the Cenozoic were fashionable in nineteenth-century thought, but their fashion declined during the twentieth century. Only those narrow corridors of land for which there is geologic evidence or those which seem plausible are acceptable today. Among these are the Bering Straits, the Panama Isthmus, Arabia, and the channel bridge between England and France.

Quite opposite to the theory of the permanence of continents is the theory of drifting continents, which more recently also has included the idea of expanding oceans. Although not originally proposed to explain animal dispersal, the theory has been seized upon by zoogeographers to help explain anomalies of plant and animal distribution.

Briefly, the theory of continental drift begins with the idea that at one time all continents were joined together in a single land mass, called Pangaea (Fig. 15–2). By the end of the Triassic (180 million years ago), Pangaea had been divided into a northern land mass—Laurasia—and a series of southern land masses, together called Gondwanaland.

If the continents did drift, and it is now generally accepted that they did, they supposedly drifted to their present positions before and during the Mesozoic, with the exception that the separation of the east coast of North America from Europe occurred later. This may not help much in the explanation of mammalian dispersal.

Simpson (1943) wrote, "The known past and present distribution of land mammals cannot be explained by the hypothesis of drifting continents. . . . The distribution of mammals definitely supports the hypothesis that continents were essentially stable throughout the whole time involved in mammalian history."

Figure 15-2 Position of continents before they drifted apart. *(After several authors.)*

But if the great Cenozoic radiation of mammals occurred after the major land masses had been defined, what effect did continental drift have on the evolution of mammals? Kurten (1969) believed that one of the results was a greater diversity of mammals. Today there are about 18 orders of mammals and four of reptiles. Kurten's belief is that mammals evolved on the several land masses under conditions of isolation and semi-isolation and therefore there has been much duplication of functional and morphologic types in separate groups. Ant- and termite-eating mammals occur in various orders on all continents. Saltatory locomotion has occurred among rodents on all continents. Reptiles, on the other hand, developed before the continents had moved apart, allowing freer interchange between evolving stocks.

CONTINENTAL FAUNAL REGIONS

The six continental faunal regions, first proposed by Sclater (1858) for birds and later confirmed by Wallace (1895) for both mammals and birds, have had minor revisions but have basically remained as originally proposed. These are:

1   The *Ethiopian region,* covering the continent of Africa south of the Atlas Mountains and including the southwest corner of Arabia and usually Madagascar. This region is mainly tropical but at the southern tip of Africa is temperate-warm. There is a large rain forest area along the equator in west Africa. From this the vegetation grades into grassland and finally desert. It has land continuity with its northern neighbor.

2   The *Oriental region,* including tropical Asia, which would be India, Indochina, the south part of China and Malaya, and the continental islands of Sumatra, Ceylon, Java, Borneo, and the Philippines. The Himalayas form the northern boundary, the Indian and Pacific Oceans border the sides, but it is difficult to draw a border in the southeast corner. Here the faunas of the Oriental and Australian regions are extremely different. The boundary first drawn by Wallace (1860) is often referred to as "Wallace's line" and has been the cause of some debate. The islands of Bali and Lombok are separated by only 24 km (15 mi) of water, yet the faunas are very different. A line further southeast, called "Weber's line," has also been proposed. The controversy has been reviewed by Mayr (1964).

The climate of the Oriental region is primarily tropical, with rain forests in the eastern part, but the region becomes relatively dry and open westward.

3   The *Palaearctic region,* including all of Eurasia above the tropics and the northern edge of Africa. This region is connected by land with the Ethiopian and Oriental regions. The climate is generally temperate, fringed with the Arctic to the north. The region includes wet forest lands, dry open steppes, large areas of coniferous forest, as well as tundra. In the wetter parts of eastern Asia are deciduous forests. The arid area of interior Asia extends through Southwestern Asia to Africa. Across Northern Asia and Europe stretches a belt of coniferous forest, and to the north of this, tundra.

4   The *Nearctic region,* including North America north of the tropics. Except for Central America, it has land connections with no other regions. The Nearctic ranges from warm temperate in the south to arctic in the north. Deciduous or mixed forests cover much of the middle part of eastern North America. In the middle part of the continent are extensive grasslands, shading

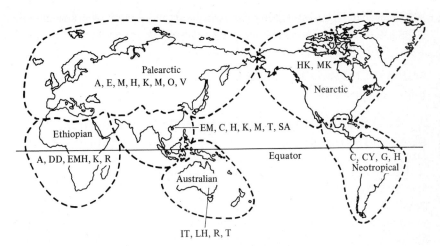

Figure 15–3 Zoogeographical realms and regional authorities.
Australian: IT, Iredale and Troughton (1936), A checklist of mammals recorded from Australia. LH, Laurie and Hill (1954), List of land mammals of New Guinea, Celebes, and adjacent islands. R, Ride (1970), A guide to native mammals of Australia. T, Troughton (1966), Furred animals of Australia.
Ethiopian: A, Allen (1939a), A checklist of African mammals. DD, Dorst and Dandelot (1970), A field guide to the larger mammals of Africa. EMH, Ellerman, Morrison-Scott, and Hayman (1935), South African mammals. K, Kingdon (1971), East African mammals. R, Roberts (1951), The mammals of South Africa.
Nearctic: HK, Hall and Kelson (1959), The mammals of North America. MK, Miller and Kellogg (1955), List of North American recent mammals.
Neotropical: C, Cabrera (1958, 1961), Catalogo de los Mamiferos de America del Sur. CY, Cabrera and Yepes (1940), Mamiferos Sud-Americanas. G, Goodwin, G. G. (1946), Mammals of Costa Rica. H, Hershkovitz, P. (1962), The recent mammals of South America.
Oriental: EM, Ellerman and Morrison-Scott (1951), Checklist of Palearctic and Indian Mammals 1758 to 1946. C, Chasen (1940), A handlist of Malaysian mammals. H, Harper (1945), Extinct and vanishing mammals of the Old World. K, Kramer (1971), Hawaiian land mammals. M, Medway (1969), The wild mammals of Malaya and offshore islands including Singapore.

into desert in the southwestern part, and into mountains with coniferous forest northwest.

5 The *Neotropical region,* including Central and South America and the tropical lowlands of Mexico; it may also include the West Indies, but the fauna there is more transitional than continental. The region is mostly tropical, with the exception of the southernmost part of South America. The Andes extend along the western border of South America, and the Amazon and its tributaries are surrounded by a huge area of rain forest. Southeastern South America is a vast grassland (pampas). There are also areas of desert and semidesert.

6 The *Australian region* includes Australia, New Guinea, Tasmania, and a few of the smaller islands of the Malay Archipelago, but not New Zealand and the islands of the Pacific. Most of the interior of Australia is arid, the "out-back" country. Southwestern Australia is moderately wet and contains some excellent forests. Tasmania is cool temperate.

FAUNA OF THE MAJOR REGIONS
Continental Faunas

The Ethiopian Region   This region is noted for its dramatic herds of big game mammals, its large carnivores (lions, cheetah, leopards), and its two large apes (chimpanzee and gorilla). A great amount of research work has been undertaken in recent years, with attendant publications and publicity, to preserve the rich mammalian fauna of this region. This mammalian fauna is the most varied of all regions, with about 47 families represented. Twelve families are exclusive to the Ethiopian region. Among them are the Potamogalidae (otter-shrews), Chrysochloridae (golden moles), Bathyergidae (mole rats), Tubulidentata (aardvarks), Trichechidae (manatees), Hippopotamidae (hippopotami), and Giraffidae (giraffes). The Ethiopian and Oriental regions share eight families. The Ethiopian region shares with the Palaearctic, dormice, jerboas, coneys, and horses, but it is without the moles, beavers, and bears found in the Palaearctic. It shares no families exclusively with the New World.

The Oriental Region   Although the mammalian fauna most closely resembles that of the Ethiopian region, only 39 families of mammals are represented in the Oriental region, including moles, bears, tapirs, and deer. The Oriental region has four endemic families, Cyanocephalidae (colugo), Tupaiidae (tree shrews), Tarsiidae (tarsiers), and the Platacanthomyidae (spiny dormice). It shares nearly a quarter of its mammalian families with Africa. It shares many of its bat families with the Australian region, and even a family, the Tapiridae, with the Neotropical region. It shares many families with the Nearctic, such as the Soricidae, Talpidae, Cricetidae, Muridae, Canidae, Ursidae, Mustelidae, Felidae, and Suidae.

The Palaearctic Region   This region has a less rich mammalian fauna than some of the other regions. There are about 33 families of mammals represented, including a complex mixture of Nearctic, Ethiopian, and Oriental faunas. Some of the families have a wide continuous range, such as the Vespertilionidae (bats), Sciuridae (squirrels), Cricetidae (cricetid rats and mice), Muridae (Old World rats and mice), Canidae (wolves, coyotes), Mustelidae (skunks, weasels), and Felidae (cats). Others have a discontinuous range, such as the Camelidae. If the two-humped camels from Central Asia are genuinely wild, then their closest wild relatives are the vicuñas and guanacos of South America. The Palaearctic shares members of the Ursidae (bears), Cervidae (deer), and Bovidae (cattle, bison) with the Nearctic, Neotropical, and Oriental regions. The Erinaceidae (hedgehogs), Hystricidae (Old World porcupines), Viverridae (civets, mongooses), Hyaenidae (Hyaenas), and Suidae (pigs) are families whose members occur in the Ethiopian and Oriental as well as the Palaearctic regions. Members of the Gliridae (dormice), Dipodidae (jerboas), Procavidae (hyraxes, conies), and Equidae (horses) occur in the Palaearctic and Ethiopian regions. Talpidae (moles), Ochotonidae (pikas), Castoridae (beaver), and Zapodidae (jumping mice) have members occurring in both the Palaearctic and Nearctic regions. Only two families are restricted to the Palaearctic. One of these is the family Spalacidae, the mole rats, fossorial mammals which dig with their incisors rather than their claws, as moles and pocket gophers do. The other family is the Seleviniidae, rodents which were not discovered until 1938. The first specimens were skeletons left by vultures in Kazakhstan, an area east of the Caspian Sea in Russia.

The Nearctic Region   This region has about 24 endemic families. Members of three families have been introduced. These are the Muridae (Old World mice), Equidae (horses), and Suidae (pigs). Four of the families are shared with the Palaearctic; these are the families already mentioned in the discussion on the Palaearctic—the Talpidae, Ochotonidae, Castoridae, and Zapodidae. Another six families are shared with the Neotropical: the Didelphidae (opossum), the Myrmecophagidae (anteater), Bradypodidae (tree sloths), Dasypodidae (armadillos), Erithizontidae (porcupine), and Tayassuidae (peccaries). Members of the family Camelidae are represented in both the neighboring regions of Neotropical and Palaeartic, but are not present anymore in the Nearctic region itself. There are four endemic families. Three are rodent families—the Aplodontidae (mountain beaver), Geomyidae (pocket gophers), Heteromyidae (kangaroo rats, kangaroo mice, pocket mice), and one is the artiodactyl family, Antilocapridae (the pronghorn). All four are primarily western North American in distribution. The pronghorn is distantly related to the bovids and the cervids.

The Neotropical Region   This region's fauna is both varied and distinctive. Forty-one families contribute to this variety. Nine of the families are of wide distribution. These are the Soricidae (shrews), Lagomorpha (rabbits), Sciuridae (squirrels), Cricetidae (cricetid mice and rats), Canidae (wolves and dogs), Procyonidae (raccoons and allies), Mustelidae (mink, weasels), Felidae (cats), and Cervidae (deer). Two Neotropical families show a discontinuous distribution, the Camelidae and the Tapiridae. The absence of Equidae in the Neotropical and the Nearctic is curious, for it is believed that members of this family were present in both regions not too many thousands of years ago. This Neotropical region has the distinction of having the highest number of endemic families, about 19. The mammalian fauna is a mixture of surviving parts of an old endemic Tertiary fauna and a newer (late Tertiary and Pleistocene) fauna received from North America. Shrews (Soricidae) extend from North America, as do the pocket gophers (Geomyidae) and pocket mice (Heteromyidae), many cricetids, and the New World porcupines (Erithizontidae). Those families extending from South America into North America are the Didelphidae (opossums), Dasypodidae (armadillos), Tayassuidae (peccaries), and Phyllostomatidae (New World leaf-nosed bats).

The Australian Region   This region is best known for its marsupial fauna, for it is here that the marsupials have diversified to parallel the diversification of the placentals elsewhere. The mammalian fauna of this region contains strikingly primitive and isolated elements. The ancestral marsupials probably reached Australia millions of years ago when the mammals of the outer world consisted mainly of small marsupial stock (Troughton, 1966). The South American marsupials are probably a "dead end" of the ancestral stock, the same stock which furnished the Australian region. The Australian marsupials provide a good example of adaptive radiation. The small dasyurid marsupials are shrewlike carnivorous and insectivorous animals. Larger dasyurids are mustelid or wolflike (Tasmanian devil and Tasmanian wolf, both now restricted to Tasmania). The marsupial counterpart of the placental anteaters is a dasyurid, *Myremecobius.* There is a marsupial mole, *Notoryctes.* Marsupial rabbits belong to the Peramelidae. The striped opossum *Dactylopsila* has rodentlike teeth with which it breaks open wood, and a long fourth finger which it uses to pull wood-boring insect larvae out of their tunnels. Not too far away on another island, Madagascar, is its ecologic counterpart among the placentals. The aye-aye, a primate, has chisellike incisors and elongated claws on its middle fingers, used for the same purpose as those of *Dactylopsila.* Some of the Phalangeridae resemble placental flying squirrels. The koala *(Phascolarctus)* is bearlike in appearance and slothlike in habits. The wombats *(Phascolomidae)* are usually given as an example of the marsupial counterparts of the woodchucks and marmots. The kangaroos are mostly terrestrial grazing mammals, although a few are arboreal. None of the marsupials are aquatic and none are aerial, but the mobility of bats precludes their living in isolation either inside or outside of Australia.

Members of about 16 families live in the Australian region. These include, in addition to marsupials, the monotreme families Tachyglossidae (spiny anteaters), and Ornithorhynchidae (platypus). Placentals include the Pteropidae (fruit bats), Emballonuridae (sheath-tailed bats), Megadermatidae (false vampires), Rhinolophidae (horseshoe-nosed bats), Hipposideridae (Old World leaf-nosed bats), Vespertilionidae (vespertilionid bats), and Molossidae (free-tailed bats). Introduced families include the Leporidae (rabbits,) Muridae (Old World rats and mice, some introduced but some, the subfamily Hydromyinae, native), and Canidae (dingo, most likely brought by ancestral man, and red fox). In addition, of course, there are whales, seals, and the dugong in the waters surrounding Australia. Only six terrestrial, placental families seem to be endemic to Australia.

Island Fauna

Invasion and distribution of mammals on islands depend a great deal on the size of the islands, their distance from continents, and the origin, both geologically and chronologically, of these islands. As yet, scientists are not always in agreement on how each island originated. As an example, Schuchert (1935) believed the Greater Antilles (Cuba, Jamaica, Hispaniola, Puerto Rico) to be continental, but they are now believed to be volcanic (Darlington, 1957). Zoogeographers also disagree among themselves concerning which of the new faunal evidence supports volcanic or continental origin.

Just as the distinction between islands and continents is arbitrary, so is the distinction between continental and oceanic islands. Wallace (1895) made a sharp distinction between continental and oceanic islands. He described continental islands as detached fragments of continents, consisting of continental rocks, and oceanic islands as originating in the ocean from volcanic material and coral. Though this is a very practical distinction, the geology of some islands may be quite complicated.

Wallace described continental islands as being always inhabited by some terrestrial mammals and amphibians, and oceanic islands as lacking these. Though this distinction may be true in some cases, it certainly is not true in all cases. An oceanic island differs in the flora, fauna, and climate from the nearest continent. Amphibia, mammals, and freshwater fish are sparse. Of the mammals, the bats, rats, and small mammals are the ones most likely to be present, and carnivores are usually absent. Because of this absence of carnivores, changes occur in the animals living there. For instance, birds may gradually change into ground-living species. The mongoose *(Herpestes auropunctatus)* was imported into the West Indies from India to curb the rat population, which was playing havoc with the sugar industry. Its influence on the fauna was described by Allen (1911), who wrote, "They had so reduced the rats that they fell upon the native ground animals and nearly annihilated certain toads, lizards, birds, and mammals." For a more recent account see Seaman (1952), and for a review of the same problem in Hawaii see Baldwin, Schwartz, and Schwartz (1952).

Examples of continental islands are the British Isles, Ireland, Borneo, Sumatra, Java, and the Falkland Islands. Examples of volcanic islands include the Aleutian, Galápagos, and Hawaiian Islands. Madagascar is difficult to place in either category at present.

A third zoogeographic grouping of islands is sometimes called "Fringing Archipelagos." Included in this group are the Philippines, the Western Pacific Islands, and the West Indies. These islands lie close to continents and receive the fringes of continental faunas.

On what makes islands important zoogeographically, MacArthur and Wilson (1967) wrote, "Many of the principles graphically displayed in the Galápagos Islands and other remote archipelagos apply in lesser or greater extent to all natural habitats." The fundamental processes of dispersal, invasion, competition, adaptation, and extinction apply to islands, as well as continents, but can perhaps be more easily isolated for study on islands, making them important as zoogeographic laboratories (MacArthur and Wilson, 1967).

For a more comprehensive review of island patterns the reader is referred to the works of Darwin (1859), Wallace (1895), Carlquist (1965), and MacArthur and Wilson (1967). Several examples of the kinds of islands and their faunas follow.

The British Isles   The British Isles are a good example of continental islands of rather recent origin. Although their geologic history is complex, they lie on the continental shelf and were not long ago connected to the mainland. The present fauna has had about 7000 (post-Pleistocene) years in which to develop. There are shrews, moles, rabbits, hares, squirrels, cricetids, murids, dormice, badgers *(Meles meles),* otters *(Lutra lutra),* marten *(Martes martes),* weasel *(Mustela erminea),* roe *(Capreolus capreolus),* red deer *(Cervus elaphus),* and bats. Many mammals which exist in a similar climate on the mainland are not found on the British Isles. These include hamsters *(Cricetus* and *Mesocricetus),* lemmings *(Lemmus lemmus),* and caribou *(Rangifer tarandus).* Mathews (1952) includes a chapter on their origin in his book on British mammals.

Ireland   Ireland is a continental island which originated from the continental island of Great Britain and thus acquired its fauna "second-hand" from Europe. Only about half the British species are represented in Ireland. Most of the differences between the Irish and British kinds are small and warrant only subspecific rank.

West Indies   The West Indies were once thought to be continental islands but are now believed to be volcanic in origin, and therefore oceanic islands. The four Greater Antilles (Cuba, Hispaniola, Jamaica, and Puerto Rico) are probably the oldest. The Lesser Antilles are a chain of smaller islands stretching from the Greater Antilles to Trinidad. With the exception of Trinidad, the Lesser Antilles are also oceanic.

Bats are numerous in the West Indies, but land mammals are few and restricted. The relict insectivore *Solenodon* is found but survives tenuously in the mountains of Cuba and Hispaniola. They presumably came from North America. Most of the land mammals are rodents, including nine species of *Capromyidae,* most likely derived from South American stock, some rice rats *(Oryzomys),* two opossums, an armadillo, agoutis, and raccoons. Of the mammals which have been introduced, the most notorious is the mongoose (Seaman, 1952).

Aleutian Islands   These islands have played an extremely important part in animal distribution. Although volcanic, they have served as a land bridge (and also as a "filter barrier") during the Tertiary and the Pleistocene. The terrestrial mammals are derived from both east and west, and of course the islands are important to some of the marine mammals (e.g., Pribiloff Islands, fur seals). For a detailed discussion see Thorne (1963).

Galápagos Islands   These islands have been called "Darwin's islands," and to biologists they have become a showcase for evolution. There are 15 islands, volcanic in origin, separated from South America by a distance of 960 km (600 mi) of deep water. They are rather arid, rising to 152 km (5000 ft). The number of species is few, thus simplifying biologic patterns. It was on these islands, which he visited in 1835, that Darwin began to crystallize his reflections on the origin of species. It was after he began analyzing the collection of "Darwin's" finches that he began to question special creation. Darwin's finches dominated the Galápagos land birds there. Lack (1947) tells in a very readable way the adaptive radiation of this natural group which derived from one ancestor.

A vast literature has sprung up concerning these islands, but for a general understanding the best volumes are perhaps Darwin's *Voyage of the Beagle* (1906), Beebe's *Galápagos—World's End* (1924) and "The Arcturus Adventure" (1926), and Eibl-Eibesfeldt's book *Galapagos—The Noah's Ark of the Pacific"* (1961). Biologists' inherent interest in the Galápagos has also resulted in the Darwin Biological Station being established there under UNESCO sponsorship. But for all their interest to biologists, the islands are more representative of the Age of Reptiles than of the Age of Mammals. Of the mammals there are only some small cricetid rodents, which Simpson (1945) includes in *Oryzomys,* and a bat of the genus *Lasiurus,* the same genus which has reached Hawaii. Since this is an American (North and South) genus of bats, the Galápagos species was probably derived from the same stock.

Madagascar   Madagascar, with an area of 620,000 km^2 (240,000 mi^2), is 418 km (260 mi) east of Africa. The channel separating Madagascar from Africa is at least 2 km (6000 ft) deep, and if they were ever connected it must have been before the Triassic. Although a land connection is not necessary to explain most of the recent fauna of the island, the presence of a fossil dinosaur presents a problem. At present it would be difficult to classify Madagascar as a continental or an oceanic island. Sedimentary rocks date back as far as the Jurassic, Cretaceous, and Tertiary. The life expectancy of purely volcanic islands is geologically short.

Madagascar is a biologically rich and curious island. There are 70 kinds of mammals on the islands, most interesting of which are the insectivores and primitive primates. There are also some native civet cats. Most or all of the rodents may have been introduced, as was the African bush pig *(Potamocho-erus porcus).* The largest land mammal to reach Madagascar was the pygmy hippopotamus *(Choreopsis liberiensis),* which was present during the Pleistocene. The mammals which have reached Madagascar by natural dispersal seem remarkably few when compared with the rich fauna of the nearest continent —Africa. Probably with enough time many of them could cross the Mozambique channel.

The lemurs are not of the main line of evolution leading to the great apes. Their lack of upright posture and a brain which apparently did not grow larger may partly explain their "down-hill" evolution. Their range today is greatly restricted (Madagascar and nearby islands, such as the Comoro Islands) from what it was when lemurs also ranged throughout Europe and North America. But what they might lack in evolutionary success (if increasing numbers are a criterion of success) they have more than made up for in interesting adaptive radiations in Madagascar. There is, for instance, a group of fruit-eating lemurs, (genus *Lemur*), which have long grasping arms and legs, long tails useful for balancing, long snouts, and foxlike faces. The ring-tailed lemur *(Lemur catta)* is the commonest and probably the most studied of the greater lemurs.

Some of the smaller fruit-eating lemurs, *Cheirogalus, Microcebus,* and *Phaner,* have developed into insect eaters, with large eyes, swift movements, and sensitive ears. *Phaner* has specialized to the extent of sharing a tree cavity with bees and feeding on the honey. But most amazing of all the lemurs, and in fact of all the nonhuman primates, is the daubentonia, aye-aye, or squirrel lemur *(Daubentonia).* It is so like a specialized rodent that it was originally thought to be a squirrel. Its face and skull are similar to those of a rodent, with large gnawing incisors. Its ears are large, facing forward. The digits are long, narrow, and provided with claws. The third finger is slimmer than the others and carries an elongated claw. The animal terminates in a long bushy tail. All these are adaptations for an arboreal life, the animal feeding on wood-boring insects. The rodentlike skull helps in the gross uncovering of the wood-boring insects, while the final spearing is done with the elongated claw of the third finger. The Australian striped opossum *(Dactylopsila)* mentioned earlier has a prominent third claw, useful for the same purpose.

A unique parallel to the adaptive radiation of the lemurs is found in the insectivore family, Tenrecidae, another primitive and endemic family in Madagascar. Tenrecs may have had the same length of time for their adaptations as did the lemurs. Among the insectivores they are believed to be quite primitive, reminding one of the relict insectivore of the West Indies, *Solenodon.* A hallmark of the insectivores that the tenrecs have is a pointed snout. Resembling somewhat the hedgehog *(Erinaceus)* of Europe is the very spiny

Echinops, which is most often found near rivers. A not quite so spiny hedgehog is the landrake, *Tenrec ecaudatus.* The specific name, "without a tail," is descriptive. The landrake is nocturnal, spending the daytime in burrows, and it probably hibernates. One of the more curious of the tenrecs is *Hemicentetes nigriceps.* It has a skunklike dark and light pattern (dark brown and yellowish white) and a pointed snout, and is hardly larger than a man's thumb. Another group of tenrecs, mouselike in size and appearance because of the long tail, is the genus *Microgale.* Most of the tenrecs are burrowing animals, and *Oryzoctes hova* has the weak eyes, strong digging claws, and soft fur of the mole. Adapted to an aquatic habitat is the tenrec *Limnogale,* which feeds extensively on aquatic plants and has webbing on its paws.

GETTING THERE

Dispersal

Dispersal occurs when an individual or a population moves to an area different from its place of origin. Mammals are *actively dispersed* by use of their own powers of locomotion—moving on foot, swimming, or flying. They can become *passively dispersed* when carried by man [either purposely—e.g., as pets (dogs) or for food (cattle)—or accidently (Old World rats and mice)] and by rafting. Rafting is a term applied to floating natural objects, from a log to a small island of debris, which can carry an animal on the water. Visher (1925) wrote:

> The floods caused by excessive rainfall associated with hurricanes, influence the dispersal of land forms. There are numerous records of the fall of more than sixty inches in three days. Under such conditions streams normally small may become great rivers and carry to sea vast quantities of driftwood. The river banks are eroded badly, and many trees are undercut and are carried out to sea. During the excessive rains, large masses of dirt and loose rock upon steep hill-sides may slip, sometimes damming valleys. If the dam breaks, the sudden rush of water does its part to contribute natural rafts of driftwood with their load of land animals and seeds.

A raft 305 m (100 ft) square with trees 10 m (30 ft) high was described by Powers (1911). It was seen off the coast of America in the Atlantic Ocean in 1892 and was known to have drifted 1600 km (1000 mi). Brewster (1924) described a raft of smaller potential on Lake Umbagog in Maine. Prescott (1959) reported the rafting of a jackrabbit 63 km (39 mi) from land off the coast of California. He did not certainly identify the rabbit but wrote that if it were *Lepus c. californicus* it would be 14.5 km (90 mi) from its closest land. The raft was 12 m (40 ft) long and "at least" 7 m (25 ft) wide.

There are, of course, limitations on dispersal, such as climate, vegetation, other mammals, and physical barriers. Climatic and geologic factors are constantly changing the environments on our planet. Mathew (1915) has defined what happens to mammals (and other living things) under these changing conditions. They may (1) change to "fit" the environment, (2) move from an unsuitable (stress-creating) environment to a more suitable (less stressful) environment, or (3) perish. The ability to disperse is very basic to mammals, and a high premium is based on dispersal ability. Udvardy (1969) wrote that, "without evolved means of dispersal most animal populations would have succumbed, over a period of time to the vicissitudes of the environment." Species that are widespread in habitat as well as in geographic range present a greater genetic flexibility than species with very specialized requirements.

The ability of terrestrial mammals to disperse depends on a great variety of factors. Certainly many have the ability to swim narrow channels. Of the land mammals, rodents have the greatest tolerance for salt water, as well as the greatest ability to move by accidental means. The largest mammals would be least able to cross wide gaps of water. Carlquist (1965) estimates 40 km (25 mi) is the greatest distance for the larger mammals. An exception to this is the semiaquatic hippopotamus, which lived on Madagascar during the Pleistocene and probably had its origin in Africa. The Madagascar civet (Viverridae) probably crossed the Mozambique Channel, about 320 km (200 mi). The murid rodents in the Philippines seemingly originated from the Celebes and Australia, and have spread throughout the Philippines; all the more remarkable if the murids did not arise until the late Miocene (Simpson, 1945). If the Hawaiian rat came to the Hawaiian Islands as a stowaway in Polynesian canoes, then surely small mammals can travel on small natural rafts. A cricetid rodent *Oryzomys,* which probably reached the Galápagos Islands 960 km (600 mi) from South America, has also reached Jamaica over a water gap of similar distance. The greatest travelers are of course those animals that can fly. They are probably helped along in their travels by winds. Bats have traveled 3200 km (2000 mi), (not in one jump), as is shown by the presence in Hawaii of a genus of North American bat, *Lasiurus.*

While land bridges loom large in any discussion of zoogeography, ice bridges must also have played a part. Small-scale evidence for this can be drawn from several recent works. The invasion or reinvasion of Isle Royale in Lake Michigan has been documented by Mech (1966). Isle Royale, not only an island, is also a National Park. The irruption of the moose population beginning in the 1930s caused concern among Park Service personnel as well as biologists and conservationists. Moose probably reached Isle Royale in the early 1900s, at a time when their numbers generally increased in the surrounding area. Although ice bridges the 64 km (40 mi) between Canada and Isle Royale occasionally, moose are believed to have reached Isle Royale by swimming this distance. As early as 1934 Murie (1934) predicted starvation. By 1945 Aldous and Krefting (1946) believed that Isle Royale's carrying capacity for moose had been reached. Natural predators such as mountain lions, timber wolves, and bears were absent. Harvesting the surplus by hunting could not be allowed in a National Park. The best solution would be to introduce a predator which might have been historically a part of the community and could now live there successfully; the timber wolf *(Canis lupus)* came closest to filling these requirements. Its presence on the island prior to 1945 was dubious, although possible. In 1952 four zoo wolves were released but the experiment was unsuccessful. The zoo animals did not recognize "meat on the hoof" and starved or were eventually disposed of. But by now it was known that wild wolves had appeared on the island, and at the time of Mech's study (1959–1961) there were 18 or 19 individuals on the island. Since it is doubtful that wolves would swim 64 km of water, or could even survive the cold waters of Lake Superior for the necessary time, Isle Royale was probably repopulated via a partial or complete ice bridge. The red fox *(Vulpes vulpes)* was first recorded on Isle Royale "about 1925" (Mech, 1966). It is interesting to note that deer mice *(Peromyscus maniculatus)* were common, and red-backed voles *(Clethrionomys gapperi)* may have been, but there were no records of the jumping mouse *(Zapus hudsonius).* The jumping mouse, of course, hibernates, and while it would be rather ludicrous to assume that other species of small mammals covered the 40 miles of ice on foot, the possibility of ice rafts does exist. Though rafts of other debris are also possibilities, the shores of Isle Royale and the shorelines of the mainland are mostly igneous rock outcrops, which decrease the possibility of anything but logs as rafts. Mech (1966) does not include bears, raccoons, skunks, or woodchucks in his list of Isle Royale mammals. Although the woodchuck is the only hibernator in this list, the others are semidormant during the winter months. For a further discussion on island dispersal, Hatt, Van Tyne, Stuart, Pope, and Grobman (1948) have studied and discussed the distribution of the land vertebrates on islands in eastern Lake Michigan.

Routes of Interchange

The ability of animals to spread from one region to another may vary from impossible to certain. Simpson (1940) has defined three main paths of inter-change—*corridors, filters,* and *sweepstakes routes.* A *corridor* is a route along which most of the animals can spread from one region to another, such as from Europe to China across Asia. A *filter* is a route which allows some groups of animals through but stops others. Examples of a filter may be a land bridge, a desert, or a mountain system. An example of a now-existent filter is the connection between North and South America. A *sweepstakes route* is one in which the odds against winning are great, but some few animals do win.

The camels of today, whose ancestors came from central North America (at least from present fossil evidence), must have crossed two filter routes (land bridges between North America and South America and North America and Asia) and one corridor route (Europe to Asia). The tapirs have also crossed land bridges to reach their present distribution, which is restricted to Central and South America and then the Malay Peninsula, Java, and Sumatra. Fossil tapirs show that tapirs must have reached these areas across Central and Northern Asia and North America.

The terrestrial fauna of Madagascar is a good example of a sweepstakes route. Very few African mammals have spread to Madagascar over a period of 50 million years. Some of these that came to the island have evolved to where Madagascar has a distinctive "endemic" fauna. The term "endemic" is certainly subjective; it does not mean that the fauna arose spontaneously in the area, but rather that a great deal of adaptation has produced many species from a few original ones. The fauna of the West Indies and of Australia as well as Madagascar and the Galápagos results from "sweepstakes" migration.

The present continents have probably varied little, except in details, during the age of mammals. The faunas now in existence have resulted from three intercontinental connections, which have not been permanent. These are the Asia–North America, the Eurasia–Africa, and the North–South America bridges. The Asia–Australia has probably always been a sweepstakes route as far as mammals were concerned.

Still other families never expanded beyond the continent of their greatest adaptation. Such is the case of the pronghorns *(Antilocapra americana).* Although the ancestor of the family may have come from Eurasia, the last survivor of the family, which radiated in North America in the Miocene, Pliocene, and Pleistocene, is the pronghorn.

The fauna of the Nearctic and Palaearctic seemed to have had a uniform composition toward the end of the Mesozoic. There were two groups of mammals, the ancestral placentals and ancestral marsupials. During the Tertiary (early Cretaceous), these evolved to presently existing mammals in three regions. The ancestral placentals radiated in the main part of the world (North America, Eurasia, Africa). The marsupials reached Australia, where they evolved into an "endemic" fauna. Marsupials and placentals also reached South America and Australia across water gaps. Monotremes are probably the remains of a group which was in Australia before the marsupials arrived. For a short period during the Paleocene, South America was joined to North America by a thin neck of land or an archipelago, across which marsupials and a few placentals moved into South America. Shortly after this but sometime in the Paleocene, South America was cut off from North America. Once isolated in the Neotropical region, the marsupials diverged widely into the mouselike and tree-living forms of the modern regions. One of the marsupial carnivores resembled the saber-toothed tiger. An extensive group of early placental herbivores also developed, probably because isolation kept out the northern herbivores and the placental carnivores. Evidently along with the few marsupials, one or several edentates and also several ungulates reached South America at the beginning of the Tertiary. A histricomorph rodent, monkeys, and procyonids all reached South America before the end of the Miocene. It is possible that all mammals reaching South America through the Tertiary came from North America. The speculation that monkeys came directly from Africa still exists.

The land bridge which exists today came into existence in the late Pliocene. Simpson (1940) has summarized the exchange of mammals across this land bridge since its establishment. Prior to this land connection, North America had probably 27 families and South America 29 families of land mammals. Members of two families, procyonids and marsupials, were found on both continents. During the Pleistocene, the two continents had 22 families in common. After a great deal of shuffling, North America ended with 23 families, South America with 29. The North American fauna was little changed, the South American greatly changed. Of the South American forms, only an opossum, a porcupine, and some cricetid rodents survived in North America. North American mammals which reached South America were a shrew, a squirrel, pocket mice, cricetid rodents, canids, an ursid, several procyonids, four mustelids, elephants, horses, tapirs, peccaries, camels, and deer. The northern (placental) carnivora replaced the marsupial carnivora. The same is true of the ungulates. A number of Old World groups have reached South America, but the reverse is not true.

Africa seems to have been part of the main world of mammals at the beginning of the Tertiary, as stated earlier, but then became isolated until the late Eocene, when it was reunited with Eurasia. Simpson (1947 and 1947a) has discussed the exchange of mammals between the Old World and North America during the Tertiary and Pleistocene. The connection between Eurasia and North America existed during much of the Tertiary, disappeared during part of the Paleocene, the middle of the Eocene, middle to late Oligocene, first half of the Pliocene, and at the present. The major selective factor for this bridge was probably climate and its effect on the vegetation. The movements from Eurasia to North America were greater than those in the opposite direction. More interesting than the fact that certain mammals crossed the bridge is the fact that certain others did not. Viverrids, Megachiroptera, "higher" Old World nonhuman primates, murids, gliroid rodents, suids, and giraffes did not reach North America. Procyonids, geomids, peccaries, and pronghorns have not reached the Old World.

FAUNAL RELATIONSHIPS OF NORTH AMERICAN MAMMALS

A logical question now would be, "Where did the North American land mammal fauna come from?" For some groups the earliest fossil records are found in North America, but for other groups the fossil history begins on other continents. The facility of interchange between continents has had a strong influence on taxonomic resemblances of North American mammals.

The criteria for determination of origin are somewhat subjective. One of the very good discussions of the subject is that of Mathew (1915). Darlington (1957), Stebbins, (1950), and Burt (1958) are among those who have discussed these criteria more recently. Savage (1958) listed five criteria for determination of origin: (1) the earliest record of the group, in the absence of contradictory evidence, (2) an earlier record of progenitors in the proposed area of origin, (3) the area of greatest taxonomic differentiation, which may possibly be the area of origin, (4) the phyletic age of the group, and (5) the vagility of the groups. Rapidly dispersing land mammals may be fossilized first far from their point of origin.

Much information on the origin and affinities of North American mammals has been added since the account of Scott (1937, originally 1913). Simpson (1947) has published an extensive and meticulous discussion of the faunal relationships of the Holarctic mammalian fauna during the Cenozoic. He has expressed taxonomic relationship between two areas by the use of an index or percentage figure based on the fraction of common species (between the two areas) in the smaller of the two faunas. The formula was $100 \ C/N$, where C equals the number of units known to be common to both faunas, and N equals the total number of taxonomic units in the smaller fauna. The indices for the Cenozoic faunal resemblance between Eurasia and North America have varied greatly, being at their highest in the early Eocene, when there were 42 for the genera and 89 for the families. These indices for the Recent were 40 for the genera and 65 for families.

Briefly, the Eurasian and North American mammalian fauna have much in common and there must have been a great deal of intermingling, though not continuously, during the Tertiary. Simpson (1947) believed that major interchanges took place during the early Eocene, late Eocene, early Oligocene, late Miocene, middle to late Pliocene and Pleistocene, and very little or none in the middle Eocene and middle and late Oligocene.

Although a direct interchange between Europe and North America as well as Asia and North America has been postulated, Simpson (1947) wrote, "All the faunal evidence is consistent with a single land route, the Bering bridge between Alaska and Siberia, as the sole means of mammalian interchange between Eurasia and America."

The mere presence of a land bridge does not ensure interchange, for, as many people have pointed out, the ecologic conditions of both the bridge and the approaches must be suitable for the animals to cross. It is believed that through Tertiary times the climate was more moderate than now. For a very readable discussion of the climate since the Late Cretaceous, see MacGinitie (1958).

Savage (1958) and Burt (1958) have discussed the affinities of the recent land mammals of western North America. Burt (1958) provides some words of caution: "Can we really talk about the origin of a fauna or a taxon without going all the way back to the origin of life itself, which may also have been a continuum?" Starting at the opposite end, i.e., at the present time, how far back must a fauna go to be called endemic? The time included in Burt's (1958) discussion is from the beginning of the Tertiary through Recent. The migrants across the Bering Straits must have been those groups "tolerant of relatively cold climates." Some mammals have a wide range of temperature tolerance, e.g., the mountain lion and the bobcat. Burt (1958) also suspects "that physiological evolution may proceed more rapidly than morphological change."

Of 70 genera of strictly terrestrial nonflying mammals of temperate North America, 30 occur in Eurasia.

In contrast, the prairie dogs *(Cynomys)* had their entire known evolutionary history in North America. There probably has been no intercontinental exchange among the raccoons, at least not since they have become recognizable as raccoons. The fossil records for the families Geomyidae and Heteromyidae indicate a strictly North American evolution. Although the pre-pronghorn ancestors may have come from Eurasia, the pronghorn *(Antilocapra americana)* is a strictly North American product.

Hooper (1949) also discussed the faunal relationships of recent North American rodents. Using the same index as Simpson (1947) as a measure of faunal relationships, he divided the Recent rodent fauna into three major units, (1) boreal fauna of the high and middle latitudes, (2) the arid western or sonoran fauna of Mexico and the Western United States, and (3) the tropical fauna of low latitudes and low elevations, excluding the interior plateau of Mexico. Hooper (1949) wrote, "If by reason of community in these features these are natural aggregations, then each has a history as a unit."

The boreal unit occupies most of Canada and areas to the north, the coastal areas and higher mountains of the Western United States, and much of the eastern part of North America. Ten of the nineteen boreal genera are shared with Eurasia, eight are restricted to North America.

The arid western or sonoran fauna occupies the arid mountains and interior plains, basins, and plateaus of most of the Western United States and Mexico. The rodent forms comprising this faunal unit are mostly indigenous forms with distant affinities to southern Eurasia and closer affinities to Central and South America. *Spermophilus* is represented in the Old World. The other 22 genera are exclusively American. Thirteen belong to the families Geomyidae and Heteromyidae. Only one genus, *Heteromys,* is represented in South America. The genera *Reithrodontomys, Peromyscus, Baiomys, Onychomys, Neotomodon, Neotoma, Nelsonia,* and *Xenomys* are endemic to North America.

The tropical unit of Hooper (1949) includes the tropical lowlands of Mexico, the Antilles, Central America, and northern South America. Twenty-five genera of rodents are represented. All components are New World and perhaps most are South American in origin. The histricomorphs are a South

American group which has moved north very recently, probably in the Pliocene or Pleistocene. The cricetid genera, found in this unit, are presumed to
be North American tropical forms that have lately moved into South America.
These include *Oryzomys, Melonomys, Neacomys, Nectomys, Rhipidomys, Ty-
lomys, Ototylomys, Nyctomys, Octonyctomys, Lygodontomys, Scotinomys, Sigmodon,* and *Rheomys.*

THE MAMMALIAN PROVINCES OF NORTH AMERICA

Interest has arisen recently in the use of statistical methods in determining
faunistic provinces. Hagmeier and Stults (1964) and Hagmeier (1966) have
made use of the distribution maps of Hall and Kelson (1959) to show that
range limits of North American terrestrial mammals were grouped, thus allowing regions of faunistic homogeneity to be identified. Hagmeier (1966)

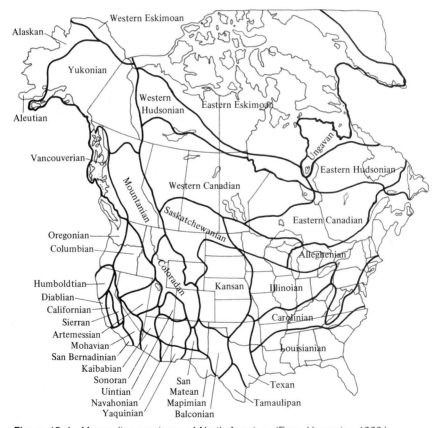

Figure 15–4   Mammalian provinces of North America. *(From Hagmeier, 1966.)*

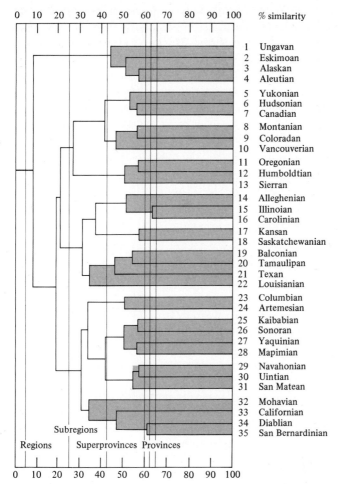

0 10 20 30 40 50 60 70 80 90 100 % similarity

1   Ungavan
2   Eskimoan
3   Alaskan
4   Aleutian

5   Yukonian
6   Hudsonian
7   Canadian

8   Montanian
9   Coloradan
10 Vancouverian

11 Oregonian
12 Humboldtian
13 Sierran

14 Alleghenian
15 Illinoian
16 Carolinian

17 Kansan
18 Saskatchewanian

19 Balconian
20 Tamaulipan

21 Texan
22 Louisianian

23 Columbian
24 Artemesian

25 Kaibabian
26 Sonoran

27 Yaquinian
28 Mapimian

29 Navahonian
30 Uintian
31 San Matean

32 Mohavian
33 Californian
34 Diablian
35 San Bernardinian

Subregions

Regions | Superprovinces Provinces

0 10 20 30 40 50 60 70 80 90 100

Figure 15–5 Dendrogram showing relationship between provinces. *(After Hagmeier, 1966.)*

provided a late Tertiary, Pleistocene, and post-Pleistocene history of the mammalian areas of the North American continent.

By separately computing the percentage of species and genera whose ranges ended within blocks of 50 by 50 mi, Hagmeier and Stults (1964) converted the ranges of 242 species of North American terrestrial mammals into a model with a working total of 2490 blocks. The number of species whose ranges ended within each block was determined, and an index of faunistic change (IFC) was computed: IFC = 100 L/n, where L is the number of range limits in the block, and n the number of species occurring there. The areas of greatest faunal homogeneity were named as zoogeographic provinces (Fig.

15–4). Degrees of faunistic affinity between provinces were also computed (Fig. 15–5). The faunal affinities were subjected to cluster analysis (Sokal and Sneath, 1963), which resulted in mean percent similarity values for each pair of the faunistically closest related provinces. Hagmeier (1966) concluded that the faunas of two geographic areas may be considered homogeneous if the mean percent of similarities is more than 65. If the mean percent is less than 65, the compared areas rank as separate faunal provinces. Mean percent similarities of unit pairs resulted in a dichotomous, hierarchical system from which higher-ranking groups, superprovinces, subregions, and regions could be formulated. Percent similarity then permitted numerical values to be given to these geographic regions. It is interesting to note that mammalian provinces of North America are strongly related to the ecogeographic units founded by Dice (1943) and revised by Kendeigh (1974).

EXTREMES OF RECENT MAMMAL RANGES

Mammalian species limited to the colder northern region but circling the globe at these higher latitudes are said to be circumpolar. These include the arctic hare *(Lepus arcticus),* arctic fox *(Alopex lagopus),* wolf *(Canis lupus),* collared lemming *(Dicrostonyx groenlandicus),* polar bears *(Ursus maritimus),* short-tailed weasel *(Mustela erminea),* least weasel *(Mustela nivalis),* caribou *(Rangifer tarandus),* and moose *(Alces americana).* These populations are probably not continuous but are a series of discrete populations, as some workers on the polar bear believe. The musk ox *(Ovibos moschatus)* is now restricted to the far north of the Western Hemisphere.

Bats of the genus *Myotis* range from the northern limit of trees, south through the tropics of both hemispheres. The bat families Vespertilionidae, Emballomuridae, and Molossidae are cosmopolitan in their distribution. The family Cricetidae is worldwide in distribution, and the family Muridae has become so because of man's activities.

Members of the insectivore genus *Crocidura* occur from Germany and eastern Siberia through Africa and the Oriental region. The leopard *(Felis)* ranges through Africa and Southern Asia to southeastern Siberia. Of the larger mammals, the most widely distributed genera are *Lutra* and *Felis.*

The monotremes *(Tachyglossus* and *Ornithorhynchus)* occur from cool-temperate Tasmania to tropical Australia. In the Western Hemisphere the opossum *(Didelphis),* a marsupial, manages to get along in a geographic range extending from a cool-temperate part of North America south through the American tropics.

In North America the deer mouse, *Peromyscus maniculatus,* can be found over an area of 5 million mi^2, from Alaska to south of Mexico City and from the Atlantic to the Pacific. In the same genus the other extreme is found in *Peromyscus floridanus,* which is restricted to the peninsula of Florida.

One of the most restricted of geographic ranges is that of *Sorex sinuosis,* which is known only from Grizzly Island, Solano County, California, and adjacent marshes. Actually this geographic restriction may be more taxonomic than real, for the shrew may in reality be only a subspecies of the wider-ranging species, *Sorex ornatus.* Surprisingly, some families of bats have a very restricted range. Members of the family Mystocinidae are restricted to New Zealand, and members of the Myzopodidae are restricted to Madagascar and vicinity.

Among the primates, the families Lemuridae, Indriidae, and Daubentoniidae are restricted to Madagascar and immediately adjacent islands.

The range of the Selevin's mouse *(Selevinia betpakdalaenis),* only member of the family Seleviniidae, is restricted to the clay and sandy deserts of Kazakhstan, U.S.S.R. The generic name derives from the name of one of the collectors of the type material, V. A. Selevin, who died before the species was described, and the specific name refers to the Betpak-Dala Desert, where the first specimens came from. This living mammal was not discovered until a 1938 expedition, when five specimens were caught by hand, none in traps.

The mammals have succeeded very well, evolutionarily, in reaching and maintaining populations which can exist in equilibrium with every climate of the world, except the extreme cold of the farthest north and farthest south.

Ecology: The External Environment

We have discussed behavioral, morphologic, and physiologic mechanisms which maintain homeostasis in the individual mammal during its lifetime. There are also mechanisms which maintain the homeostasis of a population over a long period of time, helping to maintain an equilibrium between the population and its environment. The environment is always changing, and changes in the average characteristics of a population are brought about by variations in individuals from generation to generation. Successive generations will be successful only if homeostatic readjustments to the gross and long-term environmental changes have resulted from genetic changes in the individuals of the population. Ecology is the study of this interaction between organisms and their environment. Much ecologic information can be integrated around two central themes: (1) natural communities as functional systems, and (2) populations as living systems.

This chapter will concern itself mostly with the environment and with population changes. The term environment is too often overdefined; for the purposes of this chapter it will be defined simply *as the total of the conditions surrounding an organism or group of organisms,* and includes the nonliving (abiotic) or *physical environment* and the living or *biotic environment.* There are innumerable excellent books on ecology to which the reader can refer. Among them are (excluding limnology) Allee, Emerson, Park, Park, and

Schmidt (1949); Benton and Werner (1974); Elton (1927); Farb (1963); Kendeigh (1974); Odum (1971); Ricklefs (1973); Slobodkin (1962); and Smith (1974).

The interaction between a population and its environment was originally expressed (Chapman, 1928) in a deceptively simple formula as $P = BP - ER$, where P is the total number of individuals in the population studied, BP is the biotic or reproductive potential, and ER is the environmental resistance. This formula is a summary of many factors. Just enumerating some of the factors involved will give an insight into its complexity. The biotic potential includes such categories as age at sexual maturity, number of young per litter, number of litters per year, and length of breeding activity. Environmental resistance includes such categories as restricted food and space, intra- and interspecific competition, adverse weather and climate, disease, predation, and parasites. Revisions and refinements of the population formula have added precision with the years, making it conform more closely to specific problems and usage.

Population is a loosely used word referring to a group of organisms, of the same species, which as a unit has a continuing role in the community. A *community* is an assemblage of organisms living together in an environment. A shorter definition would be a group composed of interbreeding individuals. Such a group can be the paramecia in a pond; a grove of conifers; or the red squirrels that inhabit the grove. The communities, or the living (biotic) components, together with the nonliving (abiotic) components make up the *ecosystem*.

The field of resource management consists of manipulating populations, whether for pleasure or for profit. The manipulation of a buffalo herd on a National Wildlife Refuge for pleasure (observation and photography), a herd of deer for hunting, or a white-pine forest for lumber is a process requiring a great deal of training, research, and application. For those wishing more detailed information on resource management (especially of vertebrates), the following books are recommended: Allen (1954), Dasmann (1964), Leopold (1933), and Watts (1968).

THE ENVIRONMENT

Ecologic Geography

We learn early in life that the carpeting on the earth is not everywhere the same. Along the eastern shores of North America and around the Great Lakes forests predominate. Through the central part of Canada and the United States are prairies. In the southwestern part of the United States are deserts. Along the Pacific Coast forests become more dense as one travels north. Over the centuries man has tried to organize and classify the living matter covering the surface of the earth into recognizable units. The broadest of these are the faunal regions of the world, based primarily on the past and present geologic relationships of the larger land masses. These were discussed in Chap. 15.

Other systems of classifying large communities or ecosystems are based on major terrestrial biotic communities characterized by distinctive life forms. One of the most original, best-known of the faunistic systems, and one that was in vogue for many years, was proposed by Merriam and Stejneger (1890) and came to be known as "Merriam's life zones." These authors theorized that there were two centers in North America from which animals had dispersed: (1) the boreal, in the far north, and (2) the sonoran, in the southwest. The boreal animals moved south along the higher elevations in the mountains, and as they became acclimatized to the higher temperatures, spread over the lower (and hotter) areas. Sonoran forms dispersed northward through the lowlands as they became acclimatized to the cold. Although they served a useful purpose, Merriam's life zones had their weaknesses and have to a great extent been replaced by other systems—none of which is completely satisfactory. One of the weaknesses was that too much emphasis had been placed on temperature and not enough on moisture. Life zones extended as belts across the continent, but these could not always be characterized by a uniform floral composition. Differences in vegetation, in moisture, and in geologic history may be just as important as temperature. It has also become evident that many subspecies and possibly also species have evolved from other centers than the boreal and the sonoran. In mountainous areas these life zones tend to become confused.

In traditional ecology a *community* can also be defined as an aggregate of organisms which form a distinct ecologic unit. These units may be very large, such as biomes, or smaller secondary communities, such as seral or climax communities. A climax community is a stabilized, self-maintaining community which changes very slowly, only as the climate changes. A seral community is not as stable, is not self-perpetuating, and usually contains many individuals of invading species among the dominants.

Philosophically there are two opposing ideas of communities. One is the individualistic viewpoint, which holds that the species is the basic unit and that the species responds independently to the integrated influence of the physical and biotic environment. The environment may be thought of as a pattern of gradients, e.g., from wet to dry conditions, and the species responds independently to this pattern, not with a group of associates that must be together. Exponents of this view include Gleason (1926), Whittaker (1953, 1970), and Brown and Curtis (1952). The other viewpoint is the organismic concept, which regards a community as the ultimate organism which rises from cell, to tissue, to organ, to organism, to populations, and finally to the community as a sort of superorganism. The whole is greater than the sum of its parts. Clements (1916) is usually given credit as the originator of this viewpoint.

It is doubtful that ecologists will either wholly embrace or completely discard either viewpoint. It is true that the species collectively make up the community, but a community also operates as a unit to bring about the transformation, circulation, and accumulation of matter and energy.

Three aspects of the relation of species to its environment are area, habitat, and niche. The *area* is its geographic distribution. The term *habitat* describes the kind of environment in which the species lives. Habitat can be extended to include the environment of an entire community; going to the other extreme, microdifferences in the environment cause a very localized distribution of organisms in microhabitats. Sometimes the word *niche* is used for microhabitat, but the two words do not have the same meaning. Grinnell fathered the concept of the niche (1917 and 1917a). Over the years it has been defined and redefined. It is easiest to think of the niche as a total of the organism's requirements, rather than a characteristic of the environment, with an emphasis on the relation of the animal to all components of its surroundings. Kendeigh (1974) wrote, "The restriction of a species to a particular niche depends on its structural adaptations, physiological adjustments, and developed behavior patterns." Odum (1959) described the niche as the function an animal performs in the community, or its "profession." Hutchinson (1957) has defined a niche as all interrelationships of an organism with its environment. The important thing to remember is that the niche is a functional role of the organism and is not to be thought of as a microhabitat. Every aspect of the organism's requirements helps to define its niche. The niche is ultimately an abstract concept and can never be fully measured. This concept of the niche would indicate that no two species could occupy the same niche, for the occupants would have to be identical in every respect and would therefore have to be the same species.

Species that have the same occupation in different ecosystems are defined as ecologic equivalents. The mountain lion of North America *(Felis concolor)* and the African lion *(Panthera leo)* are ecologic equivalents, just as are the kangaroo rats of North America and the gerbils of Eurasia.

Species which are found in great abundance or otherwise exert a dominant influence in a community are called *dominants* and are usually plants, but cases in which mammals exercise dominance are documented. Koford (1958) showed that prairie dogs *(Cynomys ludovicianus)* can produce a short-grass stage in the mixed prairie. Larson (1940) reported that the bison *(Bison bison)* changed areas of tall-grass prairie to short-grass prairie. Certainly ranchers and range managers know that overgrazing changes the species composition of grasses on the range. Game biologists have for years been aware of the effects of overbrowsing and overgrazing by deer (*Odocoileus* spp.) and moose *(Alces americana).*

Of the many systems of classification of large communities, or ecosystems, one which has become widely used is that of biomes. A *biome* is a major biotic community characterized by distinctive climax plant life forms (e.g., grassland, deciduous forest) which are a response to major features of the environment, including climate. Although the biome is named after the climax community, stages approaching climax communities are also included. A climax community is able to maintain itself, as long as the climate does not change. On land the most important climax species are plant dominants. If one visits a climax such as a maple-basswood forest, one finds these dominant trees in all stages of development. If one visits an aspen or birch forest (seral stages), one usually finds that all the trees are of about the same age, i.e., there is a very limited age distribution. This would be comparable to a town where all individuals were of one age group and no young people were growing up to replace them.

Biomes

The principal biomes of North America are tundra, coniferous forest, temperate deciduous forest, grassland, desert, and tropical rain forest (Odum, 1971;

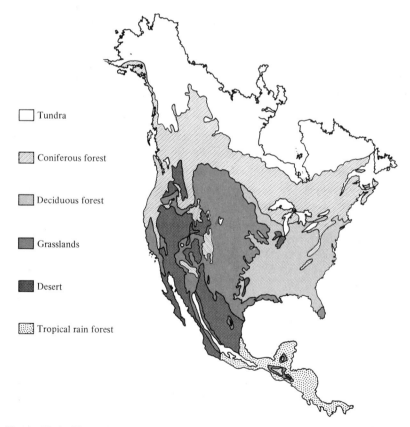

Tundra

Coniferous forest

Deciduous forest

Grasslands

Desert

Tropical rain forest

Figure 16-1 The major biomes of North America. *(From Benton and Werner, 1974, after Pitelka, 1947.)*

Oosting, 1958; Kendeigh, 1974) (Fig. 16–1). Names and classifications may vary some from author to author, and only the major North American biomes will be discussed here.

Tundra Biome The tundra biome is circumpolar in the arctic regions, and is also found farther south on mountaintops. The annual precipitation in the North American arctic tundra is about 25 cm (8 in.). The 10°C July isotherm corresponds closely to treeline. Permanently frozen soil, *permafrost,* underlies the surface and has led to statements, contrary to fact, that there are no hibernating mammals in the Arctic. During the summer the surface commonly thaws to a depth of 50 cm (20 inches), but on south-facing slopes and banks of rivers permafrost may be several feet from the surface. These are also the areas which harbor most of the scant snowfall that occurs in the Arctic. If the distribution (habitat) of the arctic ground squirrel *(Spermophilus undulatus)* were mapped, it would probably coincide with these areas of south-facing slopes with the deepest snow (Gunderson, Breckenridge, and Jarosz, 1955) (Fig. 16–2).

Although arctic plains and alpine habitats resemble each other, there are differences. Generally there is no permafrost in the alpine tundra, except in the far north, nor is there the extreme change in photoperiods. Precipitation and humidity are commonly high on the mountaintops, as are ultraviolet light intensity and evaporation. Other characteristics of alpine habitat are a low barometric pressure and oxygen concentration. The alpine environment seemingly places greater stresses on the organism than even the arctic tundra.

Many of the mammals on the tundra either remain white the year around or change to white in winter. In some instances, as in the arctic hare *(Lepus arcticus),* those farthest north remain white the year around. The most northerly form of caribou *(Rangifer tarandus pearyi)* has been described as having a long, silky, white pelage in winter, and the timber wolves *(Canis lupus)* of the arctic tundra also have a very light-colored pelage. The arctic fox *(Alopex lagopus),* the weasels *(Mustela nivalis* and *Mustela erminea),* and the collared lemming *(Dicrostonyx groenlandicus)* have a marked seasonal dimorphism. There is a great uniformity of the animal life of the tundra in Eurasia and in North America.

A major problem that tundra mammals must face is the stress of the long, cold winter, which must be either tolerated or avoided. The smaller mammals such as the voles, lemmings, and weasels stay in their nests and runways under the snow, and the ground squirrels hibernate. The larger mammals, such as the arctic hare *(Lepus arcticus),* arctic fox *(Alopex lagopus),* wolf *(Canis lupus),* and bears, have a long, dense pelage, and of course the grizzly bear *(Ursus arctos),* and the female and sometimes the male polar bears *(Ursus maritimus)* become dormant during the winter. The caribou *(Rangifer tarandus)* on the mainland migrate south even into the forests (Banfield, 1954), but those on the northern islands are restricted in their movements.

Figure 16–2 Young Arctic ground squirrel *(Spermophilus undulatus)* on tundra.

For a thorough discussion of the effects of snow cover on animals in both the tundra and the boreal forest, the reader is referred to Formozov's (1946) work, "Snow cover as an integral factor of the environment and its importance in the ecology of mammals and birds." The original is in Russian, but an English translation is available, owing to the efforts of Prychodko, Pruitt, and Fuller. Formozov describes the tundra snow as a "snowy-dune zone" and the forest snow as "fluffy forest snow." Tundra snow can be so compacted as to cause an axe to ring. Formozov wrote that the arctic fox *(Alopex lagopus)* can walk on dense tundra snow and sea ice without sinking, accomplishing long migrations, but when they get into forested regions with fluffy snow, they perish. Trappers have related to me how they are able, with snowshoes, to "run down" a bobcat *(Lynx rufus)* after a deep snowfall of light snow. This would be difficult to do with a Canada lynx *(Lynx lynx),* whose relatively larger footpads and lighter weight allow it to stay on top of loose snow. Snow cannot be ignored as a factor in mammal distribution.

Common terrestrial mammals of the tundra biome include:

Polar bear *(Ursus maritimus)*
Least weasel *(Mustela nivalis)*
Arctic fox *(Alopex lagopus)*
Wolf *(Canis lupus)*
Parry ground squirrel *(Spermophilus parryii)*
Red-backed mouse *(Clethrionomys rutilus* and *Clethrionomys c. gapperi)*
Tundra vole *(Microtus oeconomus)*
Brown lemming *(Lemmus trimucronatus)*
Collared lemming *(Dicrostonyx groenlandicus)*
Caribou *(Rangifer tarandus)*
Musk ox *(Ovibos moschatus)*

(See Banfield, 1954, 1954a; Burt and Grossenheider, 1964; Clarke, 1940; Gavin, 1945; Gunderson, Breckenridge, and Jarosz, 1955; Harper, 1961; Manning, 1946, 1948; Preble, 1902; 1908; Rausch, 1953; Shelford and Twomey, 1941; Soper, 1944.)

Coniferous Forest Biome   In North America a continent-wide belt of coniferous forest extends from New England north to the tundra and west to Alaska. There are southward extensions of this forest in the east through the Adirondacks and high Appalachians, in the west through the Rockies and Sierras into Mexico.

White spruce, black spruce, and balsam fir form the matrix of the northern, or boreal, edge of this forest, which occupies a glaciated land of bogs, cold lakes and rivers, and alder thickets (Fig. 16–3). This boreal forest encircles the globe; the Russian word "taiga" is finding increasing use as a descriptive term for it. The loosely used "spruce-moose biome" is also very descriptive.

Figure 16-3 Black spruce at boreal edge of coniferous forest biome.

Quaking aspen and paper birch occur extensively as seral stages; the aspen especially is an important facet of the coniferous forest, for it is used by the white-tailed deer *(Odocoileus virginianus)*, moose *(Alces americana)*, snowshoe rabbit *(Lepus americanus)*, and beaver *(Castor canadensis)*. Aspen groves form the ecotone between forest and grassland north and west from Minnesota to the Rocky Mountains.

Around the Great Lakes at the southern edge of the boreal forest is an area of hemlock, white pine, sugar maple, basswood, and yellow birch. Along the Pacific Coast, because of the climate and geography, the coniferous forest changes from the boreal aspect and becomes dominated by western hemlock, western red cedar, Sitka spruce, and Douglas fir—the "temperate rain forest"; still further south along the Pacific is a strip of redwood forest about 800 km (500 mi) long. In the Rockies are several coniferous forest associations, the composition of which is influenced by the elevation.

Alpine forests in the Sierras and Cascades consist of red fir, Engelmann spruce, and limber pine. At lower elevations mountain forests consist of ponderosa pine, Douglas fir, and white fir.

The Southeastern pine forests are actually seral stages of the temperate deciduous forest and unless maintained by fire and cutting, are replaced by oak, hickory, and magnolia forests.

At its northern edge, the coniferous forest borders the open tundra. At its southern edge there is no sharp line of demarcation, the coniferous forest grading into deciduous forest, aspen parkland, woodland, or chaparral.

The mammals are adapted to live in or under trees, but the coniferous forest lacks the rich humus layer of the deciduous forest. The sharp claws and opposable toes in squirrels, the prehensile tails of white-footed mice, and the bushy tails of squirrels are all adaptations to a life in the forest. The red squirrel (Tamiasciurus hudsonicus) seems more vocal than its deciduous-forest cousins. The arboreal mammals of the coniferous forest include also the marten (Martes americana) and the fisher (Martes pennanti). In winter the moose (Alces americana) feed on the tips of birch, aspen, cedar, balsam fir, and other trees, but in summer they feed on water lilies, pondweeds, sedges, and many grasses. In spite of the lack of a deep humus layer in the coniferous forest, small mammals are abundant; in winter they are well insulated under the snow, where the temperatures even in the far north may drop only a few degrees below freezing (Pruitt, 1957). Pruitt wrote:

> From the foregoing resume of the thermal environment of this sample of taiga one may make several generalizations. One feature of the forest floor environment is its stable temperature; comparatively warm in winter and cool in summer. During the two winters and one summer considered here the 22.5 centimeter (9-inch) temperature varied only 27 degrees while that of the air ranged through 152 degrees.

After the snow cover reaches a depth of 15 to 20 cm, there is a marked change in behavior of the forest-floor mammals. Before this, shrews and the red-backed mice are active on the snow surface; after this critical depth is reached, there is a marked reduction in the surface activity of these mammals.

Mammals of the coniferous forest include:

Masked shrew *(Sorex cinereus)*
Arctic shrew *(Sorex arcticus)*
Northern water shrew *(Sorex palustris)*
Star-nosed mole *(Condylura cristata)*
Red bat *(Lasiurus borealis)* (eastern third)
Hoary bat *(Lasiurus cinereus)* (eastern half only)
Black bear *(Ursus americanus)*
Marten *(Martes americana)*
Fisher *(Martes pennanti)*
Short-tailed weasel *(Mustela erminea)*
Least weasel *(Mustela nivalis)*
Wolverine *(Gulo gulo)*
Wolf *(Canis lupus)*
Lynx *(Lynx lynx)*
Red squirrel *(Tamiasciurus hudsonicus)*
Deer mouse *(Peromyscus maniculatus)*
Northern bog lemming *(Synaptomys borealis)*
Heather vole *(Phenacomys intermedius)*
Boreal red-backed vole *(Clethrionomys gapperi)*
Meadow jumping mouse *(Zapus hudsonius)*
Woodland jumping mouse *(Napaeozapus insignis)*
Porcupine *(Erethizon dorsatum)* (eastern part only)
Snowshoe hare *(Lepus americanus)*
White-tailed deer *(Odocoileus virginianus)*
Moose *(Alces alces)*
Woodland caribou *(Rangifer tarandus)*

(Adams, 1909; Buckner, 1957; Burt and Grossenheider, 1964; Cahn, 1937; Mech, 1966; Morris, 1955; Osgood, 1909; Quimby, 1944.)

The coniferous forest of the Pacific Northwest and Rocky Mountains is quite distinctive. Distributed throughout this forest are these mammals:

Grizzly bear *(Ursus arctos)*
Mountain lion *(Felis concolor)*
Mountain beaver *(Aplodontia rufa)*
Hoary marmot *(Marmota caligata)*
Yellow-bellied marmot *(Marmota flaviventris)*
Golden-mantled ground squirrel *(Spermophilus lateralis)*
Townsend chipmunk *(Eutamias townsendi)*
Yellow-pine chipmunk *(Eutamias ameonus)*
Douglas' squirrel *(Tamiasciurus douglasii)*
Bushy-tailed wood rat *(Neotoma cinerea)*
Red-backed vole *(Clethrionomys occidentalis)*
Townsend vole *(Microtus townsendii)*
Long-tailed vole *(Mirotus longicaudus)*
Water vole *(Arvicola richardsoni)*

Western jumping mouse *(Zapus princeps)*
Pika *(Ochotona princeps)*
Elk *(Cervus canadensis)*
Mule deer *(Odocoileus hemionus)*

Temperate Deciduous-Forest Biome   These forests exist as a broken belt around the world in north temperate climates. In North America this forest is best developed in the Eastern United States (Braun, 1950), although elements of it are mixed with conifers in the north, in the west, and in the mountains of Mexico. The mean annual precipitation—between 75 and 125 cm (30 to 50 in.)—falls fairly regularly throughout the year, as rain in summer and snow in winter. The climax of the deciduous-forest biome consists of broad-leaved deciduous trees, which form somewhat dense forests with a closed canopy, although conifers may occur in seral and climax forests. The seasonal aspects are well defined, and between the arrival of spring and the time of full growth of leaves there is a rich variety of flowering herbaceous plants. The annual crops of leaves produce a deep, rich litter, which, together with rotting logs and stumps, provides homes for a variety of shrews and mice. Tongues of the deciduous forest extend far into the prairie along river valleys. These are referred to as gallery forests.

The dominant trees of the deciduous-forest biome vary from east to west and north to south. Around the Great Lakes, beech and maple are common, with the maple-basswood being the climax in Minnesota and Wisconsin, south to northern Missouri. In southern Canada and the northern Appalachians is an ecotone area of beech, sugar maple, basswood, yellow birch, various northern species of pine, and eastern hemlock. Formerly the oak-chestnut forest occurred in the Appalachian Mountains and the northern Atlantic coast, but the chestnut has now largely been destroyed by blight. A mixed mesophytic forest occurs on the unglaciated Appalachian Plateau. This forest contains a rich mixture of tree species, but white basswood and yellow buckeye are the best indicators. The oak-hickory forest of the Ozark and Ouachita Mountains extends to the Gulf and out onto the prairies. Along rivers on the Piedmont Plateau and across the Southern states are found southern species of pine, as well as live oaks, magnolia, and holly, which enrich the evergreen appearance. This forest is also called the Southeastern pine forest.

The pronounced regular seasonal change in temperature and photoperiod triggers an annual reproductive season for most mammals living in the deciduous-forest biome. This is also true of the animals of the tundra and the coniferous forest. Many mammals have special adaptations for an arboreal life—sharp claws, opposable toes (squirrels), prehensile tails (opossums and white-footed mice), and bushy tails (tree squirrels.) Hearing and voice are usually well developed, as is the sense of smell, but vision is poorly developed.

A rich mammalian fauna exists in the deciduous forest, especially of tree squirrels and shrews. A deep litter layer and many rotting logs provide good habitat for shrews. Mammals present in the deciduous-forest biome include:

Opossum *(Didelphis marsupialis)*
Masked shrew *(Sorex cinereus)*
Smoky shrew *(Sorex fumeus)*
Southeastern shrew *(Sorex longirostris)*
Long-tailed shrew *(Sorex dispar)*
Short-tailed shrew *(Blarina brevicauda)*
Least shrew *(Cryptotis parva)*
Star-nosed mole *(Condylura cristata)*
Eastern mole *(Scalopus aquaticus)*
Hairy-tailed mole *(Parascolops breweri)*
Keen myotis *(Myotis keeni)*
Indiana myotis *(Myotis sodalis)*
Eastern pipistrelle *(Pipistrellus subflavus)*
Red bat *(Lasiurus borealis)*
Big brown bat *(Eptesicus fuscus)*
Hoary bat *(Lasiurus cinereus)*
Evening bat *(Nycticeius humeralis)*
Eastern big-eared bat *(Plecotus rafinesquei)*
Raccoon *(Procyon lotor)*
Long-tailed weasel *(Mustela frenata)*
Striped skunk *(Mephitis mephitis)*
Red fox *(Vulpes vulpes)*
Gray fox *(Urocyon cinereoargenteus)*
Woodchuck *(Marmota monax)*
Eastern chipmunk *(Tamias striatus)*
Eastern gray squirrel *(Sciurus carolinensis)*
Eastern fox squirrel *(Sciurus niger)*
Southern flying squirrel *(Glaucomys volans)*
Beaver *(Castor canadensis)*
Eastern harvest mouse *(Reithrodontomys humulis)*
White-footed mouse and deer mouse (*Peromyscus* spp.)
Golden mouse *(Ochrotomys nuttalli)*
Eastern wood rat *(Neotoma floridana)*
Rice rat *(Oryzomys palustris)*
Meadow vole *(Microtus pennsylvanicus)*
Pine vole *(Microtus pinetorum)*
Woodland jumping mouse *(Napaeozapus insignis)*
Snowshoe hare *(Lepus americanus)*
Eastern cottontail *(Sylvilagus floridanus)*
White-tailed deer *(Odocoileus virginianus)*

(Allen, 1938; Baumgartner, 1938; Burt and Grossenheider, 1964; D. W. and A. V. Linzey, 1968; Mohr, 1947; Pruitt, 1953; Townsend, 1935; Williams, 1936.)

Grassland Biome Grasslands are found on all continents. They are areas of low and erratic precipitation, from 25 cm to as high as 100 cm along the eastern edge, and high rates of evaporation. In Russia this biome is called the *steppe,* in Hungary, the *puszta,* in South America, the *pampas,* and in Africa, the *veld.* The climax grasses grow as sod or in bunches, and may be tall, mixed, or short; thus the grassland biome of North America has three divisions from east to west, called the tall-grass, mixed-grass, and short-grass prairie (Fi. 16–4). Sometimes the eastern portion of the biome, with its rich soil and tall grasses, is called the *prairie,* while shallow soil and short grasses in the west characterize the *plains.* A prairie peninsula extends across western Minnesota, Wisconsin, and into Illinois and Indiana, with isolated patches in Ohio. The Palouse region of eastern Washington and Oregon and western Idaho is an isolated prairie of about 60,000 km². Another isolated prairie is the Great Valley of California.

Though the prairies and the plains are sometimes despairingly described as flat, monotonous, and dull, they have a beauty, character, and history all their own. These two viewpoints were expressed by J. A. Allen (1871, 1871a), whose attempt at a scientific description of the American prairie was one of the earliest, and Ruthven (1908). Allen (1871) wrote, "With all the beauty and novelty of the primal flora of the prairies, the traveller, after a few weeks of constant wandering amid their wilds, is apt soon to experience a monotony that becomes wearisome, the full degree of which he scarcely realizes til the soft green sward and the varied vegetation of cultivated districts again meet

Figure 16–4 Grassland biome.

the eye." Ruthven (1908) wrote, "The prairie is the most interesting biotic region of North America."

If a stage setting were to be designed to display wildlife dramatically, nothing could be better than a representation of the prairie, with its simplicity of horizontal lines—one broad line, the landscape between you and the horizon; a narrow line, the horizon; then another broad line, the sky from the horizon to the zenith. This was the stage for the herds of millions of buffalo *(Bison bison),* of nearly the same number of antelope *(Antilocapra americana),* and of hundreds of millions of prairie dogs *(Cynomys ludovicianus),* of which Dobie (1962) wrote, "They were once incredibly numerous—perhaps as numerous as the extinct passenger pigeon." Merriam (1901) reported a prairie dog town north of San Angelo, Texas, 161 × 405 km (100 × 250 mi), estimated to contain 400 million animals! The prairie dog was relentlessly and successfully persecuted, and in 1947 Dobie (1962) wrote, "They have been poisoned out until now comparatively few exist." It was most interesting to read in the October 1969 issue of the *Journal of Wildlife Management* this statement, by Carpenter and Martin (1969), "and management practices have eliminated the prairie dog from much of its former range and have reduced its population to a remnant of its former numbers. There is increasing interest in transplanting these animals to other suitable areas in order to perpetuate them."

It is also safe to say that one of the deepest roots of terrestrial ecology in North America is found in the grasslands, where E. Clements and especially E. Weaver of the University of Nebraska were working at about the same time that the infant science was being nurtured by H. C. Cowles at the University of Chicago and V. E. Shelford at the University of Illinois. Visher (1916) wrote that the mammals of the northern grassland possessed two or more of this list of characteristics: ability to run swiftly, the burrowing habit, long-range vision, color adaptation of gray or tawny, ability to do without much water, daily activity confined to early morning, evening, or sometimes nighttime, ability to hibernate, and largely herbivorous diet. No mammals are exposed to any greater extremes of daily and seasonal temperature fluctuations than are the ungulates of the grasslands. The pronghorn *(Antilocapra americana)* and the bison *(Bison bison)* may live through a winter with days of −40°C and high winds, then in summer withstand temperatures of 52°C above, without shade.

For interesting accounts of the grasslands in North America the reader is referred to Allen (1967); Borchert (1950); J. R. Carpenter (1940); Costello (1969); Craig (1908); Malin (1967); Shantz (1923); Ruthven (1908); Visher (1916); Weaver (1954); Weaver and Clements (1938); and Weaver and Fitzpatrick (1934).

A characteristic of grassland and desert mammals is the development of hopping (saltatorial and/or ricochetal) locomotion. This kind of locomotion occurs in jackrabbits (*Lepus* sp.), kangaroo rats (*Dipodomys* sp.), and pocket mice (*Perognathus* sp.). This locomotion allows greater visibility. Even the mule deer *(Odocoileus hemionus)* bounds along, while the white-tailed deer *(Odocoileus virginianus)* sneaks or slinks along, reflecting the shrubbery and forest-edge habitat in which it lives and which is not conducive to hopping. Many small grassland mammals become inactive (estivate) during the dry seasons of late summer and fall and proceed from estivation to hibernation.

A list of prairie mammals would include:

Masked shrew *(Sorex cinereus)*
Least shrew *(Cryptotis parva)*
Short-tailed shrew *(Blarina brevicauda)*
Red bat *(Lasiurus borealis)*
Hoary bat *(Lasiurus cinereus)*
Raccoon *(Procyon lotor)*
Black-footed ferret *(Mustela nigripes)*
Badger *(Taxidea taxus)*
Spotted skunk *(Spilogale putorius)*
Striped skunk *(Mephitis mephitis)*
Coyote *(Canis latrans)*
Red fox *(Vulpes vulpes)*
Black-tailed prairie dog *(Cynomys ludovicianus)*
Thirteen-lined ground squirrel *(Spermophilus tridecemlineatus)*
Plains pocket gopher *(Geomys bursarius)*
Kangaroo rats (*Dipodomys* spp.)
Western harvest mouse *(Reithrodontomys megalotis)*
Prairie deer mouse *(Peromyscus maniculatus bairdii)*
Northern grasshopper mouse *(Onychomys leucogaster)*
Meadow vole *(Microtus pennsylvanicus)*
Prairie vole *(Microtus ochrogaster)*
Muskrat *(Ondatra zibethicus)*
Meadow jumping mouse *(Zapus hudsonius)*
White-tailed jackrabbit *(Lepus townsendi)*
Black-tailed jackrabbit *(Lepus californicus)*
Mule deer *(Odocoileus hemionus)*
Pronghorn *(Antilocapra americana)*
Bison *(Bison bison)*

Although they are now restricted to sanctuaries, parks, refuges, and private ranches, it is estimated that at one time there were about 35 million bison roaming the prairies and plains and extending into the eastern forests. With prolonged drought, blizzards, and prairie fires, there must have been gigantic fluctuations in their total numbers. Haley (1936), Charles Goodnight's biographer, tells how Goodnight saw dead bison near the Brazos River in Texas in the winter of 1876 so thick, "they resembled a pumpkin field." He estimated that several million bison starved that winter in an area "100 by 25 miles."

Desert Biome   Deserts vary from arid wasteland to arid regions containing trees, shrubs, and other plants especially adapted to a hot, dry climate. Deserts occur in belts at a similar latitude above and below the equator, and cover about a fifth of the earth's surface. In North America they occur in the Southwestern United States and in Mexico. The Great Basin area of the West is sometimes called the cold desert, in contrast to the "hot" deserts to the south. In North American deserts the average annual precipitation is about 5 in., and this sometimes in erratic, violent cloudbursts rather than spread out evenly. The evaporation rate is high, and relative humidity very low—20 to 30 percent at noon. Deserts are usually sunny, and the actual sunshine received of the possible annual amount is 90 percent. Differences between the daily maxima and minima temperatures are the greatest in any biome. It is with these conditions—erratic and scant rainfall, extremes of temperature, and strong winds—that desert mammals must contend.

Plants in the desert have many adaptations to retard transpiration and survive drought. Foliage can be reduced for long periods, and the photosynthesis is carried on by chlorophyll in the stems. For a discussion of these adaptations, see Weaver and Clements (1938). Although there are differences in the deserts of the Southwest, they will be treated as a single unit, the desert biome, in this discussion.

Large mammals are few on the desert; the characteristic mammals are the herbivorous rodents. The mammals that live in the desert live in burrows or otherwise seek shelter. There are few carnivores; therefore the availability of food and water, rather than predation, determines population numbers. The carnivores found on the desert are the medium-sized ones, such as coyote *(Canis latrans)*, foxes (*Vulpes* spp.), bobcats *(Lynx rufus)*, and skunks (several genera). The adjustments in grassland species to their environment are continued in desert species, and sometimes further refined. Ecologically equivalent species showing little taxonomic relationship are found throughout the deserts of the world. The genera *Dipodomys* and *Perognathus* of the family Heteromyidae are especially numerous in the North American desert. In the Eurasian deserts, the genera *Gerbillus, Meriones,* and *Dipodillus* of the family Muridae are present, and in Australia the ecologic equivalent is *Notomys,* also of the family Muridae. In South African arid regions a jackrabbit-sized rodent *Pedetes* (family Pedetidae) is said to make nightly excursions of up to 30 km from its burrow to its feeding area (Shortridge, 1934). All these animals are bipedal and exhibit behavioral and physiologic adaptations for water conservation.

A few of the mammals found in the Southwest are:

Desert shrew *(Notiosorex crawfordi)*
Fringed myotis *(Myotis thysanodes)*
Yuma myotis *(Myotis yumanensis)*

California myotis *(Myotis californicus)*
Western pipistrelle *(Pipistrellus hesperus)*
Spotted bat *(Euderma maculata)*
Pallid bat *(Antrozous pallidus)*
Mexican free-tailed bat *(Tadarida brasiliensis)*
Western mastiff bat *(Eumops perotis)*
Antelope jackrabbit *(Lepus alleni)*
Black-tailed jackrabbit *(Lepus californicus)*
Desert cottontail *(Sylvilagus audubonii)*
Spotted ground squirrel *(Spermophilus spilosoma)*
Merriam pocket mouse *(Perognathus longimembris)*
Desert pocket mouse *(Perognathus penicillatus)*
Kangaroo mice (*Microdipodops* spp.)
Kangaroo rats (*Dipodomys* spp.)
Cactus mouse *(Peromyscus eremicus)*
Canyon mouse *(Peromyscus crinitus)*
Brush mouse *(Peromyscus boylei)*
Southern grasshopper mouse *(Onychomys torridus)*
White-throated wood rat *(Neotoma albigula)*
Desert wood rat *(Neotoma lepida)*
Coati *(Nasua narica)*
Ringtail *(Bassariscus astutus)*
Badger *(Taxidea taxus)*
Hog-nosed skunk *(Conepatus leuconotus)*
Kit fox *(Vulpes macrotis)*
Peccary *(Dicotyles tajacu)*
Mule deer *(Odocoileus hemionus)*
Pronghorn *(Antilocapra americana)*
Desert bighorn sheep *(Ovis canadensis)*

(Baker, 1956; Burt, 1938; Burt and Grossenheider, 1964; Fautin, 1946; Huey, 1942; Jaeger, 1957; Linsdale, 1938; Ryan, 1968.)

POPULATIONS

A population has already been defined as a group of organisms of the same species occupying a definite area. Now we will consider a population as a self-regulating system with features over and above those of the individual. The ecologist must give some thought to the individual as a part of a population, this latter having a sex and age structure, reproductive and mortality rates, a social organization, and a density that changes with time. The population can grow, remain stable, or decline.

Different species show different population characteristics. There are many variables in population dynamics. As mentioned earlier, the population (at any time) is equal to the biotic potential, or reproductive rate minus the environmental resistance or mortality (including emigration).

Reproductive rates are greatly influenced by many characteristics of the species and the population. These include (1) litter size, (2) number of litters per year, (3) longevity of breeding age of individuals, (4) sex ratio, (5) mating habits, i.e., monogamous, polygamous, and (6) population density.

Litter size   Litter size is a character of the species, and does not vary greatly from an average figure. The litter size for bison *(Bison bison),* elk *(Cervus canadensis),* and most ungulates is one; for cottontails *(Sylvilagus floridanus)* it is four or five; for the deer mouse *(Peromyscus maniculatus)* it is five; and for the prairie vole *(Microtus ochrogaster)* it is about six.

Number of Litters Per Year   The number of litters which a female produces per year obviously influences the population and is influenced by gestation period, the length of the breeding season, and other factors. Mice and voles *(Peromyscus, Microtus)* have a gestation period of about 21 days, and breed immediately after giving birth. In *Microtus pennsylvanicus,* 10 litters per year are possible in the Temperate Zone (Hoffman, 1958). At the other extreme is the African elephant *(Loxodonta africana),* whose gestation period is 22 months.

Breeding Age   Voles and mice may begin breeding at between 1 and 2 months of age and of course have a very short life span, not much over a year. Old age is a rarity among small mammals in the wild. Bison *(Bison bison)* begin breeding at 4 years of age. Although fertile at a younger age, the bison's behavior prevents younger individuals from breeding (McHugh, 1958). African elephants do not begin breeding until 8 to 12 years of age.

Sex Ratio and Mating Habits   If a species is monogamous, an equal sex ratio would tend to favor a maximum production of young, while an unequal sex ratio would not. In a polygamous species an unequal sex ratio with more females than males would tend to favor maximum production.

Density   In a sparse population failure to find mates might keep production at a low level, while the higher densities would increase natality, except when individuals are stressed by proximity.

Maximum reproduction is seldom attained, for an optimum of conditions seldom occurs. If introduced animals find a favorable situation and the population begins to grow, it will grow slowly at first. As the size of the breeding population grows, the total numbers pyramid from year to year, and the rate of increase approaches the biotic potential. However, as the limits are approached, mortality increases! The curve of the population growth will level off at a point where the birth and death rates are in balance. This point, the maximum condition, is also referred to as the carrying capacity of the land or habitat.

Population Growth Curves

J-shaped Growth Curve   One of the first attempts to study population growth was that of Thomas Malthus, who found that populations without any environmental resistance grow in geometric fashion, and the growth curve is J shaped (Fig. 16–5). This formula is $dN/dt = rN$, where dN/dt represents the population growth per unit of time, which equals the population size N multiplied by a factor r, called the *intrinsic growth rate,* representing the maximum potential rate for increase. This maximum potential rate of increase is seldom attained but may sometimes be approximated under natural conditions during the early or accelerating phase of population growth. The value of r has been obtained for only a few species.

Sigmoid, or S-shaped, Curve   A population with unlimited resources is not a reality. To the original formula a factor representing the ability of the environment to support the population is added, or to put it another way, the factor represents the environmental resistance.

$$\frac{dN}{dt} = \frac{rN\,(K-N)}{K}$$

In this equation K is a constant representing the maximum number which the population can reach. This growth pattern is the usual one and is called the sigmoid or S curve (Fig. 16–6). This formula works well for smaller and continuously breeding species, but for species such as ungulates, which breed seasonally and mature slowly, the shape of the curve becomes slightly modified.

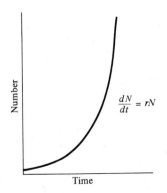

$$\frac{dN}{dt} = rN$$

Figure 16–5   Population growth pattern of J-shaped growth.

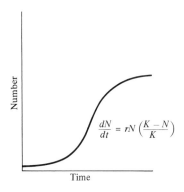

$$\frac{dN}{dt} = rN\left(\frac{K-N}{K}\right)$$

Time

Figure 16–6 Population growth pattern of S-shaped or sigmoid curve.

Stable, Irruptive, and Cyclic Growth Curves   A *stable,* or nonfluctuating, growth curve is shown by species which maintain a relatively stable population from year to year.

An *irruptive* growth curve represents a population that goes along quite regularly but suddenly reaches a high density and just as quickly returns to normal; these peaks have no regularity. Examples of irruptive populations include those of the mule deer *(Odocoileus hemionus)* of the Kaibab National Forest in Arizona (D. I. Rasmussen, 1941); mountain vole *(Microtus montanus)* in 1906 in the Humboldt Valley in California (Piper, 1909); house mice *(Mus musculus)* in California in 1927 (E. R. Hall, 1927); and shrews *(Cryptotis parva)* (W. B. Davis, 1940); *Blarina brevicauda* (Fowle and Edwards, 1955); and *Sorex cinereus* (Gunderson, 1962; Buckner, 1966). Of these the most widely known is that of the Kaibab deer herd *(Odocoileus hemionus),* whose numbers went from 4000 to 100,000 in 18 years, a population growth which was followed by a population crash. For many years the cause of the spectacular growth (and of the crash which followed) was believed to be the elimination of a natural predator on the deer, the mountain lion *(Felis concolor).* As a result, deer were believed to have reproduced far beyond the carrying capacity of the range. More recently this theory has been questioned. Caughley (1970) discredits the predator idea. In the case of the Kaibab deer herd, several factors affected the population including (1) removal of predators, (2) forest fires, and (3) removal of large numbers of cattle and sheep from the range.

A *cyclic* growth curve represents populations which vary in a somewhat uniform manner between high and low levels of density at regular intervals. Cyclic implies a recurring variation of regular timing and of constant amplitude, but with less variability than one would expect by chance (D. E. Davis, 1957). There has been a great deal of controversy over cycles. If the variability, at least in timing, is less than is to be expected by chance, then reasonably accurate predictions can be made. The argument seems to be one of terminology; the term *oscillatory* is preferred by some authorities (Fig. 16–7).

The short-term fluctuations are commonly thought of as about 4 years, but may be as long as 6 years or as short as 2 years. Short-term fluctuations were dramatized by the peculiar behavior of the lemmings which inhabit the tundra regions of the Scandinavian plateaus and mountains. Every so often they increase to enormous numbers, "eat themselves out of house and home," and start to disperse from their population or "overpopulation" centers, down the mountainside through farms, towns, cities, and into the water if any of it is on their route. These lemmings swimming in the fjords, where they had come by chance rather than intent, gave rise to the myths that they were either committing suicide or were striking out into the Atlantic following a long-lost migration route to the long-gone continent of Atlantis. Elton (1942) has discussed the subject in *"Voles, Mice and Lemmings."* Pitelka (1957) has reviewed the subject of microtine cycles in the Arctic. Hamilton (1937) believed the meadow vole *(Microtus pennsylvanicus)* exhibited a 4-year fluctuation in New York State. Siivonen (1954) has also reviewed the short-term population fluctuations.

The 10-year fluctuation is most characteristic of mammals living in the boreal forest region, but not on the tundra, where the short 4-year cycle is most prominent. The 10-year cycle has often been described, but little is known of its biology. Few researchers devote 10 to 20 years to the same project. Some of the first data were presented by Seton (1923) in his book, *The Arctic Prairies,* from a few returns of the Hudson's Bay Company. Seton presented evidence for a 10-year oscillation in the snowshoe hare *(Lepus americanus)* population and its main predator, the Canada lynx. The reports of MacLulich (1947) and Elton and Nicholson (1942) showed a remarkable picture of regularity.

Numerous factors influencing these regular oscillations have been suggested. One factor that was championed earliest was sunspots, advanced by MacLulich (1936), as well as by DeLury and O'Conner (1936). Green, Larson, and Bell (1939) have provided evidence for shock disease (hypoglycemia) as a cause for periodic decimation of the snowshoe hare *(Lepus americanus)* in

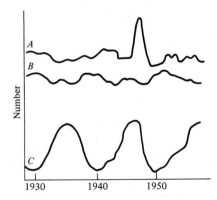

Figure 16-7 Stable, irruptive, and cyclic growth curves.

Minnesota. Shock disease was brought about by overcrowding and its accompanying stress. They also found that there was a decrease in the production of embryos, as well as a decrease in the survival of the young following the population crash. During the decade following their work on the snowshoe hare, a number of parasites and some viruses (notably the one associated with tularemia) were implicated as causative agents in population declines during cycles.

Pitelka (1958) approached the problem of population change by choosing a population living in the Arctic. He wrote:

> The total cycle is then generalized as a three-year pattern in which the main population growth of brown lemming *(Lemmus trimucronatus)* occurs in the second and third winters, the summers are intervals when the growth trend is reversed or deflected, and the decline occurs in large part if not almost entirely in the third or last summer. Winter cutting of vegetation [by the lemmings] and the resulting removal of food and cover exposes peak densities of lemmings to predation, which is the usual agent figuring in the inevitable decline of the population.

Chitty (1960 and 1967) has hypothesized that changes in social pressure causing aggressive behavior exert a regulating influence on population size. Krebs (1970), testing this hypothesis on *Microtus pennsylvanicus* and *Microtus ochrogaster,* found that aggressive behavior may act to regulate population size, but at present it is not known whether aggressive behavior is an effect of an increasing density or a cause of subsequent population regulation.

Errington (1958) has proposed that a combination of physiology and psychology associated with quality of food might be important in causing density changes in a muskrat population. Chitty (1958) tentatively suggested a rather interesting phenomenon, a hereditary hemolytic anemia, as a self-regulating mechanism in *Microtus agrestis* populations. Frank (1957) assigned the cyclic increase of *Microtus arvalis* in Germany to three factors: (1) the biotic potential of the species, (2) the carrying capacity of their habitat, and (3) a species behavior which he calls the "condensation potential," that does not occur in voles of noncyclic populations where a form of territorial behavior is more common. In the cyclic areas the females form large overwintering groups. When the condensation potential is operative, there is no behavioral limit on population increase. Under these conditions the population can increase beyond the carrying capacity. At this level the climatic stresses and conflicts between individuals cause a population crash. Under a constant climatic regime, the oscillation would follow a 3-year cycle. Frank (1957) believed the immediate cause to be shock disease, a breakdown of the adrenal-pituitary system.

An endocrinologic set of factors, the general adaptation syndrome, although not a cyclic phenomenon, has been carefully documented by Selye (1950 and 1955) and proposed by him as a cause for lowered resistance to diseases in humans in times of stress. Christian, in several papers (1953 and others) and Christian and Davis (1964) have applied this concept to nonhuman populations. In animals under stress adrenal weights increase and individuals become extremely aggressive; reproductive activity is curtailed and resistance is lowered. Eventually "shock disease" sets in. Christian (1950, 1953) and others pioneered the research along these lines on nonhuman mammals. Many students have been stimulated to do research on the phenomena loosely referred to as "stress reactions," which are often implicated as mortality factors in population decrease. Other studies have failed to show such changes in the adrenals and other endocrine glands. P. K. Anderson (1961) believed that food supply, aberrant parental behavior, and cannibalism would operate before symptoms of the general adaptation syndrome became operative. Krebs (1964) found no evidence of increase in adrenal gland size in Canadian lemmings *(Lemmus trimucronatus)* during times of high density.

Because of seemingly tenuous explanations for these "cyclic" population fluctuations and the lack of precise timing and amplitude of the oscillations, some biologists have all but discarded the whole idea. Cole (1954) has implied that populations are affected by such a variety of factors that random fluctuations are the result, but Davis's (1957) definition seems quite credible to most biologists. He wrote, "In ecological usage the term 'cycle' refers to a phenomenon that recurs at intervals. These intervals are variable in length, but it is implied that their variability is less than one would expect by chance and that reasonably accurate predictions can be made."

This brief résumé is presented not as a historical or complete survey but to emphasize the complexity of the problems associated with the study of population oscillations and the diverse ideas that workers have utilized in attempting to solve them. Such a summary can be found in Keith (1963), Hewitt (1954), or the Cold Spring Harbor Symposium on Quantitative Biology, "Population Studies: Animal Ecology and Demography" (Warren, 1958).

Some Other Factors Influencing Population Densities

If there is one area where nature is lavish, it is in the reproductive potential of most animals. Where conditions are right, the natural productivity of living things is remarkable for the wide margin of overproduction. Were it not so, geographic ranges of animals would not expand, nor would areas of catastrophic decimation become repopulated.

There are many factors which prevent the fulfillment of such productivity; they are variously called decimating factors or population depressants. Most population studies center around a search for factors that cause a change in the number of individuals.

Charles Darwin as early as 1859 in *The Origin of Species* noted what to him appeared to be four kinds of checks to overproductivity. These were (1) amount of food; (2) predation; (3) physical factors, such as climate; and (4) disease. More recently, Andrewartha and Birch (1954) persuasively argued that weather, not competition, is the most important regulating factor in insects. Lack (1954) championed the idea of competition, particularly for food. Wynne-Edwards believed that a very large part of the regulation of numbers depends on some initiative of the animals themselves, "that is to say, to an important extent it is an intrinsic phenomenon" (1965). Population growth is essentially density-dependent, and the adjustment of animal numbers is a homeostatic process. It proceeds fastest when numbers are below ceiling level and slowest when the ceiling is exceeded. Wynne-Edwards (1962 and 1965) cited such social behaviorisms as territoriality, colonialism, and the peck order as social systems which maintain population homeostasis.

Looking at various arguments we must conclude that we cannot, as yet, simplify population control to one limiting factor operating simply and directly, a concept first expressed by Liebig over a century ago.

Basically, population numbers are influenced by some force or forces outside (extrinsic) the population, or within (intrinsic) the organism, or both. Extrinsic factors are also designated as *density-dependent.* They can be either positive or negative; mortality rate is an example of a positive or direct density-dependent factor, whereas birth rate, which normally decreases with increasing density, is a negative or inverse density-dependent factor. *Density-independent* factors do not vary with populations, i.e., the same proportion of organisms is affected at any density. Climatic factors are often but not always density-independent.

High populations have been shown to be an important density-dependent factor in the control of mice and vole populations. With an increased number of individuals crowded into cages of uniform size, there was a decrease in the number of litters and the size of each litter. A social hierarchy was established, and only those near the top were able to reproduce at the normal rate; those at the bottom were very unproductive (Crowcroft and Rowe, 1957; Southwick, 1955; and Clarke, 1955).

Nutrition of the female influences her productivity by changing the age of puberty, the number of litters per year, and the number of litters per lifetime. Nutrition also affects the fertility of the male. Bodenheimer (1949) reported on the results of keeping Levant voles *(Microtus guentheri)* on various diets. He kept adult females on various diets for 100 days and counted the number of young each produced. Those maintained on wheat, barley, and carrots produced 75 young; those on a standard laboratory mouse diet averaged 73 young; wheat and water, 18; carrots alone, 18; wheat and meat and water, 0; and meat alone, 0.

In the beaver *(Castor canadensis),* Huey (1956), working in New Mexico, found a distinct relationship between kind of food and productivity. In aspen areas, average litter size was 4.20 young; in cottonwood areas, 2.1; and where willow was the principal food only 2.06 young were produced as an average.

Studies on the white-tailed deer *(Odocoileus virginianus)* in New York have shown conclusively that the ability of females to produce fawns is directly correlated with food supply. Cheatum (1947) reported that in the Adirondacks, where the soils are poor, the growing season is short, and deep snows concentrate the deer into restricted areas in winter, the deer are malnourished and many die in hard winters. Only 78 percent of the does were pregnant, one in 24 of the does had bred in the first fall season, 81 percent of the does (that were pregnant) had single fawns. By contrast, the deer in the southern tier of agricultural counties where hunting of both sexes was allowed were healthier. The females from this area produced more egg cells and a higher percentage were fertilized. Ninety-two percent of these does were pregnant, and one in three (compared with one in 24) had bred in its first fall season. In the southern agricultural area 60 percent of the does had twins (compared with 18 percent in the mountains) and seven percent had triplets (compared with only one set of triplets in the mountain sample).

In rodents the onset of reproductive activity in the spring may be greatly accelerated. Negus and Pinter (1966) have shown that small dietary supplements of either sprouted wheat or acetone-ether extracts of sprouted wheat improved reproductive performance in *Microtus montanus* by more frequent postpartum matings and lower rates of litter loss. Immature females were stimulated to immediate onset of estrus when fed sprouted wheat. Spinach extract stimulated an increased uterine weight and an increase in the number of developing follicles in 4-week-old females. The authors suggested that "hormone-like substances in plants may influence reproduction in natural populations of *Microtus montanus.*"

Predation Predation is defined by Odum (1959) as an interaction between two species, "in which one population adversely affects the other by direct attack but is dependent on the other." No phase of population interaction provokes a greater spectrum of response among a wider variety of persons than predation. Among biologists it may be considered most simply as a step in the energy transfer by a complex interaction of two or more species. The agriculturist, rancher, and sportsman most often look upon it quite differently. The argument becomes most heated when between the biologist and the sportsman or between the biologists and the applied biologists. Predator control is an issue on which a lot of light has been shed, but the darkness of ignorance is still prevalent.

Especially vulnerable were the activities of the Branch of Predator and Rodent Control of the Bureau of the Biological Survey, centered in the American West during the first third of this century. An interesting insight can be gained by the reader who might like to refer to a report of the Twelfth Annual Stated Meeting of the American Society of Mammalogists, in the *Journal of Mammalogy,* volume 11, pages 325 to 389.

The early methods of predator control usually involved the indiscriminate destruction of predators by trapping, shooting, and poisoning. Not only was the indiscriminate destruction attacked, but so were the methods. The inhumaneness of poisons and the destruction, by the use of poisons, of nonharmful or "beneficial" animals was particularly odious, even to those who had no monetary stakes involved. Although poisons are still used in rodent control (but not in predator control), their use has been severely limited, and only government officials engaged in rodent control can legally use them. Knotty problems still exist, e.g., the control of prairie dogs *(Cynomys ludovicianus),* whose burrows can also house an endangered species—the black-footed ferret *(Mustela nigripes).* For a lively, readable account of the problems involved, see McNulty (1972). These early arguments have brought results both in changes in attitude and changes in methods of control.

Although the early methods of predator control are still used, they have often been modified so that only those individuals doing the damage are destroyed or the individuals are in some way deterred from doing any damage. Imaginative research is developing new ways of population control which may eventually substitute more sophisticated methods for the trap, the gun, and poison. Lord (1961), working on gray fox *(Urocyon cinereoargenteus)* populations in northern Florida, sterilized the females by surgical hysterectomy. He wrote, "The capture and sterilization of a single female reduces the number of juveniles in the coming year's population by nearly five, a large change for a small effort." Since the gray fox is considered monogamous at least for one breeding season, an area would be occupied by only one unproductive pair. If the female were destroyed she might be replaced by a productive one. Although Lord's (1961) research was designed as a population study, his methods could easily have some application for "predator" control.

Balser (1964) has conducted field experiments with antifertility agents (diethylstilbestrol) on coyotes. Two areas in New Mexico were selected. One was treated with drop bait containing diethylstilbestrol and the other was not treated. Females from both areas were collected in April and May, and their reproductive tracts examined. Chi-square tests indicated that unsuccessful reproduction was associated with the treatment.

The literature on predator-prey relationships is voluminous. Two mathematicians, Lotka (1925) and Volterra (1926), proposed formulas attempting to show how oscillations were produced; as the predator population increased, the prey decreased to a point where the trend was reversed. Gause (1934) attempted to prove this experimentally in a closed system between *Didinium,* a ciliate predator, and *Paramecium caudatum.* However the predator destroyed the prey, and eventually itself, unless more prey was introduced. In the wild, prey is introduced (by birth of young; by immigration) and an oscillation continues indefinitely. There are plenty of examples among mammals of the starvation of prey after the destruction or removal of predators. The history of the Kaibab deer herd discussed earlier is such a case, as was the history of the moose on Isle Royale (Mech, 1966). A much more subtle effect was discussed by A. Murie (1944). In his Alaskan study of wolf predation on Dall's sheep *(Ovis dalli),* he points out that sheep discovered on the flats might be overtaken before gaining safety in the cliff. He wrote, "We have been discussing the elimination of the weak and the part this activity may play in the maintenance or improvement of the species. There is another that could be considered. As an evolutionary force the wolf may function most effectively by causing the sheep to dwell in a rocky habitat." In other words wolves *(Canis lupus)* make mountain sheep *(Ovis dalli)* live in the mountains.

Whether predation can regulate populations is a point of constant debate, and one would hope that eventually someone will have the foresight to leave a population alone long enough—maybe several human generations—to get an answer.

The moose *(Alces americana)* on Isle Royale were starving when the reintroduction of the timber wolf was unsuccessfully attempted by man. At that time the wolf *(Canis lupus)* successfully reintroduced itself (Mech, 1966). But suppose the wolf had not appeared at that time. What would then have happened? Would the moose have completely disappeared, or would their numbers have gone just low enough to let the vegetation recover? Certainly in the last 5000 years there must have been times that the moose either disappeared or nearly disappeared on Isle Royale; and the wolves disappeared and reappeared. Isle Royale would make an interesting and logical laboratory for such a long-time experiment.

A generalized predator that feeds on many kinds of prey can maintain itself under conditions in which its principal prey species are greatly reduced in numbers. The alternate prey species is called the *buffer species.* In general the greatest number of buffer species are found in the most complex communities. On Isle Royale, where the wolf fed on moose in winter, it would turn to other species in summer. The mountain lion *(Felis concolor)* is a specialized predator feeding primarily on deer *(Odocoileus)* and elk *(Cervus canadensis)* but also on bighorn sheep *(Ovis canadensis)* when available, but food alone does not limit populations of the highly territorial mountain lion in parts of Idaho (Hornocker, 1970). Intraspecific relationships manifested through territoriality acted to limit their numbers.

The reader who desires more information on predation would do well to refer to Craighead and Craighead (1946), Errington (1946), Mech (1966), A. Murie (1944), T. G. Scott (1943), Slobodkin (1962), and Hornocker (1970).

Social Behavior   Social dominance, hierarchy, territoriality, and home range (all discussed in Chap. 12) are behaviors which to a degree regulate animal populations. Wynne-Edwards (1962) and others have pointed out that social behavior limits the number of animals to a habitat, food supply, and reproductive activity. Physiology and behavior provide the regulatory machinery. Social stimulation acts on the individual through the endocrine system, and in mammals this involves especially the pituitary and the adrenal cortex. There are many studies on how social interactions inhibit population growth. One of the most comprehensive and one which the reader might want to consult is Calhoun's (1962) extensive work on the Norway rat *(Rattus norvegicus).*

ECOLOGIC ENERGETICS

Energy comes from the sun. The sun's light energy strikes the chlorophyll of green plants, leading to the formation of organic products, which is the basis of energy for all living things. Basic to an understanding of energy are two

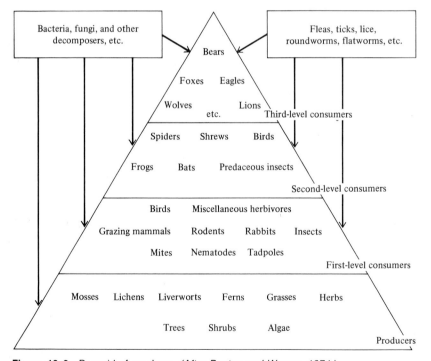

Figure 16–8   Pyramid of numbers. *(After Benton and Werner, 1974.)*

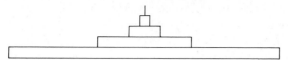

Figure 16-9 Pyramid of biomass. *(After Benton and Werner, 1974.)*

fundamental laws of thermodynamics: (1) energy cannot be created or destroyed, it can only be altered; (2) there is always degradation of energy to less and less useful forms (law of entropy).

In any ecosystem many changes of energy occur, each change causing a decrease of available energy. The noted British ecologist Charles Elton diagrammed the pyramid of numbers (Fig. 16–8). The first trophic level is known as the *producer* level. These are the plants which transform the energy from the sun into usable form for the higher, or *consumer,* levels. Herbivores are first-level consumers; than comes a series of predators. Decomposers are also important to this biotic pyramid. There is a sharp decrease in biomass at each level. Since this pyramid is actually a pyramid of numbers, another pyramid must be used to show a pyramid of biomass (weight) (Fig. 16–9).

To trace the flow of energy through this pyramid, a single *food chain* is usually isolated. When all the food chains in a community are brought together, a *food web* is the result. The shortest food chains produce the greatest amount of biomass for the least energy.

Examples of food chains are not difficult to find. Costello (1969) wrote, "An acre of tall grass, producing up to three thousand pounds of forage per year, could easily support a bison for two months or more. Several acres of the shorter grasses on the high plains were required for a month of grazing." It is often stated (although I can never find the evidence) that bison *(Bison bison)* were more efficient converters of prairie grasses to beef than are beef *(Bos taurus)* cattle themselves.

Mech (1966) wrote concerning the browse-moose-wolf ecosystem on Isle Royale, "The ratio of moose to browse is 7.7 percent; of wolves to moose 1.7 percent; and of wolves to browse .13 percent. Thus, yearly about 762 pounds of browse are consumed for each 59 pounds of moose, in turn consumed for each 1 pound of wolf." Expressed more dramatically it would take 5,823,300 lb of browse to produce 89,425 lb of moose to produce 1,512 lb of wolf. The kilocalorie loss in each trophic level seems self-evident in the browse-moose-wolf chain.

Pruitt (1968) has used biomass data in an imaginative way—to throw some light on the disagreement in the literature concerning the synchrony of cyclic fluctuation in boreal small mammals (mice, voles, lemmings, and shrews) in Alaska. Pruitt wrote, "The most interesting phenomenon exposed by this study is the general synchrony in small mammal biomass over a very large geographic area." His studies revealed synchronous fluctuations in bi-

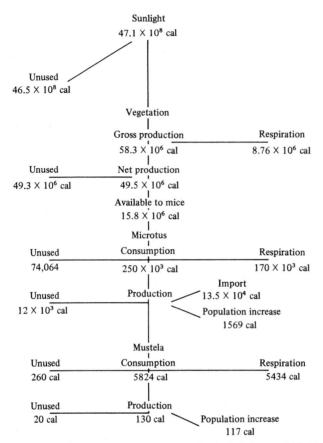

Figure 16–10 Energy flow through a food chain in an old-field community in Michigan. *(From Golley, 1960.)*

omass over an 8-year period, with peaks in 1954 and 1959 in widely separated areas of Alaska.

Energy Flow

By calculating the kilocalories at each trophic level we can begin to measure energy flow through communities and ecosystems just as physiologists have measured energy flow through individual organisms. Golley (1961), Hutchinson and Deevey (1949), Lindeman (1942), and Slobodkin (1955) are among those who have explored the energy flow of communities. Quantitative descriptions of energy flow through communities are few at best, and when they are restricted to studies of mammals, Golley's (1961) is the first. His was a study of energy flow in an old-field community in Michigan in which the primary consumers were meadow mice *(Microtus pennsylvanicus)* and the secondary consumers, or first-level carnivores, were least weasels *(Mustela nivalis)* (Fig. 16–10). The vegetation fixed about 1 percent of the solar energy into net production. The meadow mice consumed about 2 percent of the plant food, and the weasels consumed about 31 percent of the mice. In the transfer of energy through the ecosystem the energy is reduced by 100 in the step from primary producers to consumers and by 10 for each step beyond that. Food chains usually have no more than four links. The closer an animal can stay to the first trophic level the more energy will be available to it.

Although not reporting an efficiency figure for the Northern flying squirrel *(Glaucomys sabrinus),* Muul (1968) did calculate energy expenditure. The average monthly energy expenditure of the 70-g flying squirrel was 6890 kcal.

The study of energy flow through individuals, communities, and ecosystems has greatly increased in the last decade.

BIBLIOGRAPHY

Abel, A. H., ed. 1939. Tabeau's narrative of Loisel's expedition to the Upper Missouri. Translated from the French by Rose Abel Wright. University of Oklahoma Press, Norman. 272 pp.

Ables, E. D. 1969. Home range studies of red foxes *(Vulpes vulpes). J. Mammal.* **50:** 108–120.

Adams, C. C. 1909. An ecological survey of Isle Royale, Lake Superior. A report from the Univ. Mich. Mus., published by the St. Biol. Surv. as a part of the Report of the Board of the Geol. Surv. for 1908. Wynkoop Hallenbeck Crawford Co., Lansing, Mich. 468 pp.

Adams, L. 1965. Progress in ecological biotelemetry. *Biol. Sci.* **15:**83–86, 155–157.

Adolph, E. F. 1951. Response to hypothermia in several species of infant mammals. *Am. J. Physiol.* **166:**75–91.

Alcorn, J. R. 1940. Life history notes on the piute squirrel. *J. Mammal.* **21:**160–170.

Aldous, S. E. 1937. A hibernating black bear with cubs. *J. Mammal.* **18:**466–468.

———— and L. W. Krefting. 1946. The present status of moose on Isle Royale. *Trans. North Am. Wildl. Conf.* **11:**296–308.

Allee, W. C., A. E. Emerson, O. Park, T. Park, and K. P. Schmidt. 1949. Principles of animal ecology. W. B. Saunders Company, Philadelphia. 873 pp.

Allen, A. A. 1921. Banding bats. *J. Mammal.* **2:**53–57.

Allen, D. L. 1938. Ecological studies on the vertebrate fauna of a 500 acre farm in Kalamazoo County, Michigan. *Ecol. Monogr.* **8:**347–436.

————. 1943. Michigan fox squirrel management. Dep. Conserv., Lansing, Mich., Game Div. Publ. 100.

————. 1954. Our wildlife legacy. Funk & Wagnalls, New York. 422 pp.

————. 1967. The life of prairies and plains. McGraw-Hill Book Co., New York. 232 pp.

Allen, G. M. 1911. Mammals of the West Indies. *Bull. Mus. Comp. Zool. Harv. Univ.* **54**(6):175–263.

————. 1939. Bats. Harvard University Press, Cambridge, Mass. 368 pp.

————. 1939a. A checklist of African mammals. *Bull. Mus. Comp. Zool. Harv. Univ.* **83:**1–763.

————. 1940. The mammals of China and Mongolia. American Museum of Natural History, New York. pp. 621–1350.

Allen, J. A. 1871a. The flora of the prairies. *Am. Nat.* **4:**577–585.

————. 1871b. The fauna of the prairies. *Am. Nat.* **5:**4–9.

Altmann, M. 1953. Social graces in elk society. *Anim. Kingdom, N.Y. Zool. Soc.* **56:** 66–72.

————. 1956. Patterns of herd behavior in free-ranging elk of Wyoming. *Zoologica, Sci. Cont., N.Y. Zool. Soc.* **41:**65–71.

————. 1958. Social integration of the moose calf. *Anim. Behav.* **6:**155–159.

Altmann, S. A. 1959. Field observations on a howling monkey society. *J. Mammal.* **40:**317–330.

Amoroso, E. C., and J. H. Mathews. 1951. The growth of the gray seal *Halichoerus grypus* (Fabricius) from birth to weaning. *J. Anat.* **85:**427–428.

Anderson, P. K. 1961. Density, social structure and nonsocial environment in house mouse populations and implications for the regulations of numbers. *Trans. N.Y. Acad. Sci.,* ser. 11. **23:**447–451.

Anderson, S. 1967. Introduction to the rodents. pp. 206–209. *IN:* S. Anderson and J. K. Jones, Jr., eds. Recent mammals of the world. The Ronald Press Company, New York. 453 pp.

———, J. K. Doutt, and J. S. Findley. 1963. Collections of mammals in North America. *J. Mammal.* **44:**471–500.

——— and J. K. Jones, Jr. 1967. Recent mammals of the world: a synopsis of families. The Ronald Press Company, New York. 453 pp.

Andrewartha, H. G., and L. C. Birch. 1954. The distribution and abundance of animals. University of Chicago Press, Chicago. 782 pp.

Anthony, H. E. 1928. Field book of North American mammals. G. P. Putnam's Sons, New York. 625 pp.

Arlton, A. V. 1936. An ecological study of the mole. *J. Mammal.* **17:**349–371.

Aschaffenburg, R., M. E. Gregory, S. K. Kon, S. J. Rowland, and S. Y. Thompson. 1962. The composition of the milk of the reindeer. *J. Dairy Res.* **29:**324–328.

Ashworth, U. S., G. D. Ramaiah, and M. C. Keyes. 1966. Species difference in the composition of milk with special reference to the northern fur seal. *J. Dairy Sci.* **49:**1206–1211.

Audubon, J. J., and J. Bachman. 1845–1854. The viviparous quadrupeds of North America. V. G. Audubon, New York. 3 vols.

Audubon, M. R. 1897. Audubon and his journals. Charles Scribner's Sons, New York. (Paperbound in 2 vols., 1960, by Dover Publications, Inc., New York.)

Augee, M. L., and E. H. M. Ealey. 1968. Torpor in the echidna, *Tachyglossus aculeatus. J. Mammal.* **49:**466–453.

Bachop, T., and R. W. Martucci. 1968. Ruminant-like digestion of langur monkey. *Science* **161**(3842):698–699.

Bachrach, M. 1946. Fur: a practical treatise. Prentice-Hall, Inc., Englewood Cliffs, N.J. 672 pp.

Bailey, F. M. 1924. Handbook of birds of the western United States including the Great Plains, Great Basin, Pacific slopes and lower Rio Grande Valley, rev. ed. Houghton Mifflin Company, Boston. 590 pp.

Bailey, L. F., and M. Lemon. 1966. Specific milk proteins associated with resumption of development by the quiescent blastocyst of the lactating red kangaroo. *J. Reprod. Fert.* **11:**473–475.

Bailey, V. 1923. Maximilian's travels in the interior of North America, 1832 to 1834. *Nat. Hist.* **23:**337–343.

———. 1924. Breeding, feeding and other life habits of meadow mice *(Microtus). Agric. Res.* **27:**523–535.

———. 1926. A biological survey of North Dakota. *North Am. Fauna,* No. 49. 226 pp.

———. 1931. Mammals of New Mexico. *North Am. Fauna,* No. 53. 412 pp.

——— and C. C. Sperry. 1929. Life history and habits of grasshopper mice, genus *Onychomys. U.S. Dep. Agr. Tech. Bull.* 145. 19 pp.

Baird, S. F. 1857. Mammals: General report upon the zoology of the several Pacific railroad routes. Reports, exploration and surveys for railroad route from Mississippi River to Pacific Ocean. vol. 8 (pt. 1), Washington. 757 pp.

Baker, B. E., C. R. Harington, and A. L. Symes. 1963. Polar bear milk. I. Gross composition and fat constitution. *Can. J. Zool.* **41**:1035–1039.

Baker, J. R., and R. M. Ranson. 1933. Factors affecting the breeding of the field mouse *(Microtus agrestis). Proc. R. Soc. London.* **113**:486–495.

Baker, R. H. 1956. Mammals of Coahuila, Mexico. *Univ. Kans. Mus. Nat. Hist. Misc. Publ.* **9**:125–335.

Baldwin, E. 1964. An introduction to comparative biochemistry. Cambridge University Press, London. 179 pp.

Baldwin, P. H., C. W. Schwartz, and Elizabeth R. Schwartz. 1952. Life history and economic status of the mongoose in Hawaii. *J. Mammal.* **33**:335–356.

Balinsky, B. I. 1970. An introduction to embryology, 3d ed. W. B. Saunders Company, Philadelphia. 725 pp.

Balph, D. E., and A. W. Stokes. 1963. On the ethology of a population of Uinta ground squirrels. *Am. Midl. Nat.* **69**:106–126.

Balser, D. S. 1964. Management of predator populations with antifertility agents. *J. Wildl. Manage.* **28**:352–358.

Banfield, A. W. F. 1954. The role of ice in the distribution of mammals. *J. Mammal.* **35**:104–106.

———. 1954a. Preliminary investigation of the barren ground caribou. *Can. Wildl. Serv., Wildl. Manage., Bull. Ser. 1, No. 10A,* 79 pp.; *No. 10B,* 112 pp. Ottawa, Ont.

Barret, C. 1942. The echidna's secret. *Wildl. Aust.* **4**:350.

Bartholomew, G. A., Jr. 1959. Mother-young relations and the maturation of pup behavior in the Alaskan fur seal. *Anim. Behav.* **7**:163–172.

———. 1962. Hibernation, estivation, temperature regulation, evaporative water loss, and heart rate of the pigmy possum, *Cercaertus nanus. Physiol. Zool.* **35**:94–197.

——— and T. J. Cade. 1957. Temperature regulation, hibernation, and estivation in the little pocket mouse, *Perognathus longimembris. J. Mammal.* **38**:60–72.

——— and H. H. Caswell. 1951. Locomotion in kangaroo rats and its adaptive significance. *J. Mammal.* **32**:155–168.

———, W. R. Dawson, and R. C. Lasiewski. 1970. Thermoregulation and heterothermy in some of the smaller flying foxes (Megachiroptera) of New Guinea. *Vergl. Physiol.* **7**:196–209.

——— and P. G. Hoel. 1953. Reproductive behavior of the Alaska fur seal, *Callorhinus ursinus. J. Mammal.* **34**:417–435.

——— and J. W. Hudson. 1959. Effects of sodium chloride on weight and drinking in the antelope ground squirrel. *J. Mammal.* **40**:345–360.

———, P. Leitner, and J. E. Nelson. 1964. Body temperature, oxygen consumption and heart rate in three species of flying foxes. *Physiol. Zool.* **37**:179–198.

——— and R. C. MacMillen. 1961. Oxygen consumption, estivation, and hibernation in the kangaroo mouse, *Microdipodops pallidus. Physiol. Zool.* **34**:177–183.

Bassett, C. F., and L. M. Llewellyn. 1948. The molting and fur growth pattern in the silver fox. *Am. Midl. Nat.* **39**:597–601.

Baumber, J., F. E. South, L. Ferren, and M. L. Zatzman. 1971. A possible basis for periodic arousals during hibernation; accumulation of ketone bodies. *Life Sci.* **10,** part II;463–467.

Baumgartner, L. L. 1938. Population studies of the fox squirrel in Ohio. *Trans. 3d North Am. Wildl. Conf.* pp. 685–689.

Beach, F. A., and R. W. Gilmore. 1949. Responses to male dogs of urine from females in heat. *J. Mammal.* **30:**391–392.

Bear, G. D., and R. M. Hansen. No date. Food habits, growth and reproduction of white-tailed jackrabbits in southern Colorado. Colorado State Univ., Agr. Exp. Sta. Tech. Bull. 90, Fort Collins.

Beebe, W. 1924. Galápagos—worlds' end. G. P. Putnam's Sons, New York. 433 pp.

———. 1926. The Arcturus adventure. G. P. Putnam's Sons, New York.

Beecher, W. J. 1958. Chicago Academy of Science. *Science.* **128**(3339):1628–1630.

Beer, J. R., and A. G. Richards. 1956. Hibernation of the big brown bat. *J. Mammal.* **37:**31–41.

Beltrami, J. C. 1828. A pilgrimage in Europe and America. Hunt and Clarke, London. 2 vols.

Bender, M. A., and E. H. Y. Chu. 1963. The chromosomes of primates. pp. 261–310. *IN:* J. Buettner-Janusch, ed. Evolutionary and genetic biology of primates. Academic Press, Inc., New York.

Benedict, F. G.. 1938. Hibernation and marmot physiology. *Carnegie Inst. Washington Publ.* 497.

Benson, S. G. 1947. Comments on migration and hibernation in *Tadarida brasiliensis*. *J. Mammal.* **28:**407–408.

Benton, A. H., and W. E. Werner, Jr. 1974. Field biology and ecology, 3d ed. McGraw-Hill Book Company, New York. 564 pp.

Benzinger, T. H. 1964. The thermal homeostasis of man. *Symp. Soc. Exp. Biol.* **18:** 49–86.

Bewick, T. 1804. General history of quadrupeds. Printed by G. and R. Waite, 64 Maiden Lane, New York. 531 pp.

Bischoff, T. L. W. 1854. Entwicklungsgeschichte des Rehes. Giessen.

Bissonette, T. H. 1935. Relations of hair cycles in ferrets to changes in the anterior hypophysis and to light cycles. *Anat. Rec.* **63:**159–168.

——— and E. E. Bailey. 1944. Experimental modification and control of moults and changes of coat colour in weasels by controlled lighting. *Ann. N.Y. Acad. Sci.* **45:** 221–260.

——— and E. Wilson. 1939. Shortening daylight periods between May 15 and September 12 and pelt cycle of the mink. *Science* **89:**418–419.

Black, C. C. 1963. A review of the North American tertiary Sciuridae. *Bull. Mus. Comp. Zool. Harv. Univ.* **130:**111–248.

Blackwelder, R. E. 1967. Taxonomy. A text and reference book. John Wiley & Sons, Inc., New York. 698 pp.

Blair, W. F. 1940a. A study of prairie deer-mouse populations in southern Michigan. *Am. Midl. Nat.* **24:**273–305.

———. 1940. Home ranges and populations of meadow vole in southern Michigan. *J. Wildl. Manage.* **4:**149–161.

———. 1942. Size of home range and notes on the life history of the woodland deer-mouse and eastern chipmunk in northern Michigan. *J. Mammal.* **23:**27–36.

Blatt, C. N., C. R. Taylor, and M. B. Habal. 1972. Thermal panting in dogs; the lateral nasal gland, a source of water for evaporative cooling. *Science* **177**:804–805.

Blauvelt, H. 1955. Dynamics of the mother-newborn relationship in goats. pp. 221–258. *IN:* B. Schaffner, ed. Group process. Trans. 1st Conf., May Foundation, New York.

Bloedel, P. 1955. Hunting methods of fish-eating bats, particularly *Noctilis leporinus* (Noctilionidae). *J. Mammal.* **36**:390–399.

Bodenheimer, F. S. 1949. Problems of vole populations in the Middle East. Report on the population dynamics of the Levant vole *(Microtus guentheri* D. and A.). Research Council of Israel, Jerusalem. 77 pp.

Borchert, J. R. 1950. The climate of the central North American grassland. *Ann. Assoc. Am. Geog.* **40**:1–39.

Bourlière, F. 1964. The natural history of mammals. Alfred A. Knopf, Inc., New York. 387 pp.

Bovet, J. 1968. Trails of deer mice *(Peromyscus maniculatus)* traveling on the snow while homing. *J. Mammal.* **49**:713–725.

Bradshaw, G. V. 1962. Reproductive cycle of the California leaf-nosed bat, *Macrotis californicus. Science* **136**:645–646.

Bradt, G. W. 1938. A study of beaver colonies in Michigan. *J. Mammal.* **19**:139–162.

Braun, E. L. 1950. Deciduous forests of eastern North America. The Blakiston Company, Philadelphia. 596 pp.

Brazda, A. R. 1953. Elk migration patterns and some of the factors affecting movements in the Gallatin River drainage, Montana. *J. Wildl. Manage.* **17**:9–23.

Brewster, W. 1924. The birds of the Lake Umbagog region of Maine. *Bull. Mus. Comp. Zool., Harv. Univ.,* part I, **66**:1–209.

Britton, S. W., and W. E. Atkinson. 1938. Poikilothermism in the sloth. *J. Mammal.* **19**:94–99.

Bromley, P. T. 1969. Territoriality in pronghorn bucks on the National Bison Range, Moiese, Montana. *J. Mammal.* **50**:81–89.

Browman, L. G., and H. S. Sears. 1955. Mule deer milk. *J. Mammal.* **36**:473–474.

Brown, L. E. 1962. Home range in small mammal communities. *IN:* B. Glass, ed. *Surv. Biol. Prog.* **4**:131–179.

Brown, R. T., and J. T. Curtis. 1952. The upland conifer-hardwood forests of northern Wisconsin. *Ecol. Monogr.* **22**:217–234.

Brown, W. H., J. W. Stull, and L. K. Sowls. 1963. Chemical composition of the milk fat of the collared peccary. *J. Mammal.* **44**:112–113.

Bruce, H. M. 1960. A block to pregnancy in the mouse caused by proximity to strange males. *J. Reprod. Fert.* **1**:96–102.

Buckner, C. H. 1957. Population studies on small mammals in southeastern Manitoba. *J. Mammal.* **38**:87–97.

Buechner, H. K. 1961. Territorial behavior in the Uganda kob. *Science* **133**:698–699.

———, A. J. Morrison, and W. Leuthold. 1966. Reproduction in Uganda kob with special reference to behavior. *Symp. Zool. Soc. London.* **15**:69–88.

——— and R. Schloeth. 1965. Ceremonial mating behavior of Uganda kob (*Adenota kob thomasi* Neumann). *Z. Tierpsychol.* **22**:209–225.

Burbank, R. C., and J. Z. Young, 1934. Temperature changes and winter sleep of bats. *J. Physiol.* **82**:459–467.

Burlington, R. F., and J. E. Wiebers. 1965. Effect of hypoxia on cerebral anaerobic glycolysis in the hibernator and infant or adult homeotherm. *Comp. Biochem. Physiol.* **14**:201–203.

Burrell, H. 1927. The platypus. Angus and Robertson, Ltd., Sydney, Australia. 227 pp.

Burt, W. H. 1936. A study of the baculum in the genera *Perognathus* and *Dipodomys*. *J. Mammal.* **17**:145–156.

———. 1938. Faunal relationships and geographic distribution of mammals of Sonora, Mexico. *Misc. Publ. Mus. Zool. Univ. Mich.* **39**:1–77.

———. 1940. Territorial behavior and populations of some small animals in southern Michigan. *Misc. Publ. Mus. Zool. Univ. Mich.* **45**:1–58.

———. 1946. The mammals of Michigan. The University of Michigan Press, Ann Arbor. 288 pp.

———. 1958. The history and affinities of the recent land mammals of North America. pp. 131–154. *IN:* C. L. Hubbs, ed. Zoogeography. Am. Assoc. Adv. Sci. Publ. 51.

———. 1960. Bacula of North American mammals. *Misc. Publ. Mus. Zool. Univ. Mich.* **113**. 75 pp.

——— and R. P. Grossenheider. 1964. A field guide to the mammals, 2d ed. Houghton Mifflin Company, Boston. 284 pp.

Butcher, E. O. 1951. Development of the pilary system and the replacement of hair in mammals. *Ann. N.Y. Acad. Sci.* **53**:508–516.

Butsch, R. S. 1954. The life history and ecology of the red-backed vole, *Clethrionomys gapperi* Vigors in Minnesota, Ph.D. dissertation, University of Michigan, Ann Arbor. 101 pp.

Cabalka, J. L. 1952. Resting habits of the raccoon, *Procyon lotor hirtus* N. and G. in central Iowa. Unpublished M.S. thesis, Iowa State College, Iowa City.

Cabrera, A. 1957–1961. Catálogo de los mamíferos de America del Sur. Museo Argentine de Ciencias Naturales, Ciencias Zoologicas, 2 vols. 732 pp.

———. 1958. (1957). Catálogo de los Mamíferos de America del Sur. *Rev. Mus. Argentine Cien. Nat., Cien. Zool.* **4**(1):1–307.

——— and J. Yepes. 1940. Mamíferos sud-americanos. Vida, costumbres y descripción. Comp. Argentina de Editors, Buenos Aires. 370 pp.

Cade, T. J. 1963. Observation on torpidity in captive chipmunks of the genus *Eutamias*. *Ecology.* **44**:255–261.

——— 1964. The evolution of torpidity in rodents. pp. 27–30. *IN:* Paavo Soumalainen, ed. Mammalian hibernation. II. Proc. 2d In. Symp. Nat. Mammalian Hibernation, Helsinki, August 1962.

Cahalane, V. H. 1947. A deer-coyote episode. *J. Mammal.* **28**:36–39.

Cahn, A. R. 1937. Mammals of the Quetico Provincial Park of Ontario. *J. Mammal.* **18**:19–29.

Camin, J. H., and R. R. Sokal. 1965. A method for deducing branching sequences in phylogeny. *Evolution.* **19**:311–326.

Camp, C. L., and N. S. Smith. 1942. Phylogeny and functions of the distal ligaments of the horse. *Calif. Univ. Mem.,* University of California Press, Berkeley. **13**:69–124.

Carleton, M. D. 1973. A survey of gross stomach morphology in New World Cricetinae (Rodentia, Muroidea) with comments on functional interpretations, *Misc. Publ. Mus. Zoo. Univ. Mich.* **146.** 46 pp.

Carlquist, S. 1965. Island life. The Natural History Press, Garden City, N.Y. 451 pp.

Carpenter, C. R. 1934. A field study of the behavior and social relations of howling monkeys. *Comp. Psychol. Monogr.* **10:**1–168.

———. 1940. A field study in Siam of the behavior and social relations of the gibbon *Hylobates lar. Comp. Psychol. Monogr.* **16:**1–212.

———. 1942. Sexual behavior of free ranging rhesus monkeys, *Macaca mulatta. J. Comp. Psychol.* **33:**113–142.

———. 1958. Territoriality: a review of concepts and problems. pp. 224–250. *IN:* Anne Roe and G. G. Simpson, eds. Behavior and evolution. Yale University Press, New Haven. 557 pp.

———. 1962. Field studies of a primate population. pp. 286–294. *IN:* E. L. Bliss, ed. Roots of behavior. Harper & Row, Publishers, Incorporated, New York.

———. 1965. The howlers of Barro Colorado Island. pp. 250–291. *IN:* Irven De Vore, ed. Primate Behavior. Holt, Rinehart and Winston, Inc., New York.

Carpenter, J. R. 1940. The grassland biome. *Ecol. Monogr.* **10:**617–684.

Carpenter, J. W., and R. R. Martin. 1969. Capturing prairie dogs for transplanting. *J. Wildl. Manage.* **33:**1024.

Carpenter, W. B. 1845. Zoology, vol. I. Wm. S. Orr and Co., London. 566 pp.

Carr, A. 1967. The land and wildlife of Africa. Time, Inc., New York. 200 pp.

Carrick, R., S. E. Csordas, A. E. Ingham, and K. Kieth. 1962. Studies on the southern elephant seal *Mirounga leonina* (1.) III. The annual cycle in relation to age and sex. Commonwealth Science and Industrial Research Organization (Australia). *Wildl. Res.* **7:**119–160.

Carrington, R. 1963. The mammals. Life Nature Library, Time, Inc., New York. 192 pp.

Carrol, R. L. 1969. Problems of the origin of reptiles. *Biol. Rev.* **44:**393–432.

Carter, H. B. 1965. Variation in the hair follicle population of the mammalian skin. pp. 25–33. *IN:* S. G. Lyne and B. F. Short, eds. Biology of the skin and hair growth. American Elsevier Publishing Company, Inc., New York. 806 pp.

——— and D. F. Dowling. 1954. The hair follicle and apocrine gland population of cattle skin. *Aust. J. Agri. Res.* **5:**745–754.

Catesby, M. 1731–1748. The natural history of Carolina, Florida and the Bahama Islands. London. 2 vols.

Catlin, G. 1965. Letters and notes on the manners, customs, and condition of the North American Indians. Ross and Haines, Inc., Minneapolis. 2 vols. (Originally published in 1841 by Chatto and Winders, London.)

Caughley, G. 1970. Eruption of ungulate populations with emphasis on Himalayan thar in New Zealand. *Ecology.* **51:**53–72.

Chaine, J. 1925. L'os penien, étude descriptive et comparative. *Actes Soc. Linn., Bordeaux.* **78.** 195 pp.

Chapman, R. N. 1928. The quantitative analysis of environmental factors. *Ecology.* **9:** 111–122.

Chasen, F. N. 1940. A handlist of Malaysian mammals. *Bull. Raffles Mus.* **15:**1–209.

Cheatum, E. L. 1947. Whitetail fertility. *N.Y. State Conservationist.* **1:**18–32.

Chen, E. C. H., D. A. Blood, and E. H. Baker. 1965. Rocky Mountain bighorn sheep *(Ovis canadensis canadensis)* milk. *Can. J. Zool.* **43:**885–888.

Chittenden, H. 1954. The American fur trade. Stanford University Press, Stanford, Calif. 2 vols. 1029 pp.

Chitty, D. 1958. Self-regulation of numbers through changes in viability. *Cold Spring Harbor Symp. Quant. Biol.* **22:**277–280.

———. 1960. Population processes in the vole and their relation to general theory. *Can. J. Zool.,* **38:**39–113.

———. 1967. The natural selection of self-regulating behavior in animal populations. *Proc. Ecol. Soc. Aust.* **2:**51–78.

Choate, J. R., and H. H. Genoways. 1975. Collections of recent mammals of North America. *J. Mammal.* **56:**452–502.

Christensen, E. 1947. Migration or hibernation of *Tadarida mexicana. J. Mammal.* **28:**59–60.

Christian, J. J. 1950. The adreno-pituitary system and population cycles in mammals. *J. Mammal.* **31:**247–260.

———. 1953. The relation of adrenal weight to body weight in mammals. *Science* **117:** 78–80.

———. 1956. The natural history of a summer aggregation of the big brown bat, *Eptesicus fuscus fuscus. Am. Midl. Nat.* **55:**66–95.

——— and D. E. Davis. 1956. The relationship between adrenal weights and populations status in Norway rats. *J. Mammal.* **37:**475–486.

———. 1964. Endocrines, behavior and population. *Science* **146:**1550–1560.

Clark, F. H. 1938. Age of sexual maturity in mice of the genus *Peromyscus. J. Mammal.* **19:**230–234.

Clark, L. 1936. The problem of the claw in primates. *Proc. Zool. Soc., London.* **1936:** 1–24.

Clarke, C. H. D. 1940. A biological investigation of the Thelon Game Sanctuary. *Nat. Mus. Can. Bull.* 96 *Biol. Ser.* 25. 135 pp.

Clarke, J. R. 1955. Influences of members on reproduction and survival in two experimental vole populations. *Proc. R. Soc. London, Ser. (B)* **144:**68–85.

Clemens, W. 1970. Mesozoic mammalian evolution. *Ann. Rev. Ecol. Syst.,* **1:**357–390.

Clements, F. E. 1916. Plant succession. *Carnegie Inst. Washington Publ.* **242:**1–512.

Cloudsley-Thompson, J. L., ed. 1954. Biology of deserts. The proceedings of a symposium on the biology of hot and cold deserts organized by the Institute of Biology. Tavistock House, London. 224 pp.

Cockrill, W. R. 1967. The water buffalo. *Sci. Am.* **217:**118–125.

Cockrum, E. L. 1969. Migration in the guano bat, *Tadarida brasiliensis. IN:* K. Jones, ed. Contributions in mammalogy. *Univ. Kans. Mus. Nat. Hist. Misc. Publ.* **51:** 303–336.

Cohen, J. 1965. The dermal papilla. pp. 183–199. *IN:* S. G. Lyne and B. F. Short, eds. Biology of the skin and hair growth. American Elsevier Publishing Company, Inc., New York. 806 pp.

Colbert, E. H. 1961. Evolution of the vertebrates: a history of backboned animals through time. Sci. ed., John Wiley & Sons, Inc., New York, paperback. 479 pp.

Cole, H. I. 1924. Taxonomic value of hair in Chiroptera. *Philipp. J. Sci.* **24**:117–121.

Cole, L. C. 1954. Some features of random population cycles. *J. Wildl. Manage.* **18**: 2–24.

Collias, N. E. 1956. The analysis of socialization in sheep and goats. *Ecology.* **37**: 228–239.

———— and C. Southwick. 1952. A field study of population density and social organization in howling monkeys. *Proc. Am. Philos. Soc.* **96**:143–156.

Collins, H. H. 1918. Studies of normal moult and of artificially induced regeneration of pelage in *Peromyscus. J. Exp. Zool.* **27**:73–99.

Connor, P. F. 1959. The bog lemming, *Synaptomys cooperi,* in southern New Jersey. *Publ. Mus. Mich. State Univ., Biol. Ser.,* **1**:Ser.

Constantine, D. G. 1967. Activity patterns of the Mexican free-tailed bat. *Univ. N.M. Publ. Biol.* **7**:1–79.

————. 1968. An automatic bat-collecting device. *J. Wildl. Manage.* **22**:7–22.

Costello, D. F. 1969. The prairie world. Thomas Y. Crowell Company, New York. 242 pp.

Coues, E. 1875. Account of the various publications relating to the travels of Lewis and Clark, with a commentary on the zoological results of the expedition. U.S. Geol. and Geogr. Surv. of the Territories, vol. I. Washington. (Paper dated 1876 is included in this volume.)

————. 1877. Fur-bearing animals. Misc. Publ. 8, U.S. Geol. Surv. of the Territories, Washington. 348 pp.

————. 1893. History of the expedition under the command of Lewis and Clark. Francis P. Harper, New York. 4 vols.

————. 1895. The expeditions of Zebulon Montgomery Pike, 1805–1807. Francis P. Harper, New York.

————. 1896. Three subcutaneous glandular areas of *Blarina brevicauda. Science* **3**: 799–780.

Coulombe, H. N., S. H. Ridgway, and W. E. Evans. 1965. Respiratory water exchange in two species of porpoise. *Science* **149**:86–88.

Courrier, R. 1927. Étude sur le determinisme des caractères sexuels secondaires chez quelques mammifères a l'activité testiculaire périodique. *Arch. Biol.* **37**:173–334.

Couturier, M. A. J. 1954. L'ours brun. Dr. Marcel Couturier, Grenoble, France. 904 pp.

Cowie, A. T. 1972. Lactation and its hormonal control. pp. 106–143. *IN:* R. V. Short, ed. Hormones in reproduction, Book 3. Cambridge University Press, New York.

Craig, W. 1908. North Dakota life: plant, animal and human. *Bull. Am. Geol. Soc.* **40**: 321–332, 401–415.

Craighead, F. C., J. J. Craighead, and R. S. Davies. 1963. Radiotracking of grizzly bears. pp. 133–148. *IN:* L. F. Sclater, ed. Biotelemetry. Pergamon Press and The Macmillan Company, New York. 372 pp.

Craighead, J. J., G. Atwell, and B. O'Gara. 1972. Elk migrations in and near Yellowstone National Park. *Wildl. Monogr.* **29.** 48 pp.

———— and F. C. Craighead. 1946. Hawks, owls, and wildlife. Stackpole Co., Harrisburg, Pa., and Wildlife Management Institute, Washington. 443 pp.

————, M. G. Hornocker, and F. C. Craighead, Jr. 1969. Reproductive biology of young female grizzly bears. *J. Reprod. Fert.,* suppl. **6**:447–475.

Crampton, E. W., and L. E. Harris. 1969. Applied animal nutrition. W. H. Freeman and Company, San Francisco. 753 pp.

Crompton, A. W., and F. A. Jenkins, Jr. 1968. Molar occlusion in Late Triassic Mammals. *Biol. Rev.,* **43:**427–458.

Crowcroft, P. 1957. The life of the shrew. Reinhart, London. 166 pp.

────── and F. P. Rowe. 1957. The growth of confined colonies of the wild house mouse (*Mus musculus* l.) *Proc. Zool. Soc. London.* **129:**359–370.

Curasson, G. 1947. Le Chameau et ses maladies. Vigot presse, 23 Rue de l'école de médecine, Paris VI.

Currier, A., W. D. Kitts, and I. M. Cowan. 1960. Cellulose digestion in the beaver *(Castor canadensis). Can. J. Zool.* **38:**1109–1116.

Dagg, A. I., and A. Taub. 1970. Flehmen. *Mammalia.* **34:**686–695.

Darling, F. F. 1937. A herd of red deer. Originally published by Oxford University Press in 1937. The 1964 edition is a paperback edition of the Natural History Library, Anchor Books, Doubleday & Company, Inc., Garden City, N.Y.

──────. 1960. An ecological reconnaissance of the Mara plains in Kenya Colony. *Wildl. Monogr.* **5:**1–41.

Darlington, P. J., Jr. 1957. Zoogeography: the geographical distribution of animals. John Wiley & Sons, Inc., New York. 675 pp.

Darwin, C. R. 1859. On the origin of species. Murray, London. Reprinted 1950, Watts and Co., London.

──────. 1873. The expressions of emotions in man and animals. D. Appleton & Company, New York. 374 pp.

──────. 1906. Voyage of the Beagle. J. M. Dent and Sons, London. (There are many editions available of this book.)

Dasmann, R. F. 1964. Wildlife biology. John Wiley & Sons, Inc., New York, 231 pp.

Davis, D. D. 1949. The shoulder architecture of bears and other carnivores. *Fieldiana Zool.* **31:**285–305.

Davis, D. E. 1953. Rat populations. *Q. Rev. Biol.* **28:**373–401.

──────. 1957. The existence of cycles. *Ecology.* **38:**163–164.

Davis, R. B., C. F. Herreid, Jr., and H. L. Short. 1962. Mexican free-tailed bats in Texas. *Ecol. Monogr.* **32:**311–346.

Davis, W. B. 1940. Another heavy concentration of *Cryptotis* in Texas. *J. Mammal.* **21:**213–224.

──────, R. R. Ramsey, and J. M. Arendale. 1938. Distribution of pocket gophers *(Geomys breviceps)* in relation to soils. *J. Mammal.* **19:**412–418.

Davis, W. H. and H. B. Hitchcock. 1965. Biology and migration of the bat, *Myotis lucifugus* in New England. *J. Mammal.* **46:**296–313.

Dawbin, W. H. 1966. The seasonal migratory cycle of humpback whales. pp. 145–170. *IN:* K. S. Norris, ed. Whales, dolphins, and porpoises. University of California Press, Berkeley and Los Angeles.

Dawe, A. R., and W. A. Spurrier. 1969. Hibernation induced in ground squirrels by blood transfusion. *Science.* **163:**298–299.

Dawkins, M. J. R., and D. Hull. 1965. The production of heat by fat. *Sci. Am.* **213** (2):62–67.

Deanesly, R. 1966. Observations on reproduction in the mole, *Talpa europaea.* pp.

387–402. *IN:* I. W. Rowlands, ed. Comparative biology of reproduction in mammals. *Symp. Zool. Soc. London.* **15.** 559 pp.

Dearden, L. C. 1958. The baculum in *Lagarus* and related microtines. *J. Mammal.* **39:**541–553.

———. 1969. Stomach and pyloric sphincter histology in certain microtine rodents. *J. Mammal.* **50:**60–68.

DeKay, J. E. 1842. Natural history of New York. Part I. Zoology. D. Appleton & Company, and Wiley & Putnam; Gould Kendall and Lincoln, Boston; Thurlow Weed, Printer to the State, Albany. Part I includes Mammalia. 145 pp.

DeLury, R. E., and J. L. O'Conner. 1936. Regional types of response of wildlife to the sunspot cycle. *Proc.* (1st) *North A. Wildl. Conf.* **1936:**490–491. [This series is now the *Transactions of the* (X) *North American Wildlife Conference.*]

De Meijere, J. C. 1894. Über die Haare der Säugetiere besonders über ihre Anordnung. *Gegenhaur's Morphol. Jahrb.* **21:**312–424.

Denniston, R. H., II. 1956. Ecology, behavior and population dynamics of the Wyoming or Rocky Mountain moose, *Alces alces shirasi. Zoologica (N.Y.) Sci. Contrib.,* **41:**105–118.

DeVore, I., ed. 1965. Primate behavior. Field studies on monkeys and apes. Holt, Rinehart and Winston, Inc., New York. 654 pp.

——— and K. R. H. Hall. 1965. Baboon ecology. pp. 20–52. *IN:* I. DeVore, ed. Primate behavior. Holt, Rinehart and Winston, Inc., New York. 654 pp.

Días de Avila-Pires, F. 1965. The type specimens of Brazilian mammals collected by Prince Maximilian Zu Wied. *Am. Mus. Novit.* **2209:**1–21.

Dice, L. R. 1943. The biotic provinces of North America. University of Michigan Press, Ann Arbor. 78 pp.

———. 1947. Effectiveness of selection by owls of deer-mice *(Peromyscus maniculatus)* which contrast in color with their background. *Contrib. Lab. Vertebr. Biol. Univ. Mich.* **34:**1–20.

Dicker, S. E. 1970. Mechanisms of urine concentration and dilution in mammals. The Williams & Wilkins Company, Baltimore.

Dilger, W. C. 1962. Methods and objectives of ethology. pp. 83–92. *IN:* The living bird, first annual of the Cornell Laboratories of Ornithology. Cornell University Press, Ithaca, N.Y.

Dobie, J. R. 1962. The voice of the coyote. University of Nebraska Press, Lincoln. (A second printing of the Bison Book edition reproduced from the sixth printing, Little, Brown and Company, originally published in 1947.) 386 pp.

Dobzhansky, T. 1951. Genetics and the origin of species, 3d ed. Columbia University Press, New York. (1st ed., 1937.)

Dominic, G. J. 1964. Source of the male odor causing pregnancy-block in mice. *J. Reprod. Fert.* **8:**266–267.

Doolittle, R. F., and B. Blomback. 1964. Amino acid sequence investigations of fibrinopeptides from various mammals: evolutionary implications. *Nature,* **202:** 147–152.

Dorst, J., and P. Dandelot. 1970. A field guide to the larger mammals of Africa. Houghton Mifflin Company, Boston. 287 pp.

Doutt, J. K., A. B. Howell, and W. B. Davis. 1945. The mammal collections of North America. *J. Mammal.* **26:**231–272.

Downhower, J. R., and E. R. Hall. 1966. The pocket gopher in Kansas. *Univ. Kans., Mus. Nat. Hist., Misc. Publ.* **44:**1–32.

Dry, F. W. 1926. The coat of the mouse. *J. Genetics.* **16:**218–240.

Dryden, G. I., and C. H. Conaway. 1967. The origin and hormonal control of scent production in *Suncus. J. Mammal.* **48:**420–428.

Dunning, D. D., and K. D. Roeder. 1965. Moth sounds and the insect-catching behavior of bats. *Science* **147:**173–174.

DuToit, A. L. 1937. Our wandering continents. Oliver & Boyd Ltd., Edinburgh. 366 pp.

Dzuik, H. E., G. A. Fashingbauer and J. M. Idstrom. 1963. Ruminoreticular pressure patterns in fistulated white-tailed deer. *Am. J. Vet. Res.,* **24:**772–783.

Eadie, W. R. 1938. The dermal glands of shrews. *J. Mammal.* **19:**171–174.

Ealey, E. H. M. 1963. The ecological significance of delayed implantation in a population of the hill kangaroo *(Macropus robustus).* pp. 33–48. *IN:* A. C. Enders, ed., Delayed implantation. U. of Chicago Press. 318 pp.

Eaton, T. H., Jr. 1944. Modifications of the shoulder girdle related to reach and stride in mammals. *J. Morphol.* **75:**165–171.

Ecke, D. H., and A. R. Kinney. 1956. Aging meadow mice *Microtus californicus* by observation of molt progression. *J. Mammal.* **37:**249–254.

Edwards, R. L. 1946. Some notes on the life history of the Mexican ground squirrel in Texas. *J. Mammal.* **27:**105–115.

Eibl-Eibesfeldt, I. 1961. Galápagos. Doubleday & Company, Inc., New York. 102 pp.

———. 1970. Ethology, the biology of behavior. Translated by E. Klinghammer. Holt, Rinehart, and Winston, Inc., New York. 530 pp.

Eisenberg, J. F. 1963. The behavior of heteromyid rodents. *Univ. Calif. Berkeley Publ. Zool.* **69:**1–100.

———. 1964. Studies on the behavior of *Sorex vagrans. Am. Midl. Nat.* **72:**417–425.

Eisenberg, J. R., and D. G. Kleiman. 1972. Olfactory communication in mammals. pp. 1–32. *IN:* R. F. Johnston, P. W. Frank, and C. D. Michener, eds. *Annu. Rev. Ecol. Syst.* **3.**

Eisner, T. 1962. Survival by acid defense. *Nat. Hist.* **71:**10–19.

Elder, W. H. 1951. The baculum as an age criterion in mink. *J. Mammal.* **32:**43–50.

Ellenberger, W., and H. Baum. 1943. Handbuch der vergleichender Anatomie der Haustiere, 18th ed. Springer-Verlag OHG, Berlin.

Ellerman, J. R. 1940. The families and genera of living rodents, vol. I. British Museum of Natural History, London. 689 pp.

———. 1941. The families and genera of living rodents, vol. II. British museum of Natural History, London. 690 pp.

——— and T. C. S. Morrison-Scott. 1951. Checklist of Palaearctic and Indian mammals, 1758 to 1946. British Museum of Natural History. 810 pp.

———, ——— and R. W. Hayman. 1953. Southern African mammals. British Museum of Natural History, London. 363 pp.

Ellis, R. J. 1964. Tracing raccoons by radio. *J. Wildl. Manage.* **28:**363–368.

Elloff, G. 1958. The function and structural degeneration of the eye of South African rodent moles, *Cryptomys bigalke* and *Bathyergus maritimus. S. Afr. J. Sci.* **54:** 293–301.

Elsner, R., D. L. Franklin, R. L. Van Citters, and D. W. Kenney. 1966. Cardiovascular defense against asphyxia. *Science* **153**:941–949.

Elton, C. 1927. Animal ecology. The Macmillan Company, New York. 207 pp.

———. 1942. Voles, mice and lemmings. Clarendon Press, Oxford. 496 pp.

Elton, C. S., and M. Nicholson. 1942. The ten-year cycle in numbers of the lynx in Canada. *J. Anim. Ecol.* **11**:215–244.

Epple, G. 1970. Quantitative studies on scent marking in the marmoset *(Callithrix jacchus)*. *Folia Primatol.* **13**:48–62.

Erickson, A. W. 1964. An analysis of black bear kill statistics for Michigan. *Mich. State Univ. Agric. Exp. Stn. Res. Bull.* 4, part III, 68–101.

Errington, P. L. 1946. Predation and vertebrate populations. *Q. Rev. Biol.* **21**:144–177, 221–245.

———. 1958. Of population cycles and unknowns. *Cold Spring Harbor Symp. Quant. Biol.* **22**:287–300.

———. 1963. Muskrat populations. The Iowa State University Press, Ames. 665 pp.

Estes, R. D. 1967. The comparative behavior of Grant's and Thompson's gazelles. *J. Mammal.* **48**:189–209.

Evans, C. S., and R. W. Goy. 1968. Social behavior and reproductive cycles in captive ring-tailed lemurs *(Lemur catta)*. *J. Zool., London.* **156**:181–197.

Evans, D. E. 1959. Milk composition of mammals whose milk is not normally used for human consumption. *Dairy Sci. Abstr.* (rev. article no. 80). **21**:277–288.

Evans, F. G. 1942. The osteology and relationship of the elephant shrew (Macroscelididae). *Bull. Am. Mus. Nat. Hist.* **80**:85–125.

Evans, W. E., and P. F. Maderson. 1973. Mechanisms of sound production in delphinid cetaceans; a review and some anatomical considerations. *Am. Zool.* **13**:1205–1213.

——— and B. A. Powell. 1967. Proceedings of a symposium on sonic models of animal sonar systems. Frascati, Italy, 1966. pp. 363–398. Las. d' acoustique animal, Jouy-en-Josas, France.

Ewer, R. F. 1968. Ethology of mammals. Plenum Press, New York. 418 pp.

Farb, P. 1963. Face of North America. Harper & Row, Publishers, Incorporated, New York. 316 pp.

Fautin, R. W. 1946. Biotic communities of the northern desert shrub biome in western Utah. *Ecol. Monogr.* **16**:251–310.

Feldman, J. D. 1961. Fine structure of the cow's udder during gestation and lactation. *Lab. Invest.* **10**:239–255. (*J. Int. Acad. Pathol.,* The Williams & Wilkins Company, Baltimore.)

Findley, J. S. 1972. Phenetic relationships among bats of the genus *Myotis*. *Syst. Zool.,* **21**:31–52.

Finley, R. B., Jr. 1958. The wood rats of Colorado: distribution and ecology. *Univ. Kans. Mus. Nat. Hist., Publ.* **10**:213–552.

Fisher, K. C. 1964. On the mechanism of periodic arousal in the hibernating ground squirrel. *Ann. Acad. Sci. Fenn. Ser. A.* IV. **71**:141–156.

———, A. R. Dawe, C. P. Lyman, E. Schonbaum, and F. E. South, Jr., eds. 1967. Mammalian hibernation. III. Proc. 3d Int. Symp. Nat. Mammalian Hibernation, Univ. Toronto. American Elsevier Publishing Company, Inc., New York.

Fisler, G. F. 1961. Behavior of salt-marsh *Microtus* during winter high tide. *J. Mammal.* **42**:37–43.

———. 1962. Ingestion of sea water by *Peromyscus maniculatus. J. Mammal.* **43**: 416–417.

———. 1963. Effects of salt water in food and water consumption and weight of harvest mice. *Ecology* **44**:604–608.

———. 1965. Adaptations and speciation in harvest mice of the marshes of San Francisco Bay. *Univ. of Cal. Publ. in Zool.* **77**:1–108.

Fitzpatrick, T. G., P. Brunet, and A. Kukita. 1958. The nature of hair pigment. pp. 255–298. *IN:* William Montagna and Richard Ellis, eds. The biology of hair growth. Academic Press, Inc., New York. 520 pp.

Fitzwater, W. D., and W. J. Frank. 1944. Leaf nests of gray squirrels in Connecticut. *J. Mammal.* **25**:160–170.

Fleming, T. H. 1971. *Artibeus jamaicensis:* delayed embryonic development in a neotropical bat. *Science* **171**:402–404.

Florkin, M. 1966. A molecular approach to phylogeny. American Elsevier Publishing Company, New York. 176 pp.

Flyger, V. F. 1960. Movements and home range of the gray squirrel, *Sciurus carolinensis,* in two Maryland woodlots. *Ecology* **41**:365–369.

Folk, J. E., Jr. 1967. Physiological observations of subarctic bears under winter den conditions. pp. 75–85. *IN:* K. C. Fisher, A. R. Dawe, C. P. Lyman, E. Schonbaum, and F. E. South, Jr., eds. Mammalian Hibernation. III. Proc. 3d Int. Symp. Nat. Mammalian Hibernation, Univ. Toronto. American Elsevier Publishing Company, Inc., New York. 535 pp.

Folk, G. E., Jr., and R. S. Hedge. 1964. Comparative physiology of heart rate of unrestrained mammals. *Am. Zool.* **4**:297.

Forman, G. L. 1971. Histochemical differences in gastric mucosa of bats. *J. Mammal.* **52**:191–193.

Formozov, A. N. 1946. Snow cover as a integral factor of the environment and its importance in the ecology of mammals and birds. Materials for Fauna and Flora of the U.S.S.R., New Series, Zoology, 5(XX), **1946**:1–52. Moscow Society of Naturalists. English edition published by Boreal Institute, University of Alberta, Edmonton, Can. Occ. Paper No. 1. Translated from Russian by William Prychodko, William O. Pruitt, and William S. Fuller.

Fowle, C. D., and R. Y. Edwards. 1955. An unusual abundance of short-tailed shrews, *Blarina brevicauda. J. Mammal.* **36**:36–41.

Frank, F. 1957. The causality of microtine cycles in Germany. *J. Wildl. Manage.* **21**: 103–116.

Fraser, F. C., and P. E. Purves, 1954. Hearing in cetaceans. *Bull. Br. Mus. (Nat. Hist.).* **2**:103–116.

———. 1960. Hearing in cetaceans. *Bull. Br. Mus. (Nat. His.).* **7**:1–140.

———. 1960a. Anatomy and function of the cetacean ear. *Proc. R. Soc., London.* **152**: 62–77.

Frick, G. F., and R. F. Stearns. 1961. Mark Catesby, the colonial Audubon. University of Illinois Press, Urbana. 137 pp.

Gadow, H. 1913. The wanderings of animals. Cambridge University Press, Cambridge, England. 150 pp.

Gause, G. F. 1932. Ecology of populations. *Q. Rev. Biol.* **7**:27–46.

———. 1934. The struggle for existence. The Williams & Wilkins Company, Baltimore.

Gavin, A. 1945. Notes on mammals observed in the Perry River District, Queen Maude Sea. *J. Mammal.* **26**:226–230.

Geist, V. 1971. Mountain sheep: A study in evolution and behavior. The University of Chicago Press, Chicago. 383 pp.

Genest-Villard, H. 1968. L'estomac de *Lophuromys sikapusi* (Temminck). *Mammalia.* **32**:639–656.

Genoways, H. H. 1973. Systematics and evolutionary relationships of spiny pocket mice, genus *Liomys. Spec. Publ. Mus. Tex. Tech. Univ.* **5**:1–368.

George, W. 1962. Animal geography. William Heinemann, Ltd., London. 142 pp.

Getz, L. L. 1961. Home ranges, territoriality, and movement of the meadow vole. *J. Mammal.* **42**:24–36.

Gilbert, T. H. 1892. Das os priape der Säugetiere. *Morphol. Jahrb.* **18**:805–831.

Gilmore, R. M., R. L. Brownell, Jr., J. G. Mills, and A. Harrison. 1967. Gray whales near Yavaros, southern Sonora, Golfo de California, Mexico. *Trans. San Diego Soc. Nat. Hist.* **14**:198–203.

Glass, B. P. 1970. Feeding mechanisms of bats. pp. 85–92. *IN:* B. H. Slaughter, ed. About bats. A chiropteran biology symposium, Southern Methodist University Press, Dallas, with D. W. Walton. 339 pp.

Gleason, E. A. 1926. The individualistic concept of plant association. *Bull. Torrey Bot. Club.* **53**:7–26.

Gleason, K. K., and J. H. Reynierse. 1969. The behavioral significance of pheromones in vertebrates. *Psychol. Bull.* **7**:58–73.

Godfrey, G. K. 1954. Tracing field voles, *Microtus agrestis,* with a Geiger-Müller counter. *Ecology.* **35**:5–10.

Godfrey, G., and P. Crowcroft. 1960. The life of the mole. Museum Press, London. 152 pp.

Godfrey, J. 1958. The origin of sexual isolation between bank voles. *Proc. R. Phys. Soc., Edinburgh.* **27**:47–55.

Godman, J. D. 1826. American Natural History or Mastology. H. C. Carey and I. Lee, Philadelphia, Pa. 3 vols.

Goetz, R. H., and O. Budtz-Olsen. 1955. Scientific safari—the circulation of the giraffe. *S. Afr. Med. J.* **29**:773–776.

——— and E. N. Keen. 1957. Some aspects of cardiovascular system in the giraffe. *Angiology.* **8**:542–564.

———, J. V. Warren, O. H. Gauer, J. L. Patterson, Jr., J. T. Doyle, E. N. Keen, and M. McGregor. 1960. Circulation of the giraffe. *Circ. Res.* **8**:1049–1058.

Goetzmann, W. H. 1959. Army exploration in the American West 1803–1863. Yale University Press, New Haven, Conn. 508 pp.

Golley, F. B. 1960. Anatomy of the digestive tract of *Microtus. J. Mammal.* **41**:89–99.

———. 1960a. Energy dynamics of a food chain of an old-field community. *Ecol. Monogr.* **30**:187–206.

————. 1961. Interaction of natality, mortality and movement during one annual cycle in a *Microtus* population. *Am. Midl. Nat.* **66:**152–159.

Goodpaster, W. W., and D. F. Hoffmeister. 1954. Life history of the golden mouse, *Peromyscus nuttalli* in Kentucky. *J. Mammal.* **35:**16–28.

Goodwin, G. G. 1946. Mammals of Costa Rica. *Bull. Am. Mus. Nat. Hist.* **68:**1–60.

Gottschang, J. L. 1956. Juvenile molt in *Peromyscus leucopus noveboracensis. J. Mammal.* **37:**516–524.

Gould, E. 1965. Evidence for echolocation in the tenrecidae of Madagascar. *Proc. Am. Philos. Soc.* **109:**352–360.

————, N. C. Negus, and A. Novick. 1964. Evidence for echolocation in shrews. *J. Exp. Zool.* **156:**19–38.

———— and J. R. Eisenberg. 1966. Notes on the biology of the tenrecidae. *J. Mammal.* **47:**684–686.

Grasse, Pierre-P. 1955. Mammifères, vol. 16 (2 parts), 2300 pp.; 1967, part 1, 1162 pp.; 1968, part 2, 870 pp.; 1969, part 6, 1209 pp.; 1971, part 3, 1027 pp. Traité de Zoologie. Masson et Cie, Paris. 17 vols.

Gray, J. 1953. How animals move. Cambridge University Press, New York.

Green, R. C., and C. L. Larson, 1938. A description of shock disease in the snowshoe hare. *Am. J. Hyg.* **28:**190–212.

————, ———— and J. F. Bell. 1939. Shock disease as the cause of periodic decimation of the snowshoe hare. *Am. J. Hyg.* **30:**83–102.

Gregory, M. E., D. K. Kon, S. J. Rowland, and S. Y. Thompson. 1955. The composition of the milk of the blue whale. *J. Dairy Res.* **22:**108–112.

Gregory, W. K. 1912. Notes on the principles of quadripedal locomotion and on the mechanism of the limbs of hoofed animals. *Ann. N.Y. Acad. Sci.* **22:**267–294.

Griffin, D. R. 1940. Notes on the life histories of New England cave bats. *J. Mammal.* **21:**181–187.

————. 1945. Travels of cave bats. *J. Mammal.* **26:**15–33.

————. 1958. Listening in the dark. Yale University Press, New Haven, Conn. 413 pp.

————. 1968. Echolocation and its relevance to communication behavior. pp. 154–164. *IN:* T. A. Sebeok, ed. Animal communication. Indiana University Press, Bloomington. 686 pp.

————. 1970. Migrations and homing of bats. pp. 233–264. *IN:* W. A. Wimsatt, ed. Biology of bats. Academic Press, Inc., New York. 2 vols.

———— and R. Galambos. 1940. Obstacle avoidance by flying bats. *Anat. Rec.* **78:**95.

———— and ————. 1941. The sensory basis of obstacle avoidance by flying bats. *J. Exp. Zool.* **86:**481–506.

Griffiths, M. 1965. Digestion, growth and nitrogen balance in an egg-laying mammal, *Tachyglossus aculeatus* (Shaw). *Comp. Biochem. Physiol.* **14:**357–375.

————, M. A. Elliott, R. M. Leckie, and G. I. Schoefl. 1973. Observations on the comparative anatomy and ultrastructure of mammary glands and on fatty acids of triglycerides in platypus and echidna milk fats. *J. Zool.* **169:**255–279.

Grinnell, H. W. 1940. Joseph Grinnell: 1877–1939. *Condor.* **42:**3–34.

Grinnell, J. 1917. Field tests of theories concerning distributional controls. *Am. Nat.* **51:**115–128.

————. 1917a. The niche-relationships of the California thrasher. *Auk.* **34**:427–433.

Grunt, J. A., and W. C. Young. 1953. Consistency of sexual behavior patterns in individual male guinea pigs following castrations and androgen therapy. *J. Comp. Physiol. Psychol.* **46**:138–144.

Gunderson, H. L. 1950. A study of some small mammal populations at Cedar Creek Forest, Anoka County, Minnesota. *Occas. Pap. Mus. Nat. Hist. Univ. Minn.* No. 4. 49 pp.

————. 1962. An eight and one-half year study of red-backed vole (*Clethrionomys gapperi* Vigors) at Cedar Creek Forest, Anoka and Isanti Counties, Minnesota. Ph.D. dissertation, University of Michigan. 110 pp.

———— and J. R. Beer. 1953. The mammals of Minnesota. *Occas. Pap. Mus. Nat. Hist. Univ. Minn.* No. 6. 190 pp.

————, W. J. Breckenridge, and J. A. Jarosz. 1955. Mammal observations at lower Back River, Northwest Territories, Canada. *J. Mammal.* **36**: 254–259.

Gupta, B. B. 1966. Notes on the gliding mechanism in the flying squirrel. *Occas. Pap. Mus. Zool. Univ. Mich.* No. 645:1–7.

Guthrie, W. 1815. A new geographical, historical, and commercial grammar and the present state of several kingdoms of the world. Johnson and Warner.

Haan, F. W. R. 1960. Some aspects of mammalian hearing under water. *Proc. R. Soc., London.* **152**:54–62.

Hadwen, S. 1927. Color in relation to health of wild and domestic animals. *J. Hered.* **17**:450–461.

Hafez, E. S. E., ed. 1969. The behavior of domestic animals. The Williams & Wilkins Company, Baltimore. 647 pp.

————, M. W. Schein, and R. Ewbank. 1969. The behavior of cattle. pp. 235–295. *IN:* E. S. E. Hafez, ed. The behavior of domestic animals. The Williams & Wilkins Company, Baltimore, 647 pp.

———— and J. P. Signoret. 1969. The behavior of swine. *IN:* E. S. E. Hafez, ed. The behavior of domestic animals. The Williams & Wilkins Company, Baltimore. 647 pp.

Hagen, H. L. 1951. Composition of deer milk. *Calif. Fish Game.* **37**:217–218.

Hagmeier, E. M. 1966. A numerical analysis of the distributional patterns of North American mammals. II. Re-evaluation of the provinces. *Syst. Zool.* **15**:279–299.

———— and C. D. Stults. 1964. A numerical analysis of the distributional patterns of North American mammals. *Syst. Zool.* **13**:125–155.

Haines, H. 1964. Salt tolerance and water requirements in the salt-marsh harvest mouse. *Physiol. Zool.* **37**:266–272.

Hale, E. B. 1966. Visual stimulation and reproductive behavior in bulls. *J. Anim. Sci.* **25** (suppl.):36–44.

Haley, J. E. 1936. Charles Goodnight, plowman and plainsman. Houghton Mifflin Company, Boston. 485 pp.

Hall, E. R. 1927. An outbreak of house mice in Kern County, California. *Univ. Calif. Berkeley Publ. Zool.* **30**:189–203.

————. 1946. Mammals of Nevada. University of California Press, Berkeley. 710 pp.

———— and K. R. Kelson. 1959. The mammals of North America. The Ronald Press Company, New York. 2 vols.

Hall, K. R. L., and I. DeVore. 1965. Baboon social behavior. pp. 53–110. *IN:* I. Devore, ed. Primate behavior. Holt, Rinehart and Winston, Inc., New York.

Haltmeyer, G. C., and K. B. Eik-nes. 1969. Plasma level of testosterone in male rabbits following copulation. *J. Reprod. Fert.,* **19**:273–377.

Hamilton, W. J., Jr. 1929. Breeding habits of the short-tailed shrew, *Blarina brevicauda. J. Mammal.* **10**:125–134.

———. 1931. Habits of the star-nosed mole, *Condylura cristata. J. Mammal.* **12**: 345–355.

———. 1934. The life history of the rufescent woodchuck, *Marmota monax rufescens,* Howell. *Ann. Carnegie Mus.* **23**:85–178.

———. 1935. Habits of jumping mice. *Am. Midl. Nat.,* **16**:187–200.

———. 1937. Activity and home range of the field mouse, *Microtus pennsylvanicus pennsylvanicus,* Ord. *Ecology* **18**:255–263.

———. 1939. American mammals. McGraw-Hill Book Company, New York. 434 pp.

———. 1946. A study of the baculum in some North American Microtinae. *J. Mammal.* **27**:378–387.

Hansen, A., and K. Schmidt-Nielsen. 1957. On the stomach of the camel with special reference to the structure of its mucous membrane. *Acta Anat.* **31**:353–375.

Hansen, R. M. 1954. Molt patterns in ground squirrels. *Proc. Utah Acad. Sci. Arts Lett.* **31**:57–60.

Hardy, J. D. 1971. Physiology of temperature regulation. *Physiol. Rev.* **41**:521–606.

Harington, C. R. 1961. Summary—polar bear study. *Can. Wildl. Serv. Rep.* Feb., March, 1961. 14 pp.

———. 1968. Denning habits of the polar-bear (*Ursus maritimus* Phipps). *Can. Wildl. Serv. Rep. Ser.* No. 5. 30 pp.

Harlan, R. 1825. Fauna Americana. A. Finley, Philadelphia.

Harlow, H. F. 1958. The nature of love. *Am. Psychol.* **13**:673–685.

——— and M. K. Harlow. 1961. A study of animal affection. *Nat. Hist.* **70**:48–55.

——— and ———. 1962. Social deprivation in monkeys. *Sci. Am.* **207**:137–146.

Harper, F. 1945. Extinct and vanishing mammals of the Old World. Am. Comm. Int. Wildl. Protection, Spec. Publ. 12. 85 pp.

———. 1961. Land and fresh-water mammals of the Ungava Peninsula. *Univ. Kans. Mus. Nat. Hist. Misc. Publ.* No. 27. 178 pp.

Harrison, B. 1962. Orang-utan. Collins, London.

Harrison, D. L. 1964. The mammals of Arabia. 3 vols. including Insectivora, Chiroptera, Primates, Carnivora, Artiodactyla, Hyracoidea, Lagomorpha, and Rodentia. Ernst Benn Ltd., London.

Harrison, P. J., S. H. Ridgway, and P. L. Joyce. 1972. Telemetry of heart rate in diving seals. *Nature* (London). **238**(5362):280.

Hart, F. M., and J. A. King. 1966. Distress vocalization of young in two subspecies of *Peromyscus maniculatus. J. Mammal.* **47**:287–293.

Hart, G. H., S. W. Mead, and W. M. Reagen. 1946. Stimulating the sex drive of bovine males in artificial insemination. *Endocrinology* **39**:221–228.

Hartridge, H. 1920. The avoidance of objects by bats in their flight. *J. Physiol.* **54**:54–57.

Hatt, R. T. 1932. The vertebral columns of ricochetal rodents. *Bull. Am. Mus. Nat. Hist.* **63**:599–738.

———, J. Van Tyne, L. C. Stuart, C. H. Pope, and A. B. Grobman. 1948. Island life:

a study of the land vertebrates of the islands of Eastern Lake Michigan. Cranbrook Inst. Sci., Bull. No. 27. Bloomfield Hills, Mich. 179 pp.

Haugen, A. O. 1942. Home range of the cottontail rabbit. *Ecology.* **23**(3):354–367.

Hausman, L. A. 1920. Structural characteristics of the hair of mammals. *Am. Nat.* **54:** 496–523.

Hayman, R. H. 1965. Hair growth in cattle. pp. 575–590. *IN:* A. G. Lyne and B. F. Short, eds. Biology of the skin and hair growth. American Elsevier Publishing Company, Inc., New York. 806 pp.

Hayward, J. S. 1961. The ability of the wild rabbit to survive conditions of water restriction. *CSIRO Wildl. Res., Aust.* **6:**160–175.

———— and C. P. Lyman. 1967. Nonshivering heat production during arousal from hibernation and evidence for the contribution of brown fat. pp. 346–355. *IN:* K. C. Fisher et al., eds. Mammalian hibernation. III. American Elsevier Publishing Company, Inc., New York. 535 pp.

Heape, W. 1931. Emigration, migration and nomadism. W. Heffer & Sons, Ltd., London. 369 pp.

Hearne, S. 1795. A journey from Prince of Wales Fort in Hudson Bay, to the Northern Ocean in the years 1769, 1770, 1771, and 1772. A. Strahan and T. Cadell, London. 458 pp.

Hediger, H. 1950. Wild animals in captivity. Butterworth & Co. (Publishers), Ltd., London. 207 pp.

————. 1955. Wild animals in captivity. Butterworth & Co. (Publishers), Ltd., London. 650 pp.

Hellman, G. 1969. Bankers, bones and beetles. American Museum of Natural History, New York. 275 pp.

Henderson, F. R. 1960. Beaver in Kansas. *Univ. Kans. Mus. Nat. Hist., Misc. Publ.* **26:**1–85.

Henderson, J., and E. L. Craig. 1932. Economic mammalogy. Charles C Thomas, Publisher, Springfield, Ill. 550 pp.

Hennig. W. 1966. Phylogenetics systematics. University of Illinois Press, Urbana. 263 pp.

Henshaw, R. E., and G. E. Folk, Jr. 1966. Thermoregulation in bats. *IN:* B. H. Slaughter and D. W. Walton, eds. About bats. Southern Methodist University Press, Dallas. 339 pp.

Henson, O. W., Jr. 1961. Some morphological and functional aspects of certain structures of the middle ear in bats and insectivores. *Univ. Kans. Sci. Bull.* **62:**151–255.

————. 1970. The ear and audition. pp. 181–259. *IN:* W. A. Wimsatt, ed. Biology of bats, vol. 2. Academic Press, Inc., New York. 2 vols.

Herald, E. S., R. L. Brownell, Jr., F. L. Frye, E. J. Morris, W. E. Evans, and A. B. Scott. 1969. Blind river dolphin: first side-swimming cetacean. *Science* **166:** 1408–1410.

Herrick, C. L. 1892. The mammals of Minnesota. Geol. Nat. Hist. Surv. Minn., Bull. 7. 299 pp.

Herrington, L. P. 1951. The role of the pilary system in mammals and its relation to the thermal environment. *Ann. N.Y. Acad. Sci.* **53:**600–607.

Hershkovitz, P. 1962. Evolution of neotropical cricetine rodents (Muridae) with special reference to the phylotine group. *Fieldiana Zool.* **46:**1–524.

————. 1964. The recent mammals of South America. *Proc. 16th Int. Congr. Zool.* **4**:40–45. (Field Museum, Chicago.)

Herter, K. 1962. Untersuchungen an lebenden Borstenigeln (Tenricinae). I. Uber Temperaturregulierung und Aktivitätsrhythmic bei dem Igeltanreck *Echinops telfairi. Zool. Beitr.,* N.F. **7**:239–292.

Hess, E. H. 1959. Imprinting. *Science* **130**:133–141.

Hesse, R., W. C. Allee, and K. P. Schmidt. 1937. Ecological animal geography. John Wiley & Sons, Inc., New York. 715 pp.

Hewitt, O. H. 1954. Symposium on cycles in animal populations. *J. Wildl. Manage.* **18**:1–112.

Hildebrand, M. 1952. An analysis of body propulsion in Canidae. *Am. J. Anat.* **90**: 217–256.

————. 1959. Motions of the running cheetah and horse. *J. Mammal.* **40**:481–495.

————. 1961. Further studies on the locomotion of the cheetah. *J. Mammal.* **42**:84–91.

————. 1962. Walking, running, and jumping. *Am. Zool.* **2**:151–155.

Hill, J. E. 1937. Morphology of the pocket gopher mammalian genus *Thomomys. Univ. of Cal. Publ. Zool.* **42**:81–171.

Hinde, R. A. 1966. Animal behavior: A synthesis of ethology and comparative psychology. McGraw-Hill Book Company, New York. 534 pp.

Hiner, L. E. 1938. Some studies of hairs of mammals native to Minnesota. M.S. thesis, University of Minnesota.

Hinton, H. E., and A. M. S. Dunn. 1967. Mongooses: their natural history and behavior. University of California Press, Berkeley and Los Angeles. 144 pp.

Hisaw, F. L. 1923. Observations on the burrowing habits of moles *(Scalopus aquaticus machrinoides). J. Mammal.* **4**:79–88.

Hock, R. J. 1951. The metabolic rates and body temperature of bats. *Biol. Bull.* **101**: 289–299.

————. 1960. Seasonal variations in physiologic function of black bears. *IN:* C. P. Lyman and A. R. Dawe, eds. Mammalian hibernation. *Bull. Mus. Comp. Zool., Harv. Univ.* **124**:155–173.

Hock, R. J., and A. M. Larson. 1966. Composition of black bear milk. *J. Mammal.* **47**:539–540.

Hodge, P., and H. Wakefield. 1959. Echidna and its young. *Victorian Nat.* **76**:64–65.

Hoffman, R. A. 1964. Terrestrial animals in cold: Hibernators. pp. 379–403. *IN:* D. B. Dill, ed. Handbook of physiology, 2d ed., sec. 4, Adaptation to the environment. American Physiologic Society, Washington, D.C.

Hoffman, R. S. 1958. The role of reproduction and mortality in population fluctuations of voles *(Microtus). Ecol. Monogr.* **28**:79–109.

Hoffmeister, D. F. 1969. The first fifty years of the American Society of Mammalogists. *J. Mammal.* **50**:794–802.

Hooper, E. T. 1949. Faunal relationships of recent North American rodents. *Misc. Publ. Univ. Mich. Mus. Zool.* **72**:1–28.

———— 1959. The glans penis in five genera of cricetid rodents. *Univ. Mich. Occas. Pap. Mus. Zool.* **613**:1–11.

————. 1962. The glans penis in *Sigmodon, Sigmomys,* and *Reithrodon* (Rodentia, Cricetinae). *Univ. Mich. Occas. Pap. Mus. Zool.* **625**:1–11.

———— and G. G. Musser. 1964. The glans penis in Neotropical cricetines (family

Muridae) with comments on classification of muroid rodents. *Misc. Publ. Univ. Mich. Mus. Zool.* **123**:5–57.

Hopson, J. A. 1970. The classification of nontherian mammals. *J. Mammal.* **51**:1–9.

———. 1973. Endothermy, small size, and the origin of mammalian reproduction. *Am. Nat.* **107**:446–452.

——— and W. W. Crompton. 1969. Origin of mammals. pp. 15–72. *IN:* T. Dobzhansky, M. K. Hecht, and W. C. Steere, eds. Evolutionary biology. III. Appleton-Century-Crofts, Inc., New York.

Hopwood, A. T. 1947. Contribution to the study of some African mammals. III. Adaptations in the bones of the fore limb of the lion, leopard and cheetah. *J. Linn. Soc. London Zool.* **41**:259–271.

Hornaday, W. T. 1889. The extermination of the American bison with a sketch of its discovery and life history. *Smithson. Inst. Annu. Rep.* 1889.

Horner, B. E., J. M. Taylor, and H. A. Padykula. 1964. Food habits and gastric morphology of the grasshopper mouse. *J. Mammal.* **45**:513–535.

Hornocker, M. G. 1969. Winter territoriality in mountain lions. *J. Wildl. Manage.* **33**:457–464.

———. 1970. An analysis of mountain lion predation upon mule deer and elk in the Idaho primitive area. *Wildl. Monogr.* No. 21. 39 pp.

Horst, R. 1968. Observations on renal morphology and physiology in the vampire bat *Desmodus rotundus.* Ph.D. thesis, Cornell University, Ithaca, New York.

———. 1969. Observations on the structure and function of the kidney of the vampire bat *(Desmodus rotundus murinus).* pp. 78–83. *IN:* C. C. Hoff and M. L. Riedesel, eds. Physiological systems in semiarid environments. University of New Mexico Press, Albuquerque. 293 pp.

Houpt, T. R. 1959. Utilization of blood urea in ruminants. *Am. J. Physiol.* **197**:115–120.

Howard, W. E. 1949. Dispersal, amount of inbreeding and longevity in a local population of prairie deermice on the George Reserve, southern Michigan. *University of Michigan Contrib. Lab. Vert. Biol.* No. 43:1–52.

———. 1953. Growth rate of nails of adult pocket gophers. *J. Mammal.* **34**:394–396.

——— and M. E. Smith. 1952. Rate of extrusive growth of incisors in pocket gophers. *J. Mammal.* **33**:485–487.

——— and H. E. Childs, Jr. 1959. Ecology of pocket gophers with emphasis on *Thomomys bottae mewa. Hilgardia.* **29**:277–358.

Howell, A. B. 1923. The mammal collections of North America. *J. Mammal.* **4**:113–120.

———. 1928. Anatomy of the wood rat: Comparative anatomy of the subgenera of the American wood rat (genus *Neotoma*). The Williams & Wilkins Company, Baltimore. 225 pp.

———. 1930. Aquatic mammals. Charles C Thomas, Publisher, Springfield, Ill. 338 pp.

———. 1932. The saltatorial rodent *Dipodomys:* the functional and comparative anatomy of its muscular and osseous systems. *Proc. Am. Acad. Arts Sci.* **67**:377–536.

———. 1944. Speed in animals. University of Chicago Press, Chicago.

Hsu, T. C., and K. Benirschke. 1968. An atlas of mammalian chromosomes. Springer-Verlag, Inc., New York. Annual volumes are planned.

Hudson, J. W. 1962. The role of water in the biology of the antelope ground squirrel *Citellus leucurus leucurus. Univ. Calif. Berkeley Publ. Zool.* **64**:1–56.

Huey, L. M. 1942. A vertebrate faunal survey of the Organ Pipe Cactus National Monument, Arizona. *Trans. San Diego Soc. Nat. Hist.* **9**:355–375.

Huey, W. S. 1956. New Mexico beaver management. *N. M. Dep. Game Fish Bull.* No. 4. 49 pp.

Huibgreste, W. H. 1966. Some chemical and physical properties of bat milk. *J. Mammal.* **47**:551–554.

Hume, E. E. 1942. Ornithologists of the United States Army Medical Corps. The Johns Hopkins Press, Baltimore. 583 pp.

Hungate, R. E. 1966. The rumen and its microbes. Academic Press, Inc., New York. 533 pp.

Hutchinson, G. E. 1957. Concluding remarks. *Cold Spring Harbor Symp. Quant. Biol.* **22**:415–427.

———— and E. S. Deevey, Jr. 1949. Ecological studies on animal populations. *Surv. Biol. Progr.* **1**:325–359.

Iredale, T., and E. Troughton. 1934. A checklist of the mammals recorded from Australia. *Aust. Mus. Sydney Mem.* **6**:1–122.

Jaeger, E. C. 1957. The North American deserts. Stanford University Press, Stanford, Calif. 308 pp.

Jameson, E. W., Jr. 1947. Natural history of the prairie vole (mammalian genus *Microtus*). *Univ. Kans. Mus. Nat. Hist. Misc. Publ.* **1**:125–151.

Jenness, R. 1974. The composition of milk. *IN:* B. L. Larson and V. L. Smith, eds. Lactation, vol. III, Nutrition and biochemistry maintenance. Academic Press, Inc., New York. 425 pp.

———— and S. Patton. 1959. Principles of dairy chemistry. John Wiley & Sons, Inc., New York. 446 pp.

————, E. A. Regehr, and R. E. Sloan. 1964. Comparative biochemical studies of milks. II. Dialyzable carbohydrates. *Comp. Biochem. Physiol.* **13**:339–352.

———— and R. E. Sloan. 1970. The composition of milks of various species: a review. *Dairy Sci. Abstr.* **32**:599–612, review article No. 158.

Johansen, K. 1961. Temperature regulation in the ninebanded armadillo *(Dasypus novemcinctus mexicanus). Physiol. Zool.* **34**:126–144.

————. 1962. Buoyancy and insulation in the muskrat. *J. Mammal.* **43**:74–78.

———— and J. Krog. 1959. Diurnal body temperature variations and hibernation in the birch mouse *Sicista betulina. Am. J. Physiol.* **196**:1200–1204.

Johnsgard, P. A. 1967. Animal behavior. Wm. C. Brown Company Publishers, Dubuque, Iowa. 156 pp.

Johnson, G. E. 1917. The habits of the thirteen-lined ground squirrel *(Citellus tridecemlineatus),* with special reference to the burrows. *Q. J., Univ. N.D.* **7**:261–271.

————. 1931. Hibernation in mammals. *Q. Rev. Biol.* **6**:439–461.

Johnson, L. A. 1968. Rainbow's end: the quest for an optimal taxonomy. *Proc. Linn. Soc. N.S.W.* **93**:1–45. Reprinted 1970, *Syst. Zool.* **19**:203–239.

Jonsgard, A. 1959. Recent investigations concerning sound production in cetaceans. *Nor. Hvalfangst-Tid.* **10**:501–509.

————. 1966. The distribution of Balaenopteridae in the North Atlantic Ocean. pp. 114–124. *IN:* K. S. Norris, ed. Whales, dolphins, and porpoises. University of California Press, Berkeley and Los Angeles. 789 pp.

Joslin, P. W. B. 1967. Movements and home sites of timber wolves in Algonquin Park. *Am. Zool.* **7**:279–288.

Kahmann, H., and K. Ostermann. 1951. Warnehmen und Hervorbrigen hoher tone bei Kleiner Säugetieren. *Experientia.* **7**:268–269.

Karlson, P., and M. Lüscher. 1959. 'Pheromones': a new term for a class of biologically active substances. *Nature.* **183**:55–56.

Katz, H., and M. Katz. 1965. Museums, U.S.A. A history and guide. Doubleday & Company, Inc., Garden City, N.Y. 395 pp.

Kaufman, J. H. 1962. Ecology and social behavior of the coati, *Nasua narica,* on Barro Colorado Island, Panama. *Univ. Calif. Berkeley Publ. Zool.* **60**:95–222.

Kaye, S. 1961. Gold—198 wires used to study movements of small mammals. *Science.* pp. 131–824.

Kayser, C. 1939. Les échanges respiratoires des hibernants. *Ann. Physiol.* **15**:1087–1219.

———. 1965. Hibernation, pp. 179–296. *IN:* W. Mayer and R. W. Van Gelder (eds.). Physiological mammology, Vol. III. Academic Press, Inc., N.Y.

Keating, W. 1825. Narrative of an expedition to the source of the St. Peter's River. Cox and Baylis, London. 2 vols.

Keeler, C. E., T. Mellinger, E. Fromm, and L. Wade. 1970. Melanin, adrenalin and the legacy of fear. *J. Hered.* **61**:81–89.

Keith, L. B. 1963. Wildlife's ten year cycle. The University of Wisconsin Press, Madison. 216 pp.

Kellogg, Remington. 1946. A century of progress in Smithsonian biology. *Science.* **104**:132–141.

Kelsall, J. P. 1968. The migratory barren-ground caribou of Canada. Dept. of Indian Affairs and Northern Development. *Can. Wildl. Serv. Monogr.,* No. 3. 340 pp.

Kendeigh, S. C. 1974. Ecology. Prentice-Hall, Inc., Englewood Cliffs, N.J. 468 pp.

Kennerly, T. E., Jr. 1959. Contact between the ranges of two allopatric species of pocket gophers. *Evolution.* **8**:247–263.

———. 1964. Microenvironmental conditions of the pocket gopher burrow. *Tex. J. Sci.* **14**:397–441.

Kennicott, R. 1857. The quadrupeds of Illinois, injurious and beneficial to the farmers. pp. 52–110. U.S. Patent Office, Dep. (Agr.) for 1856.

———. 1858. The quadrupeds of Illinois, injurious and beneficial to the farmers. pp. 52–110. U.S. Patent Office, Dep. (Agr.) for 1857.

———. 1859. The quadrupeds of Illinois, injurious and beneficial to the farmers. pp. 241–256. U.S. Patent Office, Dep. (Agr.) for 1858.

Kenyon, K. W. 1960. Territorial behavior in the Alaskan fur seal. *Mammalia.* **24**:431–444.

——— and F. Wilke. 1953. Migration of the northern fur seal, *Callorhinus ursinus.* *J. Mammal.* **34**:86–98.

King, J. A. 1955. Social behavior, social organization and population dynamics in the black-tailed prairie dog town in the Black Hills of South Dakota. *Contrib. Lab. Verteb. Biol. Univ. Mich.* **67**:1–123.

———. 1963. Maternal behavior in *Peromyscus.* pp. 58–93. *IN:* Harriet L. Rheingold, ed. Maternal behavior in animals. John Wiley & Sons, Inc., New York. 349 pp.

King, J. E. 1964. Seals of the world. British Museum of Natural History, London. 154 pp.

Kingdon, J. 1971. East African mammals, an atlas of evolution in Africa. Academic Press, Inc., New York. 446 pp.

Kingston, B. H. 1964. The chemistry and olfactory properties of musk, civet and castoreum. pp. 209–214. Proc. 2d Int. Congr. Endocrinol., London. Excerpta Medica Foundation, Amsterdam.

Kirmiz, J. P. 1962. Adaptation to desert environment—a study on the jerboa, rat, and man. London Butterworths. 154 pp.

Kirstin, A. D. 1929. The mole clavicle. *J. Mammal.* **10**:305–313.

Kitts, W. D., I. Cowan, J. Bandy, and A. J. Wood. 1956. The immediate post-natal growth in the Columbian black-tailed deer in relation to the composition of the milk of the doe. *J. Wildl. Manage.* **20**:212–214.

Kleiber, M. 1932. The linear relation of the logarithm of body weight to logarithm of energy metabolism. *Hilgardia.* **6**:315–353.

Kleiman, D. G. 1966. Scent marking in the Canidae. *Symp. Zool. Soc. London.* **18**: 167–177.

———. 1968. Some aspects of the social behavior in the Canidae. *Am. Zool.* **7**:365–372.

——— and J. F. Eisenberg. 1973. Comparisons of canid and feline social systems from an evolutionary perspective. *Anim. Behav.* **21**:673–689.

Kleinenberg, S. E., A. V. Yablakov, B. M. Bel'kovich, and M. N. Tarasevich. 1969. Beluga *(Delphinopterus leucas),* investigation of the species. Translated from Russian by the Israel Program for the USSR. Moscow (1964). 376 pp.

Knappe, H. 1964. Zur Funktion des Jacobson Organs (*Organon vomeronasale* Jacobsoni). *Zool. Gart.* **28**:188–194.

Knowlton, F. 1960. Food habits, movements and populations of moose in the Gravelly Mountains, Montana. *J. Wildl. Manage.* **24**:162–170.

Knudsen, G. J. 1962. Relationship of beaver to forests, trout, and wildlife in Wisconsin. Wisc. Conserv. Dep. Tech. Bull. No. 25, Madison. 52 pp.

Knudsen, V. O. 1931. The effect of humidity upon the absorption of sound in a room, and determination of the coefficients of absorption of sound in air. *J. Acoust. Soc. Am.* **3**:126–138.

Koford, C. B. 1958. Prairie dogs, white faces, and blue grama. *Wildl. Monogr.* No. 3. 78 pp.

Komarek, E. V. 1939. A progress report on southeastern mammal studies. *J. Mammal.* **20**:292–299.

Kon, S. D., and A. T. Cowie. 1961. Milk: the mammary gland and its secretion, vol. I, 515 pp.; vol. II, 423 pp. Academic Press, Inc., New York.

Kooyman, G. L. 1963. Milk analysis of the kangaroo rat. *Dipodomys merriami. Science.* **142**:1467.

Kramer, R. J. 1971. Hawaiian land mammals. Charles E. Tuttle Co., Rutland, Vt. 347 pp.

Krause, H. 1957. The mammals of the Major Long Expedition, 1823. *Naturalist.* **8**:1–5.

Krebs, C. J. 1964. The lemming cycle at Baker Lake, Northwest Territories, during 1959–62. Arctic Inst. North Am. Tech. Pap. No. 15. 104 pp.

———. 1970. *Microtus* population biology; behavioral changes associated with the population cycle in *M. ochrogaster* and *M. pennsylvanicus. Ecology.* **51**:35–52.

Kristofferson, R., and A. Saivio. 1964. Hibernation in the hedgehog *(Erinaceus europaeus L.)*. *Ann. Acad. Sci. Fenn. Ser. A.4.* No. 82.

Krutsch, P. H. 1955. Observations on the Mexican free-tailed bat, *Tadarida mexicana. J. Mammal.* **36**:236–242.

Kruuk, H. 1972. The spotted hyena. The University of Chicago Press, Chicago, 335 pp.

Kulzer, E. 1956. Flughunde erzeugen Orientierung durch Zungenschlag. *Naturwissenschaften.* **43**:117–118.

———. 1958. Untersuchungen über die Biologie von Flughunden der Gattung *Rousettus* Gray. *Z. Morph. Ökol. Biere.* **47**:374–402.

Kuroda, N. 1940. A monograph of the Japanese mammals. The Sansiedo Co., Ltd., Tokyo and Osaka. 311 pp.

Kurten, B. 1969. Continental drift and evolution. *Sci. Am.* **220(3)**:54–64.

———. 1971. The age of mammals. Weidenfeld and Nicolson, London. 250 pp.

Lack, D. 1947. The Galapagos finches (Geospizinae). *Occas. Pap. Calif. Acad. Sci.* **21**: 1–151.

———. 1954. The natural populations of animal numbers. Clarendon Press, Oxford. 343 pp.

———. 1968. Ecological adaptations for breeding in birds. Methuen & Co., Ltd., London. 401 pp.

Larson, F. 1940. A role of bison in maintaining the short grass plains. *Ecology.* **21**: 113–121.

Larson, P. 1970. Deserts of America. Prentice-Hall, Inc., Englewood Cliffs, N.J. 340 pp.

Laurie, E. M. O., and J. E. Hill. 1954. List of land mammals of New Guinea, Celebes and adjacent islands, 1758–1952. British Museum of Natural History, London. 175 pp.

Lawrence, B., and W. E. Schevill. 1956. The functional anatomy of the delphinid nose. *Bull. Mus. Comp. Zool. Harv. Univ.* **114**:103–151.

Lechleitner, R. R. 1954. Age criteria in mink. *J. Mammal.* **35**:496–503.

———. 1958. Certain aspects of behavior of the black tailed jackrabbit. *Am. Mid. Nat.* **60**:145–155.

Lee, S., van der, and L. M. Boot. 1955. Spontaneous pseudo-pregnancy in mice. *Acta Physiol. Pharmacol. Neerl.* **4**:442–444.

Lehausen, P., and R. Wolff. 1959. Das Revier einer Haskatze. *Z. Tierpsychol.* **16**: 666–670.

Lent, P. C. 1965. Rutting behavior in a barren-ground caribou population. *Anim. Behav.,* **13** (nos. 2 and 3):259–264.

Leopold, A. 1933. Game management. Charles Scribner's Sons, New York. 481 pp.

Leopold, A. S. (ed.) 1969. The desert. Time-Life Books, New York. Revised. 192 pp.

Lesley, L. B. 1929. Uncle Sam's camels. Harvard University Press, Cambridge, Mass. 298 pp.

Liggins, G. C. 1969. The foetal role in the initiation of parturition in the ewe. pp. 218–231. *IN:* G. E. W. Wolstenholme and M. O'Connor, eds. Foetal anatomy. Ciba Foundation Symposium. J. and A. Churchill, Ltd., London. 326 pp.

Lilly, J. C. 1966. Sonic-ultrasonic emissions of the bottle-nose dolphin. *IN:* K. S. Norris, ed. Whales, dolphins, and porpoises. University of California Press, Berkeley and Los Angeles. 503 pp.

Lindeman, R. L. 1942. The trophic-dynamic aspect of ecology. *Ecology* **23**:399–418.

Linduska, J. P. 1950. Ecology and land-use relationships of small mammals on a Michigan farm. Game Div. Dep. Conserv., Lansing, Mich. 144 pp.

Ling, J. K. 1965. Hair growth and moulting in the southern elephant seal, *Mirounga leonina* (Linn.). pp. 525–544. *IN:* A. G. Lyne and B. F. Short, eds. Biology of the skin and hair growth. American Elsevier Publishing Company, Inc., New York. 806 pp.

———. 1970. Pelage and moulting in wild animals with special reference to aquatic forms. *Q. Rev. Biol.* **45**:16–54.

Linnaeus, C. 1758. Systema naturae per regna tria naturae, secundum classes, ordines, genera, species cum characteribus, differentiis, synonymis, locis. 10th ed., rev. Laurentii Salvii, Stockholm.

Linsdale, J. M. 1938. Environmental responses of vertebrates in the Great Basin. *Am. Midl. Nat.* **19**:1–206.

Linzey, D. W., and A. V. Linzey. 1968. Mammals of the Great Smoky Mountains National Park. *J. Elisha Mitchell Sci. Soc.* **84**:384–414.

Livezey, R., and F. Evenden, Jr. 1943. Notes on the western red fox. *J. Mammal.* **24**: 500–501.

Lockley, R. M. 1954. The seals and the curragh; introducing the natural history of the grey seal of the North Atlantic. J. M. Dent, London. 149 pp.

———. 1961. Social behavior of the European wild rabbit. *J. Anim. Ecology* **30**: 385–423.

Long, C. A. 1972. Two hypotheses of the origin of lactation. *Am. Nat.* **106**:141–144.

Longley, W. H., and J. B. Moyle. 1963. The beaver in Minnesota. Minn. Dep. Conserv., Div. Game Fish, Tech. Bull. No. 6, St. Paul. 87 pp.

Lord, R., Jr. 1961. A population study of the gray fox. *Am. Midl. Nat.* **66**:87–109.

———. 1963. The cottontail rabbit in Illinois. State Ill., Dep. Conserv., Tech. Bull. No. 3, Springfield. 94 pp.

Lorenz, K. Z. 1935. Der Kumpan in der Umwelt des Vogels. *J. Ornithol.* **83**:137–213.

———. 1937. The companion in the birds' world. *Auk.* **54**:254–273. (An English version of Der Kumpan in der Umwelt des Vogels.)

———. 1952. King Solomon's ring. (Paperback edition of his German book, published by Thomas Y. Crowell Company, New York.)

———. 1966. On aggression. Harcourt, Brace and Company, Inc., New York. 306 pp.

Lotka, A. J. 1925. Elements of physical biology. The Williams & Wilkins Company, Inc., Baltimore. 465 pp.

Lovaas, A. L. 1958. Mule deer food habits and range use, Little Belt Mountains, Montana. *J. Wildl. Manage.* **22**:275–283.

Lucas, F. A. 1926. Thomas Jefferson—paleontologist. *Nat. Hist.* **26**:328–330.

Lydekker, R. 1896. A geographical history of mammals. Cambridge University Press, Cambridge, England.

Lyman, C. P. 1942. Control of coat color in the varying hares by daily illumination. *Proc. New England Zool. Club.* **19**:75–78.

———. 1958. Oxygen consumption, body temperature and heart rate of woodchuck entering hibernation. *Am. J. Physiol.* **194**:83–91.

———. 1970. Thermoregulation and metabolism in bats. pp. 301–330. *IN:* W. A. Wimsatt, ed. Biology of bats, vol. I. Academic Press, Inc., New York. 406 pp.

MacArthur, R. H., and E. O. Wilson. 1967. The theory of island biogeography. Princeton University Press, Princeton, N.J. 203 pp.

MacGinitie, H. D. 1958. Climate since the Late Cretaceous. pp. 61–79. *IN:* Carl L. Hubbs, ed. Zoogeography. Am. Assoc. Adv. Sci. Publ. No. 51. 509 pp.

Mackintosh, N. A. 1966. The distribution of southern blue and fin whales. pp. 125–144. *IN:* K. S. Norris, ed. Whales, dolphins, and porpoises. University of California Press, Berkeley and Los Angeles. 789 pp.

MacLulich, D. A. 1936. Sunspots and abundance of animals. *J. R. Astron. Soc. Can.* **30:**233–246.

———. 1947. Fluctuations in the numbers of varying hare *(Lepus americanus). Univ. Toronto Stud. Biol. Ser.* 43. 136 pp.

MacMillen, R. E. 1965. Aestivation in the pocket mouse, *Peromyscus eremicus. Comp. Biochem. Physiol.* **16:**227–248.

———. 1972. Water economy of nocturnal desert rodents. Symp. Zool. Soc., London. **31:**147–173.

Malin, J. C. 1967. The grassland of North America. Prolegomena to its history, with addenda and postscript, 5th printing. Peter Smith, Gloucester, Mass. 490 pp.

Manning, T. H. 1946. Bird and mammal notes from the east side of Hudson Bay. *Can. Field Nat.* **60:**71–85.

———. 1948. Notes on the country, birds and mammals west of Hudson Bay between Reindeer and Baker Lakes. *Can. Field Nat.* **62:**1–28.

Marler, P. R. 1965. Communicating in monkeys and apes, pp. 544–584. *IN:* I. Devore (ed.). Primate behavior. Holt, Rinehart and Winston, Inc., N.Y. 654 pp.

Marsden, H. M., and N. R. Holler. 1964. Social behavior in confined populations of the cottontail and the swamp rabbit. *Wildl. Monogr.* No. 13. 39 pp.

Marston, H. R. 1926. The milk of the monotreme—*Echidna aculeata. Abstr. J. Exp. Biol. Med. Sci.* **3:**217–220.

Mathew, W. D. 1915. Climate and evolution. *Ann. N.Y. Acad. Sci.* **24:**171–318. (Reprinted in 1939 as Spec. Publ., N.Y. Acad. Sci. 1.)

Mathews, L. H. 1952. British mammals. William Collins Sons & Co., Ltd. London. 410 pp.

Mathiak, H. A. 1938. A key to the hairs of mammals of southern Michigan. *J. Wildl. Manage.* **2:**251–268.

Matson, J. R. 1946. Notes on dormancy in the black bear. *J. Mammal.* **37:**203–212.

Mayer, W. W. 1953. Some aspects of the ecology of the Barrow ground squirrel, *Citellus parryi barrowensis.* Current biologic research in the Alaska Arctic, *Stanford Univ. Publ. Univ. Ser. Biol. Sci.* **11:**48–55.

———. 1960. Histological changes during the hibernating cycle in the Arctic ground squirrel. pp. 131–152. *IN:* C. P. Lyman and A. R. Dawe, eds. Mammalian hibernation. *Bull. Mus. Comp. Zool. Harv. Univ.* **124.** 549 pp.

Mayr, E. 1942. Systematics and the origin of species. Columbia University Press, New York. 344 pp.

———. 1963. Animal species and evolution. The Belknap Press of the Harvard University Press, Cambridge, Mass. 767 pp.

―――. 1964. Wallace's line in the light of recent zoogeographic studies. *Rev. Biol.* **19**:1–14.

―――. 1969. Principles of systematic zoology. McGraw-Hill Book Company, New York. 428 pp.

―――, E. G. Linsley, and R. L. Usinger. 1953. Methods and principles of systematic zoology. McGraw-Hill Book Company, New York. 328 pp.

McCarley, H. 1966. Annual cycle, population dynamics and adaptive behavior of *Citellus tridecemlineatus. J. Mammal.* **47**:294–316.

McClintock, M. K. 1971. Menstrual synchrony and suppression. *Nature.* **229**:244–245.

McCullough, D. R. 1969. The tule elk, its history, behavior and ecology. *Univ. Calif. Berkeley Publ. Zool.* **88**:1–209.

McFarland, W. N. 1961. Water metabolism in temperate and tropical climates. 10th Proc., Pac. Sci. Congr. Univ. Hawaii, Honolulu.

――― and W. A. Wimsatt. 1965. Urine flow and composition in the vampire bat. *Am. Zool.* **5**:662(abstr.).

McGrady, H. 1938. The embryology of the opossum. *Wistar Inst. Anat. Biol.* **16**: 161–233.

McHugh, T. 1958. Social behavior of the American buffalo *(Bison bison bison). Zoologica.* **43**:1–40.

―――. 1972. The time of the buffalo. Alfred A. Knopf, Inc., N.Y. 339 pp.

McKay, C. M. 1949. Nutrition of the dog, 2d ed. Comstock Publishing Associates, Ithaca, New York. 337 pp.

McKeever, S. 1966. Reproduction in *Citellus beldingi* and *Citellus lateralis* in northeastern California. Symp. Zool. Soc. London. No. 15: 363–385.

McManus, J. J. 1969. Temperature regulation in the opossum, *Didelphis marsupialis virginiana. J. Mammal.* **50**:550–558.

McNab, B. L. 1969. The economics of temperature regulation in neotropical bats. *Comp. Biochem. Physiol.* **31**:227–268.

McNulty, F. 1972. Must they die? National Audubon Society, New York, and Ballantine Books, Inc., New York. 124 pp.

Mead, J. G. 1972. On the anatomy of the external nasal passages and facial complex in the family Delphinidae of the order Cetacea. Ph.D. thesis, The University of Chicago.

Mearns, E. A. 1907. The mammals of the Mexican boundary of the United States. *Bull. U.S. Nat. Mus.* No. 56. Part V. 530 pp.

Mech, L. D. 1966. The wolves of Isle Royale. Fauna of the National Parks of the United States, Fauna Series 7, Washington. 210 pp.

―――. 1970. The wolf: the ecology and behavior of an endangered species. The Natural History Press, Garden City, N.Y. 384 pp.

―――, J. R. Tester, and D. W. Warner. 1966. Fall daytime resting habits of raccoons as determined by telemetry. *J. Mammal.* **47**:450–466.

Medway, G. G. 1969. The wild mammals of Malaya and offshore islands including Singapore. Oxford University Press, London. 127 pp.

Merriam, C. H. 1887. Do any Canadian bats migrate? Evidence in the affirmative. *Trans. R. Soc. Can.,* sec. 4, pp. 85–87.

―――. 1901. The prairie dog of the Great Plains. *Yearb. U.S. Dep. Agri.* pp. 259–270.

――― and L. Stejneger. 1890. Results of a biological survey of the San Francisco

Mountain Region and desert of the Little Colorado, Arizona. *North Am. Fauna.* No. 3. 416 pp.

Messer, M., and K. R. Kerry. 1973. Milk carbohydrates of the echidna and platypus. *Science.* **180**:203.

Metzger, L. H. 1967. An experimental comparison of screech owl predation on resident and white-footed mice *(Peromyscus leucopus). J. Mammal.* **48**:387–391.

Miller, G. S., Jr. 1912. Catalogue of the mammals of western Europe. British Museum of Natural History, London. 1019 pp.

———— and R. Kellogg. 1955. List of North American recent mammals. *Bull. U.S. Natl. Mus.* **205**:1–954.

Miller, M. D., G. C. Christensen, and H. E. Evans. 1964. Anatomy of the dog. W. B. Saunders Company, Philadelphia. 941 pp.

Milne, L. J., and M. Milne. 1958. Paths across the earth. Harper & Brothers, New York. 216 pp.

Mitchell, G. C., and J. R. Tigner. 1970. The route of ingested blood in the vampire bat *(Desmodus rotundus). J. Mammal.* **51**:814–817.

Moen, A. N. 1973. Wildlife ecology: an analytical approach. W. H. Freeman and Company. San Francisco. 458 pp.

Mohr, C. O. 1947. Tables of equivalent populations of North American small mammals. *Am. Midl. Nat.* **37**:223–249.

Mohres, F. P. 1953. Über die Ultraschallorientierung der Hufeisennasen (Chiroptera-Rhinolophidae). *Z. Vgl. Physiol.* **34**:547–588.

Moir, R. J. 1968. Ruminant digestion and evolution. Handbook of physiology. **5:** 2673–2694.

————, M. Somers, G. Sharman, and H. Waring. 1954. Ruminant digestion in a marsupial. *Nature.* **173**:269–270.

———— and H. Waring. 1956. Studies on marsupial nutrition. *IN:* Ruminant-like digestion in a herbivorous marsupial *(Setonix brachyurus). Aust. J. Biol. Sci.* **9**:293–304.

Montagna, W. 1956. The structure and function of skin. Academic Press, Inc., New York. 356 pp.

Moody, P. D. 1958. Serological evidence on the relationship of musk ox. *J. Mammal.* **39**:554–559.

Morris, D. 1964. The mammals, a guide to the living species. Harper & Row, Publishers, Incorporated, New York. 448 pp.

Morris, R. F. 1955. Population studies in some small forest mammals in eastern Canada. *J. Mammal.* **36**:21–35.

Morrison, F. B. 1949. Feeds and feeding, 21st ed. Morrison Publishing Co., Ithaca, N.Y. 1027 pp.

Morrison, P. R. 1959. Body temperatures of some Australian mammals. I. *Chiroptera. Biol. Bull.* **116**:484–497.

————. 1962. Metabolism and body temperature in a small hibernator, the meadow jumping mouse, *Zapus hudsonius. J. Cell. and Comp. Physiol.* **60**:169–180.

———— and F. A. Ryser. 1951. Temperature and metabolism in some Wisconsin mammals. *Fed. Proc.* (Wisc. Acad. Sci.) **10**:93–94.

Morse, M. A. 1937. Hibernation and breeding of the black bear. *J. Mammal.* **18:** 460–465.

————. 1942. Wildlife restoration and management planning project. *U.S. Dep. Inter., Fish Wildl. Serv., Pittman-Robertson Q.* **2**:24–25.

Moss, M. L. 1969. Evolution of mammalian dental anatomy. *Am. Mus. Novit.* **2360:** 1–39.

Mossman, H. W. 1953. The genital system and fetal membranes as criteria for mammalian phylogeny and taxonomy. *J. Mammal.* **34:**289–298.

Moulton, D. G. 1967. Olfaction in mammals. *Zoologist.* **7:**421–429.

Müller-Schwarze, D. 1971. Pheromones in black-tailed deer *(Odocoileus hemionus columbianus). Anim. Behav.* **19:**141–152.

Murie, A. 1934. The moose of Isle Royale. *Univ. Mich. Mus. Zool. Misc. Publ.* No. 25. 44 pp.

———. 1936. Following fox trails. *Univ. Mich. Mus. Zool. Misc. Publ.* No. 32. 45 pp.

———. 1944. The wolves of Mt. McKinley. U.S. Dep. Inter. Natl. Park Serv. Fauna, Ser. 5. 238 pp.

Murie, O. J. 1935. Alaska-Yukon caribou. *North Am. Fauna,* No. 54. 93 pp.

———. 1951. The elk of North America. The Stackpole Co., Harrisburg, Pa., and the Wildlife Management Institute, Washington. 376 pp.

Murphy, D. A. 1960. Rearing and breeding of white-tailed fawns in captivity. *J. Wildl. Manage.* **24:**439–440.

Mutere, F. E. 1967. The breeding biology of equatorial vertebrates: reproduction in the fruit bat, *Eidolon helvum* at latitude 0° 20' N. *J. Zool.* **153:**153–161.

Muul, T. 1968. Behavior and physiological influences on the distribution of the flying squirrel, *Glaucomys volans. Univ. of Mich. Mus of Zool. Misc. Publ.* No. 134. 66 pp.

Muybridge, E., L. S. Brown, eds. 1957. Animals in motion. Republished by Dover Publications, Inc., New York. 77 pp. (Originally published in 1899 by Chapman J. Hall, L & D, London.)

Myers, G. S. 1938. Fresh water fishes and West Indian zoogeography. Smithsonian Report 1937. pp. 339–364.

Mykytowycz, R. 1965. Further observations on the territorial function and histology of the submandibular cutaneous (chin) gland in the rabbit, *Oryctolagus cuniculus* (L.). *Anim. Behav.* **13:**400–412.

Nalbandov, A. J. 1958. Reproductive physiology; comparative reproductive physiology of domestic animals, laboratory animals and man. W. H. Freeman and Company, San Francisco. 271 pp.

Nason, E. S. 1948. Morphology of the hair of eastern North American bats. *Am. Midl. Nat.* **39:**345–372.

Negus, N. C. 1958. Pelage stages in the cottontail rabbit. *J. Mammal.* **39:**246–252.

——— and A. J. Pinter. 1966. Reproductive responses of *Microtus montanus* to plants and plant extracts in the diet. *J. Mammal.* **47:**596–601.

Nissen, H. W. 1958. Axes of behavioral comparison. pp. 183–205. *IN:* Anne Roe and G. G. Simpson, eds. Behavior and evolution. Yale University Press, New Haven.

Noirot, D. 1969. Change in responsiveness to young in the adult mouse. V. Priming. *Anim. Behav.* **17:**542–546.

Norris, K. S. 1964. Some problems in echolocation in cetaceans. pp. 317–336. *IN:* W. N. Tavolga, ed. Marine bio-acoustics. Pergamon Press, New York. 413 pp.

———. 1968. The evolution of acoustic mechanisms in odontocete cetaceans. pp. 297–324. *IN:* Ellen T. Drake, ed. Evolution and environment. Peabody Museum of Natural History. Yale University Press, New Haven. 470 pp.

————. 1969. The echolocation of marine mammals. pp. 391–423. *IN:* H. T. Andersen, ed. The biology of marine mammals. Academic Press, Inc., New York. 511 pp.

———— and J. H. Prescott. 1961. Observations on Pacific cetaceans of California and Mexican waters. *Univ. Calif. Berkeley Publ. Zool.* **63:**291–402.

Odum, E. F. 1959. Fundamentals of ecology, 2d ed. W. B. Saunders Company, Philadelphia. 546 pp.

————. 1971. Fundamentals of ecology, 3d ed. W. B. Saunders Company, Philadelphia. 574 pp.

Ognev, S. I. 1947. The mammals of Russia (U.S.S.R.) and adjacent countries. Moscow. (9 volumes have been translated so far.)

Ogren, H. A. 1965. Barbary sheep. New Mexico, Dep. Game Fish. Bull. No. 13.

Olsen, R. W. 1969. Agonistic behavior of the short-tailed shrew *(Blarina brevicauda).* *J. Mammal.* **50:**494–500.

Olson, H. F. 1938. Deer tagging and population studies in Minnesota. *Trans. 3d North Am. Wildl. Conf.* pp. 280–286.

Osborn. H. F. 1910. The age of mammals in Europe, Asia and North America. The Macmillan Company, New York. 635 pp.

Osgood, W. H. 1909a. Biological investigations in Alaska and Yukon Territory. 1. East-central Alaska; 2. Ogilvie River, Yukon; 3. Macmillan River, Yukon. *North Am. Fauna,* No. 30. 96 pp.

————. 1909b. Revision of the mice of the American genus *Peromyscus. North Am. Fauna,* No. 28. 285 pp.

————. 1943. Clinton Hart Merriam, 1885–1942. *J. Mammal.* **24:**421–436.

Oosting, H. J. 1958. The study of plant communities, 2d ed. W. H. Freeman and Company, San Francisco. 440 pp.

Oxford, A. E. 1958. Rumen microorganisms and other products. *N.Z. Sci. Rev.* **16:** 38–44.

Pack, J. C., H. S. Mosby, and P. B. Siegel. 1967. Influence of social hierarchy on gray squirrel behavior. *J. Wildl. Manage.* **31:**720–728.

Palmer, R. S. 1954. The mammal guide, mammals of North America north of Mexico. Doubleday & Company, Inc., Garden City, N.Y. 384 pp.

Palmer, T. S. 1954. In memoriam: Clinton Hart Merriam. *Auk.* **71:**130–136.

Panuska, J. A. 1959. Weight patterns and hibernation in *Tamias striatus. J. Mammal.* **40:**554–566.

———— and N. J. Wade. 1956. The burrow of *Tamias striatus. J. Mammal.* **37:**23–31.

Park, H., and E. R. Hall. 1951. The gross anatomy of the tongues and stomachs of eight New World bats. *Trans. Kans. Acad. Sci.* **54:**64–72.

Parkes, A. S., and H. M. Bruce. 1962. Pregnancy-block in female mice placed in boxes soiled by males. *J. Reprod. Fert.* **4:**303–308.

Parrington, F. R. 1967. On the Upper Triassic mammals. *Philos. Trans. R. Soc. London Ser. B.* **261:**231–272.

Payne, R. S. 1970. Songs of the humpback whale. (An LP record by CRM Records with explanatory booklet.) Del Mar, Calif.

———— and S. McVay. 1971. Songs of the humpback whales. *Science* **173:**585–597.

Pearson, O. P. 1942. On the cause and nature of a poisonous action produced by the bite of a shrew, *Blarina brevicauda. J. Mammal.* **23:**159–166.

————. 1946. Scent glands of the short-tailed shrew. *Anat. Rec.* **94:**615–625.

————. 1955. Shrews. pp. 100–107. *IN:* Twentieth century bestiary, by *Sci. Am.,* publ. by Simon and Schuster, New York.

————, M. R. Koford, and A. K. Pearson. 1952. Reproduction of the lump-nosed bat *(Corynorhinus rafinesqui)* in California. *J. Mammal.* **33:**273–320.

Pedersen, T. 1952. The milk fat of sperm whale. *Norw. Whaling Gaz.* **41:**375–378.

Pengelley, E. T., and K. C. Fisher. 1961. Rhythmical arousal from hibernation in the golden-mantled ground squirrel. *Can. J. Zool.* **39:**105–120.

———— and J. Asmundson. 1972. An analysis of the mechanisms by which mammalian hibernators synchronize their behavioral physiology with the environment. pp. 637–661. *IN:* F. E. South, J. P. Hannon, J. R. Willis, E. T. Pengelley, and N. R. Alpert, eds. Hibernation and hypothermia, perspectives or challenges. American Elsevier Publishing Company, Inc., New York. 743 pp.

Perry, R. 1967. The world of the walrus. Taplinger Publishing Co., New York. 162 pp.

Peterka, H. E. 1936. A study of the myology and osteology of three sciurids with regard to adaptation to arboreal, glissant and fossorial habits. *Trans. Kans. Acad. Sci.* **34:**313–332.

Peterson, R. L. 1955. North American moose. University of Toronto Press, Toronto. 280 pp.

Peterson, R. S., and G. A. Bartholomew. 1967. The natural history and behavior of the California sea lion. Spec. Publ. No. 1, Am. Soc. Mammal., Stillwater, Oklahoma. 79 pp.

Petter, J., Jr. 1962. Ecological and behavioral studies of Madagascar lemurs in the field. *Ann. N.Y. Acad. Sci.* **102:**267–287.

Peyer, B. 1968. Comparative odontology. The University of Chicago Press, Chicago. 347 pp. (Original in German; translated by Rainer Zangerl.)

Picton, H. D. 1960. Migration patterns of the Sun River elk herd, Montana. *J. Wildl. Manage.* **24:**279–290.

Pike, G. C. 1962. Migration and feeding of the gray whale *(Eschrichtius gibbosus).* *J. Fish Res. Board Can.* **19:**815–838.

Pilson, M. E. Q. 1965. Absence of lactose from the milk of the Otarioidea, a superfamily of marine mammals. Abstr. *Am. Zool.* **5:**220–221.

———— and A. L. Kelly. 1962. Composition of the milk from *Zalophus californianus,* the California sea lion. *Science.* **135:**104–105.

———— and D. W. Walker. 1970. Composition of milk from spotted and spinner porpoises. *J. Mammal.* **51:**74–79.

Piper, S. E. 1909. The Nevada mouse plague of 1907–1908. U.S. Dep. Agri. Farmer's Bull. No. 352. 23 pp.

Pitelka, F. A. 1947. Distribution of birds in relation to major biotic communities. *Am. Midl. Nat.* **25:**113–137.

————. 1957. Some characteristics of microtine cycles in the Arctic. 18th Biol. Colloq. Oregon State College Proc. pp. 73–88.

————. 1958. Some aspects of population structure in the short-term cycles of the brown lemming in northern Alaska. *Cold Spring Harbor Symp. Quant. Biol.* **22:** 237–251.

Pocock, R. 1914. On the facial vibrissae of mammalia. *Proc. Zool. Soc. London.* **40:** 889–912.

Poulter, T. C. 1968. Marine mammals. pp. 405–465. *IN:* T. A. Sebeok, ed. Animal communication. Indiana University Press, Bloomington. 686 pp.

Powers, S. 1911. Floating islands. *Pop. Sci. Monthly.* **79**:303–307.

Preble, E. A. 1902. A biological investigation of the Hudson Bay region. *North Am. Fauna,* No. 22. 140 pp.

———. 1908. A biological investigation of the Athabaska-Mackenzie region. *North Am. Fauna,* No. 27. 574 pp.

Prell, H. 1930. Die verlangerte Tragzeit der einheimischen Martes-arten: Ein Erklärungsversuch. *Zool. Anz.* **74**:122–128.

Prescott, J. H. 1959. Rafting of jackrabbit on kelp. *J. Mammal.* **40**:443–444.

Prosser, C. L. ed. 1973. Comparative animal physiology, 3d ed. W. A. Saunders Company, Philadelphia. 966 pp.

Pruitt, W. O., Jr. 1953. An analysis of some physical factors affecting the local distribution of the shorttail shrew *(Blarina brevicauda)* in the northern part of the lower peninsula of Michigan. *Misc. Publ. Mus. Zool. Univ. Mich.* No. 79. 39 pp.

———. 1957. Observation on the bioclimate of some taiga mammals. *Arctic.* **10**: 131–138.

———. 1960. Behaviour of the barren-ground caribou. Univ. Alaska Biol. Pap. No. 3. 44 pp.

———. 1968. Synchronous biomass fluctuation of some northern mammals. *Mammalia.* **32**:172–191.

Purves, P. E. 1966. Anatomy and physiology of the outer and middle ear in cetaceans. pp. 320–330. *IN:* K. W. Norris, ed. Whales, dolphins, and porpoises. University of California Press, Berkeley and Los Angeles. 788 pp.

Quay, W. B. 1953. Seasonal and sexual differences in the dorsal skin gland of the kangaroo rat *(Dipodomys). J. Mammal.* **34**:1–14.

———. 1954. The Meibomian glands of voles and lemmings (Microtinae). *Misc. Publ. Univ. Mich. Mus. Zool.* No. 82. 24 pp.

———. 1965. Integumentary modifications of North American desert rodents. pp. 59–74. *IN:* S. G. Lyne and B. F. Short, eds. Biology of the skin and hair growth. American Elsevier Publ. Company, Inc., N.Y. 806 pp.

Quimby, D. C. 1942. *Thomomys* in Minnesota. *J. Mammal.* **23**:216–217.

———. 1944. A comparison of overwintering populations of small mammals in a northern coniferous forest for two consecutive years. *J. Mammal.* **25**:86–87.

———. 1951. The life history and ecology of the jumping mouse, *Zapus hudsonius. Ecol. Mono.* **21**: 61–95.

Rabb, G., J. H. Woolpy, and B. E. Ginsburg. 1967. Social relationships in a group of captive wolves. *Am. Zool.* **7**:305–311.

Ralls, K. 1971. Mammalian scent marking. *Science* **171**:443–449.

Rasmussen, A. T. 1917. Seasonal changes in the interstitial cells of the testis in the woodchuck *(Marmota monax). Am. J. Anat.* **22**:475–515.

Rasmussen, D. I. 1941. Biotic communities of Kaibab Plateau. *Ecol. Monogr.* **3**:229–275.

Raun, G. G., and B. J. Wilks. 1964. Natural history of *Baiomys taylori* in southern

Texas and competition with *Sigmodon hispidus* in a mixed population. *Tex. J. Sci.* **16**:28–49.

Rausch, R. 1953. On the status of some arctic mammals. *Arctic.* **6**:91–148.

Read, B. E. 1925. Chemical constituents of camel's urine. *J. Biol. Chem.* **64**:615–617.

Reeder, W. G., and R. B. Cowles. 1951. Aspects of thermoregulation in bats. *J. Mammal.* **32**:389–403.

Reinhardt, R. 1967. Out west on the Overland Train. Across-the-continent excursion with Leslie's Magazine in 1877 and the overland trip in 1967. The American West Publishing Co., Palo Alto, Calif. 207 pp.

Reynolds, V., and F. Reynolds. 1965. Chimpanzees of the Budongo Forest. pp. 368–424. *IN:* Irven DeVore, ed. Primate behavior. Holt, Rinehart and Winston, Inc., New York. 654 pp.

Reysenback de Haan, F. W. 1957. Hearing in whales. *Acta Oto-Laryngol. Suppl.* **134**: 1–114.

———. 1966. Listening under water: thoughts on sound and cetacean hearing. pp. 583–596. *IN:* K. W. Norris, ed. Whales, dolphins and porpoises. University of California Press., Berkeley. 789 pp.

Rice, D. W., and A. A. Wolman. 1971. The life history and ecology of the gray whale *(Eschrichtius robustus)*. Am. Soc. Mammal., Spec. Publ. No. 3. pp. 142.

Richardson, Sir J. 1937. Fauna boreali: Americana, or The zoology of the northern parts of British America: containing descriptions of the objects of natural history collected on the northern land expedition under command of Sir John Franklin, R.N. John Murray (Publishers), Ltd., London. 4 vols.

Richardson, K. C. 1949–1950. Contractile tissue in the mammary gland, with special reference to myoepithelium in the goat. *Proc. R. Soc. London, Ser. B.* **136**:30.

Richmond, M., and C. H. Conaway. 1969. Induced ovulation and oestrus in *Microtus ochrogaster. J. Reprod. Fert., Suppl.* **6**(1969):357–376.

Ricklefs, R. E. 1973. Ecology. Chiron Press, Inc., Newton, Mass. 816 pp.

Ride, W. D. L. 1970. A guide to native mammals of Australia. Oxford University Press, London and New York. 249 pp.

Roberts, A. 1951. The mammals of South Africa. Central News Agency, Capetown, S.A. 700 pp.

Robinson, D. T., and I. M. Cowan. 1953. An introduced population of the gray squirrel, *Sciurus carolinensis* Gmelin in British Columbia. *Can. J. Zool.* **32**:261–282.

Roe, F. G. 1970. The North American buffalo, 2d ed. University of Toronto Press, Toronto. 991 pp.

Roeder, K. D. 1964. Night fighters in a sonic duel. *Nat. Hist.* **73**:33–39.

———. 1965. Moths and ultrasound. *Sci. Am.* **212**:94–102.

Rollings, C. 1945. Habits, foods and parasites of the bobcat in Minnesota. *J. Wildl. Manage.* **9**:131–145.

Romer, A. S. 1956. The vertebrate body. W. B. Saunders Company, Philadelphia, 486 pp.

———. 1966. Vertebrate paleontology, 3d ed. University of Chicago Press, Chicago. 468 pp.

Roosevelt, T. 1910. African game trails. Charles Scribner's Sons, New York. 583 pp.

Rosenzweig, M. R., D. A. Riley, and K. Krech. 1955. Evidence for echolocations in the rat. *Science* **121**:600.

Rouk, C. S., and B. P. Glass. 1970. Comparative gastric histology of five North and Central American bats. *J. Mammal.* **51**:455–472.

Ruff, F. J. 1938. The white-tailed deer of the Pisgah National Game Preserve. U.S. Dep. Agri. Forest Service (mimeo.). 249 pp.

Runcorn, S. K. 1962. Continental drift. Academic Press, Inc., New York. 338 pp.

Rust, C. C., and R. K. Meyer. 1972. Hair, color, molt, testis size in male short-tailed weasels treated with melatonin. *Science,* **165**:921–922.

Rutherford, W. H. 1964. The beaver in Colorado: its biology, ecology, management and economics. Colorado Game, Fish Parks Dep., Denver, Tech. Bull. No. 17. 49 pp.

Ruthven, A. G. 1908. The faunal affinities of the prairie region of central North America. *Am. Nat.* **42**:388–394.

Ryan, R. M. 1968. Mammals of Deep Canyon, Colorado desert, California. The Desert Museum, Palm Springs, Calif. 137 pp.

Ryder, M. L. 1962. Structure of rhinoceros horn. *Nature (London).* **193**:1199–1201.

———. 1973. Hair. Edward Arnold (Publishers) Ltd., London. 56 pp.

Sanborn, C. C. 1952. Philippine zoological expedition, 1947–48: Mammals. *Fieldiana, Zool.* **33**:89–158.

Sanderson, G. C. 1961. Techniques for determining age of raccoons. Ill. Nat. Hist. Surv. Biol. Notes No. 45. Urbana, Ill. 16 pp.

———. 1966. The study of mammal movements—a review. *J. Wildl. Manage.* **30**: 215–235.

Sandoz, M. 1954. The buffalo hunters. The story of the hide men. Hastings House, New York. 372 pp.

Savage, D. E. 1958. Evidence from fossil land mammals on the origin and affinities of the Western Nearctic fauna. *IN:* C. L. Hubbs, ed. Zoogeography. *Am. Assoc. Adv. Sci. Publ.* **51**:97–129.

Savile, D. B. O. 1962. Gliding and flight in the vertebrates. *Am. Zool.* **2**:161–166.

Schaller, G. B. 1961. The orang-utan in Sarawak. *Zoologica.* **46**:73–82.

———. 1963. The mountain gorilla: ecology and behavior. University of Chicago Press, Chicago.

———. 1965. The behavior of the mountain gorilla. pp. 324–367. *IN:* I. DeVore, ed. Primate behavior. Holt, Rinehart and Winston, Inc., New York. 654 pp.

———. 1972. The Serengeti lion. University of Chicago Press, Chicago. 480 pp.

Scheffer, V. B. 1958. Seals, sea lions, and walruses. Stanford University Press, Stanford, Calif. 179 pp.

——— and D. W. Rice. 1963. A list of marine mammals of the world. *U.S. Fish Wildl. Serv. Spec. Sci. Rep. Fish.* **431**:1–12.

Schenkel, R. 1947. Expression studies of wolves. *Behaviour.* **1**:81–129. Translated from German by Agnes Klasson.

Schevill, W. E., and B. Lawrence. 1949. Underwater listening to the white porpoise *(Delphinapterus leucas). Science.* **109**:143–144.

Schmidt, D. W., L. E. Walker, and K. E. Ebner. 1971. Lactose synthesase activity in northern fur seal. *Biochim. Biophys. Acta.* **252**:439–442.

Schmidt, G. H. 1971. Biology of lactation. W. H. Freeman and Company, San Francisco. 317 pp.

Schmidt-Nielsen, K. 1952. Water metabolism of desert animals. *Physiol. Rev.* **32:** 135–166.

————. 1953. The desert rat. *Sci. Am.* **189**(1):73–78.

————. 1957. Urea excretion in the camel. *Am. J. Physiol.* **188:**477–484.

————, 1964. Desert animals, physiological problems of heat and water. Oxford University Press, Fairlawn, N.J.

————, T. J. Dawson, and E. C. Crawford. 1966. Temperature regulation in the echidna *(Tachyglossus aculeatus). J. Cell. Comp. Physiol.* **67:**63–71.

————, F. R. Hainsworth, and D. E. Murrish. 1970. Counter-current heat exchange in the respiratory passages: effects on water and heat balance. *Resp. Physiol.* **9:** 263–276.

———— and B. Schmidt-Nielsen. 1950. Do kangaroo rats thrive when drinking sea water? *Am. J. Physiol.* **160:**291–294.

————, ————, S. A. Jornum, and T. R. Haupt. 1956. Body temperature of the camel and its relation to water economy. *Am. J. Physiol.* **188:**103–112.

Schneider, K. M. 1931. Das Flehmen. *Zool. Gart.* **4:**349–362.

Schoen, A. 1969. Water conservation and the structure of the kidneys of tropical bovids. *J. Physiol.* **204:**143–144.

Scholander, P. F. 1940. Experimental investigations on the respiratory function in diving mammals and birds. *Hvalradets Skr.* **22:**1–131.

Schuchert, C. 1935. Historical geology of the Antillian-Caribbean region. John Wiley & Sons, Inc., New York. 811 pp.

Schultze-Westrum, T. G. 1969. Social communication by chemical signals in flying phalangers *(Petaurus breviceps papuanua). IN:* C. Pfaffman, ed. Olfaction and taste. Rockefeller University Press, New York. pp. 268–277.

Schwartz, C. W. 1941. Home range of the cottontail in central Missouri. *J. Mammal.* **22:**386–392.

Sclater, P. L. 1858. On the general geographical distribution of the members of the class Aves. *Proc. Linn. Soc. London, Zool.* **2:**130–145.

Scott, J. P. 1958. Animal behavior. The University of Chicago Press, Chicago. Paperback edition by the Natural History Library, Anchor Books, Doubleday & Company, Inc., Garden City, N.Y. 331 pp.

————. 1969. Introduction to animal behavior. pp. 3–21. *IN:* E. S. E. Hafez, ed. The behavior of domestic animals. The Williams & Wilkins Company, Baltimore. 647 pp.

Scott, T. G. 1943. Some food coactions of the Northern Plains red fox. *Ecol. Monogr.* **13:**427–479.

Scott, W. B. 1937. A history of land mammals in the western hemisphere, rev. ed. (1st ed., 1913). The Macmillan Company, New York. 786 pp.

Seaman, G. A. 1952. The mongoose and Caribbean wildlife. *Trans. 17th North Am. Wildl. Conf.* pp. 188–197.

Searle, A. G. 1968. Comparative genetics of coat color in mammals. Academic Press, Inc., New York. 308 pp.

Sears, P. B. 1947. Deserts on the march. University of Oklahoma Press, Norman. 178 pp.

Selye, H. 1950. Stress. The physiology and pathology of exposure to stress. A treatise based on the concepts of the general-adaptation syndrome. Acta Inc., Montreal, Canada. 1025 pp.

Seton, E. T. 1909. Life histories of northern mammals. Charles Scribner's Sons, New York. 2 vols., 1267 pp.

———. 1923. The Arctic prairies. Charles Scribner's Sons, New York. 308 pp.

———. 1929. Lives of game animals. Doubleday, Doran & Company, Inc., Garden City, N.Y. 4 vols.

Severaid, J. H. 1945. Pelage changes in the snowshoe hare. *J. Mammal.* **26**:41–63.

Sewell, G. D. Ultrasonic signals from rodents. *Ultrasonics.* **8**:26–30.

Shahani, K. M. 1965. Milk enzymes: Their role and significance. *J. Dairy Sci.* **49**: 907–920.

Shanks, C. E. 1948. The pelt-primeness method of aging muskrats. *Am. Midl. Nat.* **39**: 179–187.

Shantz, H. L. 1923. The natural vegetation of the Great Plains region. *Ann. Assoc. Am. Geog.* **13**:81-107.

Sharman, G. B. 1970. Reproductive physiology of marsupials. *Science.* **167**:1221–1228.

———, J. H. Calaby, and W. E. Poole. 1966. Patterns of reproduction in female diprotodont marsupials. *IN:* I. W. Rowlands, ed. Comparative biology of reproduction in mammals. *Symp. Zool. Soc. London,* No. **15**:208–232.

Shaul, D. M. B. 1962. The composition of milk of wild animals. *Int. Zoo Yearb.* **4**: 333–342. Published for the Zoological Society of London, by Hutchison of London.

Shaw, W. T. 1925. Duration of the estivation and hibernation of the Columbian ground squirrel, *Citellus columbianus,* and sex relations of same. *Ecology.* **6**:75–81.

Sheets, R. G., R. L. Linder, and R. B. Dahlgren. 1971. Burrow systems of prairie dogs in South Dakota. *J. Mammal.* **52**:450–453.

Sheldon, C. 1934. Studies on the life histories of *Zapus* and *Napaeozapus* in Nova Scotia. *J. Mammal.* **15**:290–300.

———. 1938. Vermont jumping mice of the genus *Zapus. J. Mammal.* **19**:324–332.

Sheldon, W. G. 1950. Denning habits and home range of red foxes in New York State. *J. Wildl. Manage.* **14**:33–42.

Shelford, V. E., and A. C. Twomey. 1941. Tundra animal communities in the vicinity of Churchill, Manitoba. *Ecology.* **22**:47–69.

Sherman, H. B. 1937. Breeding habits of the free-tailed bat. *J. Mammal.* **18**:176–187.

Short, R.V. 1972. Species differences. pp. 1–33. *IN:* C. R. Austin and R. V. Short, eds. Reproduction in mammals. Book 4. Reproductive patterns. Cambridge University Press, Cambridge, England. 5 books.

——— and M. F. Hay. 1966. Delayed implantation in the roe deer *Capreolus capreolus. Symp. Zool. Soc. London,* No. **15**:171–194.

Shortridge, C. C. 1934. The mammals of southwest Africa, vol. 1. William Heinemann, Ltd., London. 437 pp.

Shull, A. F. 1907. Habits of the short-tailed shrew. *Am. Nat.* **41**:495–522.

Shutt, F. T. 1932. Milk of North American buffalo. *Analyst.* **56**:454.

Siivonen, L. 1954. Some essential features of short-term population fluctuation. *J. Wildl. Manage.* **18**:33–45.

Sikes, S. K. 1971. The natural history of the African elephant. Weidenfeld and Nicolson, London. 397 pp.

Silver, H. 1961. Deer milk compared with substitute milk for fawns. *J. Wildl. Manage.* **25**:66–70.

Simpson, G. G. 1940. Mammals and land bridges. *J. Wash. Acad. Sci.* **30**:137–163.

———. 1943. Mammals and the nature of continents. *Am. J. Sci.* **241**:1–31.

———. 1945. The principles of classification and a classification of mammals. *Bull. Am. Mus. Nat. Hist.* **85**:1–350.

———. 1947. Holarctic mammalian faunas and continental relationships during the Cenozoic. *Bull. Geol. Soc. Am.* **58**:613–688.

———. 1947a. Evolution, interchange, and resemblance of the North American and Eurasian Cenozoic mammalian fauna. *Evolution* **1**:218–220.

———. 1953. Evolution and geography. Condon Lectures, Oregon State System of Higher Education, Eugene. 74 pp.

———. 1959. Mesozoic mammals and the polyphyletic origin of mammals. *Evolution* **13**:405–414.

———. 1962. Principles of animal taxonomy. Columbia University Press, New York. 247 pp.

Sink, J. D. 1967. Theoretical aspects of sex odor in swine. *J. Theor. Biol.* **17**:174–180.

Sisson, S., and J. D. Grossman. 1969. The anatomy of domestic animals, 4th ed., rev. W. B. Saunders Company, Philadelphia. 972 pp.

Slijper, E. J. 1962. Whales. Hutchinson Publishing Group, Ltd., London. 475 pp.

Sloan, R. E., R. Jenness, A. L. Kenyon, and E. A. Regehr. 1961. Comparative biochemical studies of milks. I. Electrophoretic analysis of milk proteins. *Comp. Biochem. Physiol.* **4**:47–62.

Slobodkin, L. B. 1955. Conditions for population equilibrium. *Ecology* **36**:530–533.

———. 1962. Growth and regulation of animal populations. Holt, Rinehart and Winston, Inc., New York. 184 pp.

Smalley, R., and R. Dryer. 1966. Brown fat in hibernation. pp. 324–344. *IN:* K. C. Fisher, A. R. Dawe, C. P. Lyman, Edward Schonbaum, and Frank E. South, Jr., eds. Int. Symp. Natural Mammalian Hibernation. III. American Elsevier Publishing Company, Inc., New York. 535 pp.

Smith, C. F. 1948. A burrow of the pocket gopher *(Geomys bursarius)* in eastern Kansas. *Trans. Kans. Acad. Sci.* **51**:313–315.

Smith, H. M. 1960. Evolution of chordate structure: an introduction to comparative anatomy. Holt, Rinehart and Winston, Inc., New York. 529 pp.

Smith, H. W. 1951. The kidney: structure and function in health and disease. Oxford University Press, New York.

Smith, J. C. 1972. Sound production by infant *Peromyscus maniculatus* (Rodentia: Myomorpha). *J. Zool.* **168**:279–369.

Smith, J. D. 1972. Systematics of the Chiroptera family Mormoopidae. *Univ. Kans. Mus. Nat. Hist. Misc. Publ.* **56**:1–132.

Smith, J. M., and R. J. G. Savage. 1956. Some locomotory adaptations in mammals. *J. Linn. Soc. London. Zool.* **42**(288):603–627.

Smith, R. E. 1958. Natural history of the prairie dog in Kansas. *Univ. Kans. Mus. Nat. Hist., Misc. Publ.* No. 16, 36 pp.

Smith R. F. 1964. Thermoregulatory and adaptive behavior of brown adipose tissue. *Science* **146**:1686–1689.

Smith, R. L. 1974. Ecology and field biology, 2d ed. Harper & Row, Publishers, Incorporated, New York. 850 pp.

Sneath, P. H. A., and R. R. Sokal. 1973. Numerical taxonomy. W. H. Freeman and Company, San Francisco. 573 pp.

Sokal, R. R., and P. H. Sneath. 1963. Principles of numerical taxonomy. W. H. Freeman and Company, San Francisco. 359 pp.

Soper, J. D. 1941. History, range and home life of the northern bison. *Ecol. Monogr.* **11**:347–412.

―――. 1944. The mammals of southern Baffin Island, Northwest Territories, Canada. *J. Mammal.* **25**:221–254.

Sorokin, S. P. 1965. On the cytology and cytochemistry of the opossum's bronchial glands. *Am. J. Anat.* **177**:311–338.

Southwick, C. H. 1955. The population dynamics of confined house mice supplied with unlimited food. *Ecology* **36**:212–225.

Sowls, L. K., V. E. Smith, R. Jenness, R. E. Sloan, and E. Regehr. 1961. Chemical composition and physical properties of the milk of the collared peccary. *J. Mammal.* **42**:245–251.

Sperber, I. 1944. Studies on the mammalian kidney. *Zool. Bidr. Uppsala.* **22**:240–431.

Stains, H. J. 1956. The raccoon in Kansas, natural history, management and economic importance. *Univ. Kans. Mus. Nat. Hist. Misc. Publ.* **10**:1–76.

―――. 1962. Game biology and game management. A laboratory manual. Burgess Publishing Company, Minneapolis, Minn. 141 pp.

Stebbins, G. L., Jr. 1950. Variation and evolution in plants. Columbia University Press, N.Y.

Stehlia, H. G., and Schaub, S. 1951. Die Trigonodontidae der Simplicidenten Nager. *Palaeont. Abhand.* **67**:200–236.

Stephenson, A. B. 1969. Temperature within a beaver lodge in winter. *J. Mammal.* **50**:134–136.

Sterling, K. B. 1974. Last of the naturalists: the career of C. Hart Merriam. Arno Press, New York.

Steuwer, F. W. 1943. Raccoons: their habits and management in Michigan. *Ecol. Monogr.* **13**:203–257.

Stevens, P. G., and J. L. E. Erickson. 1942. The chemical composition of the musk of the Louisiana muskrat. *J. Am. Chem. Soc.* **64**:144–147.

Stevens, S. A. 1959. The first ninety years. *Nat. Hist.* (Am. Mus.), **68**:186–219.

Stieve, H. 1950. Anatomische-biologische Untersuchungen ueber die Fortpflanzungstätigkeit des europaischen Rehes (*Capreolus capreolus capreolus* L.) *A. Mikr.-anat. Forsch.* **55**:427–530.

Stokes, A. W., and D. F. Balph. 1965. The relation of animal behavior to wildlife management. *Trans. North Am. Wildl. and Nat. Resources Conf.* **30**:401–410.

Stones, R. C. 1964. Thermal regulatory responses in the bat, *Myotis lucifugus.* Ph.D. dissertation, Purdue University, Lafayette, Ind.

―――– and J. E. Wiebers. 1965. A review of temperature regulations in bats (Chiroptera). *Am. Midl. Nat.* **74**(1):155–167.

Storer, T. I. 1969. Mammalogy and the American Society of Mammalogists. *J. Mammal.* **50**:785–793.

―――, F. C. Evans, and F. G. Palmer, 1944. Some rodent populations in the Sierra Nevada of California. *Ecol. Monogr.* **14**:165–192.

Storer, T. I., R. L. Usinger, and J. W. Nybakken. 1968. Elements of zoology. McGraw-Hill Book Company, New York. 552 pp.

Straile, W. E. 1965. Root sheath–dermal papilla relationships and the control of hair growth. pp. 35–37. *IN:* S. G. Lyne and B. F. Short, eds. Biology of skin and hair growth. American Elsevier Publishing Company, Inc., New York.

Strumwasser, F., J. J. Gilliam, and J. L. Smith. 1964. Long term studies on individual hibernating animals. pp. 401–414. *IN:* P. Soumalainon, ed. Proc. 2d Int. Symp. Nat. Mammalian Hibernation, Helsinki. *Ann. Acad. Sci. Fenn., Ser. AIV,* Biol. 71.

Sikes, S. K. 1971. The natural history of the African elephant. Weidenfeld and Nicolson, London. 397 pp.

Tate, G. H. H. 1941. Results of the Archbold Expeditions. No. 36. Remarks on some Old World long-nosed bats. *Am. Mus. Novit.* **1140:**1–11.

———. 1947. Mammals of eastern Asia. The Macmillan Company, N.Y. 366 pp.

Taylor, C. R. 1966. The vascularity and possible thermoregulatory function of the horns in goats. *Physiol. Zool.* **39:**127–139.

———. 1969. The eland and oryx. *Sci. Am.* **220:**89–95.

———. 1970. Strategies of temperature regulation: effect on evaporation in East African ungulates. *Am. J. Physiol.* **219:**1131–1135.

———. 1972. The desert gazelle: a paradox resolved. *IN:* G. M. O. Maloiy, ed. Comparative physiology of desert animals. *Symp. Zool. Soc. London.* **31:**215–227.

Taylor, E. H. 1934. Philippine land mammals. Philipp. Bur. Sci. Monogr. No. 30.

Tembrock, G. 1968. Land mammals. pp. 338–404. *IN:* T. A. Sebeok, ed. Animal communication. Indiana University Press, Bloomington.

Tener, J. S. 1956. Gross composition of muskox milk. *Can. J. Zool.* **34:**569–571.

Thompson, R. D., G. C. Mitchell, and R. J. Burns. 1972. Vampire bat control by systemic treatment of livestock with an anticoagulant. *Science.* **177:**806–807.

Thomson, A. L. 1926. Problems of bird migration. Houghton Mifflin Company, Boston.

Thorne, R. F. 1963. Biotic distribution patterns in the tropical Pacific. pp. 311–350. *IN:* J. L. Gressitt, ed. Pacific Basin biogeography. Bishop Museum Press, Honolulu. 563 pp.

Throckmorton, L. H. 1968. Concordance and discordance of taxonomic characters in *Drosophila* classification. *Syst. Zool.* **17:**355–387.

Thwaites, R. G. 1904. The original journals of the Lewis and Clark expeditions. Dodd, Mead & Company, Inc., New York. 8 vols.

Tobias, P. V. 1956. Chromosomes, sex cells and evolution in a mammal: based mainly on studies of the reproductive glands of the gerbil and a new list of chromosome numbers of mammals. Published for the South African Council for Scientific and Industrial Research by Lund, Humphries, London. 420 pp.

Toepfer, K. 1891. Die morphologie des magens der Rodentia. *Morphol. Jahrb., Leipzig* **17:**380–407.

Toldt, K., Jr. 1935. Aufbau und natürliche Färbung des Haarkledes der Wildsäugetiere. *Dtsch. Ges. Kleintier-und Peltztierzucht.* Leipzig. 291 pp.

Townsend, M. T. 1935. Studies on some of the small mammals of central New York. *Bull. N.Y. State Coll. Forestry.* **8:**1–120. Also as *Roosevelt Wildl. Ann.* **4:**1–120.

Trapido, H. 1946. Observations of the vampire bat with special reference to longevity in captivity. *J. Mammal.* **27:**217–219.

Troughton, E. 1966. Furred animals of Australia. Livingston Publishing Co. 376 pp.

Tucker, V. A. 1965. Relations between torpor cycle and heat exchange in California pocket mouse, *Perognathus californicus. J. Cell. Comp. Physiol.* **65**:405–414.

Tullberg, T. 1899. Uber das System der Nagethiere, eine phylogenetische studie. *Nova Acta Regiae Soc. Sci. Ups.* **18**:329–541.

Turner, C. D. 1966. General endocrinology, 4th ed. W. B. Saunders Company, Philadelphia. 579 pp.

Twente, J. W., 1956. Ecological observations on a colony of *Tadarida mexicana. J. Mammal.* **37**:42–48.

Udvardy, M. D. F. 1969. Dynamic zoogeography. D. Van Nostrand-Reinhold Co., New York. 445 pp.

Ulrich's International Periodicals Directory, 15th ed. 1973–1974. R. R. Bowker Company, N.Y. 2706 pp.

Urquhart, F. A. 1958. A discussion of the use of the word "migration" as it relates to a proposed classification for animal movements. *Contrib. R. Ontario Mus., Div. Zool. Paleontol.,* No. 50:3–11.

Valenta, J. G., and M. K. Rigby. 1968. Discrimination of the odor of stressed rats. *Science.* **1968**:599–601.

Van Citters, R. L., W. S. Kemper, and D. L. Franklin. 1966. Blood pressure responses of wild giraffes: Studies by radiotelemetry. *Science.* **152**:384–387.

Vandenbergh, J. G. 1969. Male odor accelerates female sexual maturation in mice. *Endocrinology.* **84**:658–660.

Van Gelder, R. 1956. Echolocation failure in migratory bats. *Trans. Kans. Acad. Sci.* **59**:220–222.

Van Tienhoven, A. 1968. Reproductive physiology of vertebrates. W. B. Saunders Company, Philadelphia. 498 pp.

Vaughan, T. A. 1959. Functional morphology of three bats, *Eumops, Myotis, Macrotus. Univ. Kans. Mus. Nat. Hist. Misc. Publ.* **13**:1–53.

———. 1970. Adaptations for flight in bats. pp. 127–143. *IN:* B. H. Slaughter and D. W. Walton, eds. About bats. Southern Methodist University Press, Dallas.

———. 1972. Mammalogy. W. B. Saunders Company, Philadelphia. 463 pp.

Vershchagin, N.Y. 1959. The mammals of Caucasus, Acad. Sci. U.S.S.R. Translated from the Russian Israel Program for Scientific Translation, Jerusalem, 1967. 816 pp.

Villa, R. B. 1956. *Tadarida brasiliensis mexicana,* el murcielago guanero, es una subspecies migratoria. *Acta Zool. Mex.* **1**:1–11.

Vimtrup, B. J., and B. Schmidt-Nielsen. 1952. The histology of the kidney of kangaroo rats. *Anat. Rec.* **114**:515–528.

Visher, S. S. 1916. The biogeography of the northern Great Plains. *Geogr. Rev.* **2**:89–115.

———. 1925. Tropical cyclones of the Pacific. *Univ. Hawaii, Bernice P. Bishop Mus. Bull.* **20**:1–163.

Volterra, V. 1926. Variazione e fluttazioni de numero d' individiu in specie animali conviventi. Ma. Accad. Lincie, translated. pp. 31–113. *IN:* R. N. Chapman, Animal Ecology, 1931, McGraw-Hill Book Company, New York, Vol. 2.

Vorhies, C. T. 1945. Water requirements of desert animals in the Southwest. *Univ. Ariz. Tech. Bull.* **107**:487–525.

Wade, O. 1930. The behavior of certain spermophiles with special reference to estivation and hibernation. *J. Mammal.* **11**:160–188.

———. 1950. Soil temperatures, weather conditions, and emergence of ground squirrels from hibernation. *J. Mammal.* **31**:158–161.

Waites, G. M. H., and G. R. Moule. 1961. Relation of vascular heat exchange to temperature regulation in the testis of the ram. *J. Reprod. Fert.* **2**:213–224.

Walker, E. P., F. Warnick, K. I. Lange, H. E. Uible, S. E. Hamlet, M. A. Davis, and P. R. Wright. 1964. Mammals of the world. The Johns Hopkins Press, Baltimore. 3 vols.

———. 1968. Mammals of the world, 2d ed. Revised by J. L. Paradiso. The Johns Hopkins Press, Baltimore. 2 vols.

Wallace, A. R. 1860. On the zoological geography of the Malay Archipelago. *J. Proc. Linn. Soc. London.* **4**:172–184.

———. 1876. The geographical distribution of animals. The Macmillan Company, London. 2 vols.

———. 1895. Island life, or the phenomena and causes of insular faunas and floras including a revision and attempted solution of the problem of geological climates. The Macmillan Company, London and New York. 563 pp.

Walls, G. L. 1942. The vertebrate eye and its adaptive radiation. Cranbrook Inst. Sci. Bull. No. 19. 785 pp.

Walther, F. 1958. Zum kampf- und paarungsverhalten einiger Antilopen. *Z. Tierpsychol.* **15**:340–382.

———. 1965. Verhalten studien an der Grant gazell (*Gazella granti* Brooke, 1872) in Ngorongoro-Krater. *Z. Tierpsychol.* **22**:167–208.

Waring, S. H. 1970. Sound communication of black-tailed, white-tailed and Gunnison's prairie dog. *Am. Midl. Nat.* **83**:167–185.

Warren, K. B., exec. ed. 1958. Population studies: animal ecology and demography. *Cold Spring Harbor Symp. Quant. Biol.* **22**. 437 pp.

Wasserman, R. H., and F. W. Lengemen. 1960. Further observation on lactose stimulation of the gastrointestinal absorption of calcium and strontium in the rat. *J. Nutri.* **70**:377–384.

Watkins, W. A., and W. E. Schevill. 1968. Underwater playback of their own sounds to *Leptonychotes* (Weddell seals). *J. Mammal.* **49**:287–296.

Watson, D. M. S., 1915. The monotreme skull: a contribution to mammalian morphogenesis. *Philos. Trans. R. Soc. London B,* **207**:311–374.

Watt, K. E. F. 1966. Systems analysis in ecology. Academic Press, Inc., New York and London. 273 pp.

———. 1968. Ecology and resource management. McGraw-Hill Book Company, N.Y. 450 pp.

Weaver, J. E. 1954. North American prairie. Johnson Publishing Co., Lincoln, Nebr. 348 pp.

——— and F. E. Clements. 1938. Plant ecology. McGraw-Hill Book Company, New York. 520 pp.

——— and T. J. Fitzpatrick. 1934. The prairie. *Ecol. Monogr.* **4**:111–295.

Weber, M. 1927. Die Säugetiere, vol. I. Anatomischer Teil, Gustav Fischer, Jena, 2d ed. 444 pp.

———. 1927. Die Säugetiere vol. II. Systematischer Teil, Gustav Fischer, Jena, 2d ed. 898 pp.

Webster, D. B. 1961. The ear apparatus of the kangaroo rat, *Dipodomys. Am. J. Anat.* **108:**123–1248.

———. 1962. A function of the enlarged middle ear cavities of the kangaroo rat, *Dipodomys. Physiol. Zool.* **35:**248–255.

———. 1966. Ear structure and function in modern mammals. *Am. Zool.* **6:**451–466.

———, R. F. Ackerman, and G. C. Longa. 1968. Central auditory system of the kangaroo rat, *Dipodomys merriami. J. Comp. Neurol.* **133:**477–494.

Welker, W. I., and S. Seidenstein. 1959. Somatic sensory representation in the cerebral cortex of the raccoon *(Procyon lotor). J. Comp. Neurol.* **111:**469–501.

Welty, J. C. 1962. The life of birds. W. B. Saunders Company, Philadelphia. 546 pp.

Whiteman, E. E. 1940. Habits and pelage changes in captive coyotes. *J. Mammal.* **21:** 435–438.

Whitney, L. E., and A. D. Underwood. 1952. The raccoon. Practical Science Publishing Co., Orange, Conn. 147 pp.

Whittaker, R. H. 1953. A consideration of climax theory: the climax as a population pattern. *Ecol. Monogr.* **23:**41–78.

———. 1970. Communities and ecosystems. The Macmillan Company, New York. 162 pp.

Whitten, W. K. 1959. Occurrence of anestrus in mice caged in groups. *J. Endocrinol.* **18:**102–107.

———. 1966. Pheromones and mammalian reproduction. *Adv. Reprod. Physiol.* **1:** 155–177.

———, R. H. Bronson, and J. A. Greenstein. 1968. Estrus-inducing pheromone of male mice: transport by movement of air. *Science.* **161:**584–585.

Wied, M. 1839. Reise in das innere Nord-Amerika in den jahren 1832 bis 1834. J. Hoelscher, Coblenz.

Wilks, B. J. 1963. Some aspects of the ecology and population dynamics of the pocket gopher *(Geomys bursarius)* in southern Texas. *Tex. J. Sci.* **15:**241–283.

Williams, A. B. 1936. The composition and dynamics of a beech-maple climax community. *Ecol. Monogr.* **6:**318–408.

Williams, C. S. 1938. Aids to the identification of mole and shrew hairs with general comments on hair structure and hair determination. *J. Wildl. Manage.* **2:**239–250.

Willis, R. 1847. The works of William Harvey. Sydenheim Society, London.

Wilson, E. O. 1963. Pheromones. *Sci. Am.* **208:**100–114.

———. 1968. Chemical systems. pp. 75–102. *IN:* T. Seobeok, ed. Animal communication. Indiana University Press, Bloomington.

Wimsatt, W. A. 1944. Further studies on the survival of spermatozoa in the female reproductive tract of the bat. *Anat. Rec.* **88:**193–204.

———. 1945. Notes on breeding behavior, pregnancy and parturition in some vespertilionid bats of eastern United States. *J. Mammal.* **26:**23–33.

———. 1960. Some problems of reproduction in relation to hibernation in bats. *Bull. Mus. Comp. Zool. Harv. Univ.* **124:**249–263.

——— and A. L. Gurriere. 1962. Observations on the feeding capacities and excretory functions of captive vampire bats. *J. Mammal.* **43:**17–46.

Winkelmann, R. K. 1965. Innervation of the skin: Notes on a comparison of primate and marsupial nerve-endings. pp. 171–182. *IN:* S. G. Lyne and B. F. Short, eds. Biology of the skin and hair growth. American Elsevier Publishing Company, New York. 806 pp.

Wislocki, G. B. 1956. Further notes on antlers in female deer of the genus *Odocoileus. J. Mammal.* **37**:231–235.

Wohlbach, S. B. 1952. The hair cycle of the mouse and its importance in the study of sequences of experimental carcinogenesis. *Ann. N.Y. Acad. Sci.* **53** (art. 3): 517–536.

Wood, A. E. 1955. A revised classification of rodents. *J. Mammal.* **36**:165–187.

———. 1965. Grades and classes among rodents. *Evolution* **19**:115–130.

Woolpy, J. H., and B. E. Ginsburg. 1967. Wolf specialization: a study of temperament in a wild species. *Am. Zool.* **7**:357–363.

Wright, B. S. 1960. Predation on big game in East Africa. *J. Wildl. Manage.* **24**:1–15.

Wynne-Edwards, V. C. 1962. Animal dispersion in relation to social behavior. Oliver & Boyd Ltd., Edinburgh and London. 653 pp.

———. 1965. Self-regulating systems in populations of animals. *Science.* **147**:1543–1548.

Young, S. P., and H. H. T. Jackson. 1951. The clever coyote, parts I and II. Wildlife Management Institute, Washington. 411 pp.

Ziegler, L. 1843. Beobachtungen ueber die Brunst und den Embryo der Rehe. Hanover.

Zwerew, M. D. 1953. Zur Biologie des Wüstenschlafers. *Z. Säugetierk.* **17**:158–159. (This note is difficult to find, for on the page on which the note appears is this: *Bd. 17, 1942–1949.* On the cover of the journal is "Berlin 1952," but the "2" has been struck out and "3" substituted.)

Index

Index